MW01505144

Christian Menn

Prestressed Concrete Bridges

Translated and edited by Paul Gauvreau

Birkhäuser Verlag
Basel · Boston · Berlin

Originally published in 1986 under the title "Stahlbetonbrücken" by Springer-Verlag, Wien.
© 1986 by Springer-Verlag, Wien.

Author's address:
Christian Menn
Professor of Civil Engineering
Swiss Federal Institute of Technology
Zurich (Switzerland)

Library of Congress Cataloging-in-Publication Data

Menn, Christian, 1927–
[Stahlbetonbrücken. English]
Prestressed concrete bridges / Christian Menn ; translation into
English and edited by Paul Gauvreau.
p.   cm.
Translation of: Stahlbetonbrücken.
Includes bibliographical references (p.   ) and index.
ISBN-13:978-3-0348-9920-8
(U.S.).
1. Bridges, Concrete–Design and construction. 2. Reinforced concrete construction.
I. Gauvreau, Paul.   II. Title.
TG340.M4613   1990
624'.257–dc20

Deutsche Bibliothek Cataloging-in-Publication Data

Menn, Christian:
Prestressed concrete bridges / Christian Menn. Transl. into
Engl. and ed. by Paul Gauvreau. – Basel ; Boston ; Berlin :
Birkhauser, 1990
Einheitssacht.: Stahlbetonbrücken ⟨engl.⟩
ISBN-13:978-3-0348-9920-8   e-ISBN-13:978-3-0348-9131-8
DOI:10.1007/978-3-0348-9131-8

© 1990 Birkhäuser Verlag AG Basel
Softcover reprint of the hardcover 1st edition 1990

ISBN-13:978-3-0348-9920-8

Preface

This book was written to make the material presented in my book, *Stahlbetonbrücken*, accessible to a larger number of engineers throughout the world. A work in English, the logical choice for this task, had been contemplated as *Stahlbetonbrücken* was still in its earliest stages of preparation. The early success of *Stahlbetonbrücken* provided significant impetus for the writing of *Prestressed Concrete Bridges*, which began soon after the publication of its predecessor.

The present work is more than a mere translation of *Stahlbetonbrücken*. Errors in *Stahlbetonbrücken* that were detected after publication have been corrected. New material on the relation between cracking in concrete and corrosion of reinforcement, prestressing with unbonded tendons, skew-girder bridges, and cable-stayed bridges has been added. Most importantly, however, the presentation of the material has been extensively reworked to improve clarity and consistency. *Prestressed Concrete Bridges* can thus be regarded as a thoroughly new and improved edition of its predecessor.

This book is guided by the same philosophy as its German-language counterpart: quality in the design and construction of bridges is achieved through the application of a small number of fundamental principles. In the last decade, the issue of durability in bridges has grown in importance, giving new urgency to the need for quality. The deterioration of bridges is often due to deficiencies in design, building materials, detailing, or construction. An awareness of quality and how it is achieved is thus crucial to preventing a repetition of past mistakes in the rehabilitation of existing structures and the design of new ones. Analytical refinement for its own sake continues to be the primary obstacle to quality in design. The preference found throughout *Stahlbetonbrücken* for clear, simple, case-specific calculations over more general analyses of greater complexity has therefore been maintained in the present work.

Stahlbetonbrücken was largely based on direct experience gained from the design and construction of bridges in Switzerland over the past thirty years. As a result, much of the material discussed in the present work appears in a decidedly Swiss context. The main differences between this book and practise in other countries can be summarized as follows:

1. Loadings, assumptions of material behaviour, and rules for checking safety and serviceability have been taken directly from Swiss SIA Standards 160 and 162. No attempt has been made to adapt these aspects of the book to the standards or local practise of other countries, since, armed with an understanding of the fundamental principles of bridge design, the reader should have little difficulty in doing so for himself.

2. A unified theory of concrete design, which is now well established in Swiss practise, underlies this book. This theory provides one consistent set of rules for design, valid for structures in plain, reinforced, or prestressed concrete. One of the most important consequences of this approach is partial prestressing, by which the engineer is freed from arbitrary restrictions on tensile stresses in concrete at service load levels. Prestressing can thus be chosen to achieve certain specific objectives under service conditions, while contributing fully to resistance at ultimate limit state. These objectives are referred to in this book as *prestressing concepts*, a powerful tool which makes possible a rational use of prestressing. Engineers in countries where only full prestressing is allowed may find some of the material in this book somewhat frustrating, since many of the solutions presented are not workable without partial prestressing. It is hoped that a wider recognition of the value of partial prestressing will facilitate the adoption of the unified theory of concrete design throughout the world.

3. Loads, prestressing, and restrained or imposed deformations are treated identically in design in many parts of the world. Consistent with current design standards in Switzerland, this book makes a clear distinction between the three, which are denoted collectively as *actions*. An awareness of the differences among the three types of actions is essential to a proper understanding of structural behaviour.

Thanks are due to Guido Göseli for relettering the figures into English and drawing the new figures, and to Heinrich Schnetzer for assistance with coordination during the final stages of production. The support and advice provided by the staff of Birkhäuser Verlag is also greatly appreciated.

Finally, I would like to pay special tribute to my collaborator Paul Gauvreau, who transformed *Stahlbetonbrücken* into the present work. He brought to the task his skill as a translator, his good judgement as an editor, and his insight as a practising engineer. This rare combination of talents was instrumental in producing a book that is a significant improvement over its predecessor. His dedication, curiosity, and patience made possible the most cordial of professional collaborations, which was not only rewarding in itself, but is also reflected in the quality of this work.

Christian Menn
Zurich 1989

Preface to the German Edition

> "Everything should be made as
> simple as possible, but not simpler."
>
> Albert Einstein

This book provides engineers with a comprehensive overview of the fundamental principles governing the design and construction of concrete bridges. Experience has shown that safety and quality are the direct result of the consistent and rational application of these principles. Refinement and volume of calculations, on the other hand, have no significant influence on quality or economy.

The book has its origins in lectures on the design of concrete bridges given to undergraduate and graduate students at the Department of Civil Engineering of the Swiss Federal Institute of Technology, Zurich. Its eight chapters cover the fundamentals of conceptual design, analysis, and detailed design of bridge superstructures and substructures.

Conceptual design is of primary importance to quality and economy. The principles and objectives of conceptual design are therefore thoroughly discussed in the first chapters. In accordance with modern design standards, the concepts of safety and serviceability are clearly distinguished in discussions of analysis and detailed design. The proposed analytical models and methods of calculation are simple, clear, practical, and sufficiently accurate for design.

Truss models are used extensively throughout the book to establish the flow of forces in structural components. This method, known for decades yet "rediscovered" in recent years, was used consistently in my former design office and has proven itself in the analysis of many complex problems. Truss models are often more reliable than finite-element calculations, which are usually based on the assumption of isotropic, elastic material behaviour.

Reserves of structural resistance are considered in the verification of safety at ultimate limit state, in accordance with the theory of plasticity. It is recommended as a general rule, however, that designs be based on sectional forces obtained from the elastic solution. The greater the deviation of the design sectional forces from the elastic solution, the more important it becomes to ensure that the structure is capable of sufficient plastic deformations. Behaviour under service conditions and fatigue must also be given greater consideration as the difference between the design sectional forces and the elastic solution increases.

Experience has shown that quality and durability of concrete structures cannot be achieved with careful analysis and design alone. The arrangement of reinforcement and detailing of individual structural components play a decisive role in this regard. For this reason, recommended details are presented and discussed throughout the book.

The most important diagrams for the calculation of slender compression members have been collected in the Appendix. The interaction diagrams can be used to calculate ultimate resistance of cross-sections and flexural stiffness at ultimate limit state. The latter is necessary for the calculation of second-order effects. A small number of important reference works has been listed, giving additional information and guidance for in-depth treatment of special problems.

Engineers today are faced with an explosion of technical information. Although this information will have beneficial effects on future developments, it has largely distracted engineers from the most fundamental principles. It is the purpose of this book to make engineers aware of these principles once again.

Acknowledgements

The extensive preparatory work for this book was supported financially by the Swiss Federal Institute of Technology, Zurich, the Swiss Federal Department of Transportation and Energy (Highway Research), Berne, and the Foundation for Scientific, Systematic Research in Concrete and Reinforced Concrete (*Stiftung für wissenschaftliche, systematische Forschung auf dem Gebiet des Beton- und Eisenbetonbaus*), Zurich. Their support is gratefully acknowledged.

I would like to express my gratitude to members of the Institute of Structural Engineering at the Swiss Federal Institute of Technology and my personal collaborators, in particular Bruno Zimmerli for coordinating the work, Martin Käser, Thomas Keller, Christoph Künzli, Susanna Schenkel, Silvio Toscano, and Michael Wagner for their assistance in preparation of calculations, examples, and sketches, Guido Göseli for drafting the figures, and Sybille Burki and Rita Feusi for typing the manuscript. I am especially grateful to Springer-Verlag in Vienna for their excellent and understanding cooperation.

Christian Menn
Zurich, November 1986

Contents

1 Historical Overview

Prior to the nineteenth century, the two most important materials used in bridge construction were stone and timber. Bridges of either material were built as far back as prehistoric times. Although drawings and written descriptions are the only remaining evidence of the timber bridges of antiquity, several impressive stone bridges from Roman times have survived to this day. These include the Pons Fabricius (fig. 1.1) and the Ponte Sant'Angelo, both in Rome, the bridge over the Tagus at Alcántara, Spain, and the Pont du Gard near Nîmes, France (fig. 1.2).

From Roman times to the twentieth century, the forms used for stone bridges have been based on arches, retaining walls, and piers. These systems are effective in compression only and are thus well suited to masonry construction. The arch

Figure 1.1
Pons Fabricius, Rome

Figure 1.2
Pont du Gard, near Nîmes, France

was the main spanning element; piers were used for the intermediate supports of multiple-span bridges. Fill retained by walls constructed along both edges of the arch made possible level roadways and, in the case of aqueducts, allowed piping to be placed at constant slope.

Roman arch bridges are characterized by semicircular vaults. The timber centering used in their construction was often supported by stones protruding from the piers. (These are clearly visible, for example, in figure 1.2.) Although they had no purpose in the completed structure, these supports were often incorporated with special care into the final visual form. A similar careful treatment of such details can also be observed in some of the most recent stone bridges of the early twentieth century (see, for example, figure 1.5).

The semicircular vault remained the preferred form for arch bridges until the Middle Ages, when new arch shapes first began to be adopted. A particularly elegant example of a Renaissance bridge with elliptical arches is the Santa Trinità Bridge over the Arno in Florence, built in 1560 (fig. 1.3).

The Pont de Neuilly, completed in 1773, exemplifies the refinements in masonry bridge construction achieved in eighteenth century France (fig. 1.4). Designed by Jean Perronet (1708–1794), this structure is remarkable for its shallow elliptical arches, narrow piers, and harmonious proportions.

The bridges of the Rhätische Bahn railway, built in Switzerland at the beginning of the twentieth century, represent the last major milestone in the history of masonry

Figure 1.3
Santa Trinità Bridge, Florence

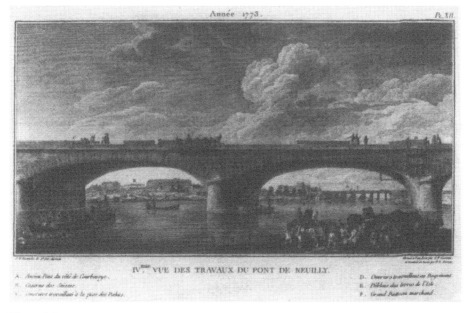

Figure 1.4
Pont de Neuilly, Paris

Figure 1.5
Landwasser Viaduct, Switzerland

bridge construction (figs. 1.5 and 1.6). These structures, although remarkable for their technical and aesthetic achievements, differ little from traditional stone bridges with regard to overall concept. Contrary to tradition, however, the proportions of the load-carrying members were designed using the theory of structures.

The simplest timber bridges of antiquity were little more than wooden beams supported on stone piers. Arches and simple inclined-leg frames were used for the construction of longer spans. These systems were followed by compound inclined-leg frames of increasing complexity, culminating in the intricate systems devised by the Swiss master builder Hans Ulrich Grubenmann (1709–1783), who achieved spans of up to 70 m (fig. 1.7).

Bridge building underwent major transformations as a result of the social, technological, and scientific developments that occurred in Europe during the Industrial Revolution. Bridges had heretofore been designed by master builders using empirical guidelines and built using simple tools and traditional materials. By the early nineteenth century, however, bridges were being designed by engineers using the newly developed theory of structures and built from stronger, lighter, industrially produced materials.

The impetus for change was provided by a major restructuring of the European transportation system. Many new bridges were required as early as the eighteenth

Figure 1.6
Wiesener Viaduct, Switzerland

Figure 1.7
Bridge over the Rhine at Schaffhausen, Switzerland

century to accommodate the growth of a postal coach network. Of much greater significance, however, was the subsequent appearance of the railway and the dramatic increase in demand for new bridges that occurred as a result. Factors such as economy, construction time, and strength assumed much greater importance in the construction of railway bridges than in previously built bridges. Although it was possible in some cases to adapt the traditional materials and methods of construction to these stricter requirements, the inherent advantages of new materials and techniques became increasingly evident.

Industrialized iron, the product of newly developed manufacturing processes, quickly became the pre-eminent material for railway bridge construction. The world's first iron bridge was built in England by Abraham Darby III (1750–1791) in 1779. This structure, which crossed the Severn River at the town of

Figure 1.8
Iron Bridge at Coalbrookdale, England

Coalbrookdale, consisted of parallel cast-iron arches spanning 30 m (fig. 1.8). The further development of cast-iron bridge technology was prevented by the brittleness of the material, which limited its application to compression members only. The dramatic growth of iron bridge construction in the nineteenth century was rather the result of wrought iron, a material with sufficient tensile strength and ductility to be used in tension and bending. The rolling of standardized iron sections, pioneered in England in the 1820s, enabled wrought iron to be produced economically.

At the same time, the development of a new and rational basis for the design of structures was underway. The experience acquired in the construction of timber and stone bridges was of little help in the design of iron bridges, since the advantages of iron could only be fully exploited using structural systems that differed radically from their predecessors. A rational basis for the design of these new systems was provided by a theory of structures based on the science of mechanics, developed by the French engineers Charles Coulomb (1736–1806) and Louis Navier (1785–1836). Practical application of the theory was greatly facilitated by the method of graphic statics, formulated by Carl Culmann (1821–1881).

The development of wrought iron (and its successor steel), together with graphic statics, made possible the design and construction of a great variety of truss-based bridges spanning up to 500 m. These structures made a profound visual impression on the built environment of the late nineteenth century. The most notable bridges of this era are remarkable for their daring dimensions and highly

Figure 1.9
Firth of Forth Bridge, Scotland

Figure 1.10
Bridge over the Elbe at Hamburg, Germany

functional character. Symbols of technical excellence in their day, many are still the objects of great admiration (figs. 1.9 through 1.12).

Industrialized iron transformed the hanging cable, a structural system that dated back to prehistoric times, into the modern suspension bridge. This development lead to the construction of spans longer than had ever before been considered possible. One of the earliest masterpieces in modern suspension bridge design was the 116 m long Conway Bridge in Wales. Designed by Thomas Telford (1757–

Figure 1.11
Garabit Viaduct, France

Figure 1.12
Eads Bridge, St. Louis, U.S.A.

1834) and completed in 1826, its main supporting element is a chain of wrought iron links (fig. 1.13).

Significant contributions to the development of suspension bridge construction on the European continent were made by Guillaume Henri Dufour (1787–1875) and Joseph Chaley (1795–1861). Dufour was responsible for the first major wire cable

Figure 1.13
Conway Bridge, Wales

bridge in Europe. The structure, completed in 1823, had two main spans of roughly 40 m. Chaley designed the most daring bridge of its day, the 1834 Grand Pont Suspendu over the River Sarine in Fribourg, Switzerland, whose main span measured 273 m (fig. 1.14).

American innovations in suspension bridge construction date back to the early nineteenth century. The world's first wire cable suspension bridge was built in the United States in 1816. American leadership in suspension bridge design and construction was firmly established by the achievements of John Roebling (1806–1869). These include the Niagara River Bridge, the first major suspension bridge designed to carry a railway. Its 250 m main span crossed a deep gorge at a height of 60 m (fig. 1.15). Roebling's most famous work is the Brooklyn Bridge over the East River in New York (fig. 1.16). Its main span of 488 m was without precedent; at the time of its completion in 1883, it was the most daring work of structural engineering yet achieved. The building of the bridge lasted 13 years and posed immense challenges, especially the construction of the tower foundations using compressed air caissons. Roebling did not live to see the completion of this great structure, which would become a monument to his extraordinary talents as bridge builder.

Roebling's great triumph in iron was echoed in the twentieth century by the steel suspension bridges of Othmar H. Ammann (1897–1965). One of Ammann's most outstanding works is the 1932 George Washington Bridge over the Hudson River, which links the American states of New York and New Jersey (fig. 1.17). Its main span, the first to exceed 1000 m, was twice as long as the previous record holder,

Figure 1.14
Grand Pont Suspendu, Fribourg, Switzerland

Figure 1.15
Niagara River Bridge, U.S.A.-Canada

Figure 1.16
Brooklyn Bridge, New York

Figure 1.17
George Washington Bridge, New York-New Jersey, U.S.A.

making it one of the boldest leaps in the history of structural engineering. Ammann's masterpiece was rendered even more daring by its elimination of the stiffening truss. (The truss that now forms part of the bridge was built in 1962. Provision was made for its addition in Ammann's original design, not to improve structural behaviour, but to double the traffic capacity of the bridge.)

Iron and steel were not, however, unqualified successes. Their strength, light weight, and speed of construction were offset by high maintenance costs and a high ratio of live load to dead load. The latter factor was responsible for the costly strengthening of many nineteenth-century railway bridges, which could not carry the heavier, faster trains that appeared soon after their construction. These disadvantages of iron and steel helped bring about the application of another new material, Portland cement concrete, to bridge construction. Compared to metal, concrete was regarded as maintenance-free; the heavier weight of concrete structures reduced the ratio of live load to dead load. The first plain concrete bridges were arches, built in France and Spain in the 1860s. These structures appeared identical to contemporary masonry structures, since plain concrete and masonry have essentially the same structural behaviour.

The discovery of reinforced concrete can be traced back to several originators. In 1850, the American Thaddeus Hyatt (1816–1901) conducted investigations of reinforced concrete beams and developed structural details that were fifty years ahead of their time. In France, a patent for a reinforced concrete boat was taken out by A. Lambot in 1855. His countryman François Coignet (1814–1888) obtained a patent in 1860 for concrete slabs with embedded iron mesh. The French gardener Joseph Monier (1823–1906), who patented a wire-reinforced concrete flowerpot in 1867, was the first to recognize the vast potential of reinforced concrete as a material for major structures. Following an additional patent awarded in 1873, he built the world's first reinforced concrete bridge in Chazelet, France in 1875 (fig. 1.18).

Based on Monier's patents, reinforced concrete technology soon spread to other European countries. The first theoretical basis for calculating the strength of reinforced concrete sections was developed in Germany in the 1880s by M. Koenen (1849–1924). Koenen neglected the tensile strength of concrete and computed steel and concrete stresses due to bending by locating the neutral axis at the centroidal axis of the section. The development of his theory relied heavily on tests conducted under the supervision of local building officials in Berlin in 1886. These tests established that, under live load applied to one half of the span, the capacity of a reinforced concrete arch was roughly three times that of an unreinforced concrete arch of identical dimensions. This result encouraged the construction of extremely flat, thin arches.

Additional investigations in Germany, France, and Austria confirmed that internal tensile forces must be resisted by reinforcement and revealed that reinforcing steel embedded in concrete is adequately protected against corrosion.

Figure 1.18
Reinforced concrete bridge at Chazelet, France

The most significant contributions to the expansion of reinforced concrete design and construction at the turn of the century were made by the French engineer François Hennebique (1842–1931) and by the German engineer Emil Mörsch (1872–1950).

Hennebique was a gifted designer who took full advantage of the monolithic behaviour of reinforced concrete and recognized that concrete members can safely fulfill several different structural functions. He developed effective reinforcement details, many of which are still used today. His 1892 patent for T-beams considered the compressive resistance of the slab in computing flexural resistance. In 1893, he patented a complete system for the construction of monolithic, composite members using stirrups and bent-up longitudinal reinforcement. The reinforced concrete bridges designed by Hennebique were considerably cheaper to construct than traditional masonry or iron bridges. In spite of considerable resistance from building officials against reinforced concrete, his central office in Paris and his concessionaires in several European countries designed and built more than 600 bridges. These bridges served as examples for designers for many years. The Bridge over the Ourthe in Liège, Belgium (1904, span length 55 m) and the Risorgimento Bridge in Rome (1911, span length 100 m) are two of Hennebique's most famous bridges (figs. 1.19 and 1.20). In both structures, the vault, spandrel walls, and deck slab were integrated into a box cross-section.

Figure 1.19
Bridge over the Ourthe, Liège, Belgium

Mörsch's principal contribution was to the theory of reinforced concrete. His thinking was always guided, however, by practical knowledge acquired through his experience as a designer. He developed methods of calculating sectional forces according to elastic theory and based the design of cross-sections on allowable stresses under service loads. These procedures ensured an adequate margin of safety and restricted the formation of cracks. Of far greater importance, however, was his development of truss models for reinforced concrete, which facilitated the design and detailing of reinforcement and provided designers with a clear, simple, and reliable tool for visualizing structural behaviour. Mörsch's ideas were published in 1902 as *Der Betoneisenbau, seine Anwendung und Theorie* (Reinforced concrete, application and theory), one of the most important reference works of its generation. The soundness of Mörsch's truss analogy has been confirmed by subsequent investigations. This powerful tool is still in use today.

Figure 1.20
Risorgimento Bridge, Rome

Figure 1.21
Wildegg Bridge, Switzerland

In 1903, the first provisional design standard for reinforced concrete, *Bauten in armiertem Beton* (Structures in reinforced concrete), was published by the Swiss Society of Engineers and Architects. Its purpose was to ensure a sufficient and uniform margin of safety in reinforced concrete construction. This example was soon followed by Germany, France, Italy, and Austria. The first definitive Swiss standard followed in 1909. The allowable stresses were specified as follows:

1. Compressive stress in concrete at extreme fibre: 4.5 N/mm^2
2. Compressive stress in concrete at centroid: 3.5 N/mm^2
3. Tensile stress in reinforcement: 120 N/mm^2

Switzerland's first reinforced concrete bridge was a 37.2 m arch span built in 1890 in the town of Wildegg (fig. 1.21). It consisted of an arch vault which supported monolithically cast retaining walls along its edges. The space between the walls was backfilled up to the level of the roadway. The structural system was thus similar to systems used in traditional masonry construction. The proportions of the arch, however, could only have been realized in reinforced concrete: the vault was only 0.2 m thick at the crown; the span to rise ratio was 10.6 to 1. The slenderness of this bridge is even more astounding when it is realized that the stiffening effect of the retaining walls was not considered by its designers.

The 1907 Allier River Bridge at Le Veurdre, France, designed by Eugène Freyssinet (1879–1962), is one of the most daring reinforced concrete bridges ever to be built (fig. 1.22). The crossing is achieved by three arch spans of 72.5 m. The span to rise ratio of 15 to 1 is so large that it can hardly be considered reasonable from a structural point of view. The young, insufficiently experienced engineer had ventured into unknown and dangerous territory with this bridge. Breaking with the tradition of stiffening flat arches with solid spandrel walls (e.g. fig. 1.20), Freyssinet supported the deck from the arch using thin columns. The choice of inclined columns was clever, since the resulting truss action significantly increased the buckling strength of the system.

Figure 1.22
Allier River Bridge, Le Veurdre, France

Contrary to Hennebique, who eliminated hinges wherever possible, Freyssinet designed the bridge at Le Veurdre as a series of three-hinged arches, in accordance with the wishes of his superiors. This proved to be a blessing in disguise. The removal of formwork was greatly facilitated by jacking the two arch halves apart at the crown. Downward deflections due to elastic deformations of the arch could thus be largely eliminated. Soon after completion of the bridge, however, deflections at the crown of approximately 130 mm were observed. This was a consequence of creep in concrete, a little-known phenomenon at that time. Freyssinet was able to reproduce the original profile of the bridge by jacking apart both halves of the arch at the crown. The hinge at midspan was subsequently concreted, transforming the spans into two-hinged arches.

The design of reinforced concrete bridges in the early twentieth century was strongly influenced by contemporary steel design practise. This resulted in bridges with little or no structural interaction among the individual, visually distinct components. Bridges designed in this way were relatively easy to analyse. Two examples of this tendency are the Swiss arch bridges shown in figures 1.23 and 1.24. Mörsch's Gmündertobel Bridge (1909) has an arch span of 79 m; the Langwieser Viaduct (1914) was the first railway bridge to have an arch span of 100 m.

Robert Maillart (1872–1940) did not follow the conventions of steel structures but rather pioneered new concrete bridge forms remarkable for their efficiency, economy, and beauty. Maillart was fascinated by the ease with which concrete could be formed into surfaces of arbitrary shape and recognized the advantages offered by monolithic connections and two-dimensional structural systems. He designed structural elements to carry out more than one function simultaneously and exploited the interactive structural behaviour of the entire system. Maillart believed that emphasizing the overall system rather than the individual components resulted both economic and aesthetic advantages. His 41 bridges, built between 1899 and 1940, are ample proof that his belief was correct.

Maillart's first bridge as an independent engineer was a 30 m arch built in 1901 over the Inn River at Zuoz, Switzerland (fig. 1.25). Its visual form is similar to that of the Wildegg Bridge. The bridge at Zuoz, however, is not merely a masonry system built in concrete; the thin spandrel walls, deck slab, and arch are monolithically connected into a box cross-section. Due to the high span to rise ratio of 10 to 1, concrete hinges were provided at the springing lines and at the

Figure 1.23
Gmündertobel Bridge, Switzerland

Figure 1.24
Langwieser Viaduct, Switzerland

Figure 1.25
Inn River Bridge, Zuoz, Switzerland

crown. Difficulties such as large cracks or large deformations, which plagued many other flat arch bridges, were avoided by Maillart's choice of structural system and cross-section.

Maillart later achieved his most visually compelling three-hinged arch forms by eliminating the spandrel walls near the springing lines, where they were not required for structural reasons. The most famous example of this type is the Salginatobel Bridge, a 90 m arch completed in 1930 in Switzerland (fig. 1.26). Falsework costs in the steep mountain canyon were substantially reduced by the choice of a three-hinged arch system with thin arch slab and spandrel walls. The falsework was designed to carry the weight of the arch slab only. After hardening, the arch could carry the additional weight of the walls and deck slab by itself; the falsework was required only to stiffen the system.

Maillart also used thin arch slabs to reduce falsework costs of deeper arches. The deck-stiffened arch was his preferred system for these bridges. The interaction of the entire structure, especially the combined action of arch and deck, is clearly visible in figure 1.27. The arch resists only axial force from arch thrust; bending moments due to live load are resisted entirely by the girder. Removing all bending from the arch as Maillart did is not completely justifiable from a statical point of view, however, since arches in compression can resist substantial moments without additional reinforcement. Maillart may have done so mainly to achieve the aesthetically pleasing form of the thin arch and to ensure simple structural behaviour.

Figure 1.26
Salginatobel Bridge, Switzerland

Figure 1.27
Landquart Bridge, Klosters, Switzerland

Many impressive reinforced concrete bridges were built during the 1930s and early 1940s. Arch spans of over 100 m became common; girder bridges reached spans of over 70 m.

A major crossing of the Mosel in Germany was designed by Franz Dischinger (1887–1953) and completed in 1934 (fig. 1.28). The bridge consisted of three spans of length 100 m, 105 m, and 119 m; each span was designed as two parallel three-hinged arches. Since the rise was constant over all three spans, the span to rise ratio varied from 12:1 in the shortest span to 14:1 in the longest. Solid slabs were used for the arches of the 100 m and 105 m spans; a hollow-box section was used for the arch of the 119 m span. This reduced dead load in the longest span and balanced the arch thrust due to dead load from the 105 m and 119 m spans, thus enabling the use of symmetrical deep-water caisson foundations. The arch spans were constructed in two longitudinal halves by displacing the falsework sideways after completion of one half.

Many of the large bridges constructed for the Autobahns in Germany were built as political monuments rather than as works of structural engineering art. Two noteworthy exceptions, however are the bridge over the Danube at Leipheim (fig. 1.29) and the Teufelstal Bridge at Jena (fig. 1.30). The Leipheim Bridge, built in 1936, had four spans of 73 m, 80 m, 85 m, and 78 m; its careful attention to visual design is reminiscent of the bridges of Maillart. Its relatively high span to rise ratio of 7 to 1 led to the choice of three-hinged arches. The Teufelstal Bridge,

Figure 1.28
Mosel River Bridge, Koblenz, Germany

Figure 1.29
Bridge over the Danube, Leipheim, Germany

built in 1938, had a record arch span of 138 m and a span to rise ratio of 5.3 to 1. A somewhat stiffer, fixed arch was therefore chosen for this bridge. The Teufelstal Bridge is remarkable for the design of its deck girder, which is free of joints over the entire 250 m length of the bridge between abutments.

Many important innovations in bridge design and construction were pioneered in France between the two world wars. The most significant French bridge of this period, the Plougastel Bridge, was designed by Freyssinet and completed in 1930 (fig. 1.31). It crosses the mouth of the Elorn River with three spans of 186 m. The deck girder was designed as a truss; highway and railway traffic were carried by the upper and lower chords, respectively. The falsework for the arch was a free-

Figure 1.30
Teufelstal Bridge, Jena, Germany

standing timber tied arch, constructed on land and floated into position. The same falsework was reused for all three spans.

The Bridge over the Lot at Castelmoron (1933) is remarkable not only for its structural system and its span length of 120 m, but for the way it was constructed. The slender arch and thin suspended deck girder, which acted as a tension tie, were stiffened using inclined hangers, prestressed by the self-weight of the deck. The arch consisted of precast core elements which were preloaded by jacking the two halves of the arch apart at the crown. The centering, unloaded in this way, would then be required to support only the remainder of the arch dead load in a subsequent construction phase.

Figure 1.31
Plougastel Bridge, France

During this period, French engineers also achieved record spans for girder bridges. The construction of long-span girders in reinforced concrete was difficult, not only because of the unwieldy arrangement and splicing of reinforcement, but also due to deformations and cracking, which were unavoidable and often severe. These bridges were rarely successful visually, since adequate structural behaviour could only be ensured with span to depth ratios of less than 10 to 1.

Albert Caquot (1881–1976) designed the Rue Lafayette Bridge, a concrete truss bridge over the railway yards of the Gare du Nord in Paris, which was completed in 1928 (fig. 1.32). The bridge consisted of two simply supported spans of 72 m and 77 m. The parallel-chord, 10.4 m deep trusses were spaced transversely at 20.4 m apart. The lower panel points were connected to floor beams, also designed as trusses. The deck was located at roughly the lower quarter point of the main trusses. The arrangement of reinforcement at the nodal points was extremely challenging. Although this structure must be considered as an extraordinary pioneering achievement in bridge construction, it is strong evidence of the inherent limitations of conventional reinforced concrete.

The span length of Caquot's Rue Lafayette Bridge was slightly surpassed in 1939 by the Villeneuve-St. Georges Bridge over the Seine. The span lengths were 39 m, 78 m, and 39 m. A cantilever system was used, resulting in a more favourable distribution of moments. Since a hinge was provided in each of the side spans 11 m from each end of the bridge, the structural system was statically determinate. This was presumably done to prevent uplift reactions at the abutments. The cross-

Figure 1.32
Rue Lafayette Bridge, Paris

section was designed as a three-cell box girder, the depth of which varies from 7.75 m at the intermediate supports to 2.52 m at the midpoint of the main span. This bridge still holds the record for the longest span of a conventionally reinforced concrete girder bridge.

The state of the market for reinforced concrete bridges was less favourable in the U.S.A. Concrete was often not competitive against steel, which could be inexpensively produced and quickly erected. Long-span reinforced concrete bridges were therefore much less common than in Europe during the 1930s. Of particular note, however, are concrete arch bridges built during this period for the Pacific coast highway in California.

Brazil was the site of the first application of the cantilever method to concrete bridge construction (fig. 1.33). The 68.5 m main span of the Rio do Peixe Bridge at Herval was built out from both piers to eliminate the risk of damage from flash floods, which could raise the water level more than 10 m in the space of a few hours. Segments of length 1.5 m were used. The reinforcement consisted of 45 mm diameter bars, which were spliced using threaded sleeves. Although cantilever construction had long been used for steel bridges and would later become an important method for the construction of long-span prestressed concrete bridges, the achievement of the Rio do Peixe Bridge was neither recognized nor appreciated at that time.

Figure 1.33
Rio do Peixe Bridge, Herval, Brazil

The last major conventionally reinforced concrete girder bridge was built in 1941 in Santa Fé, Argentina. The bridge is continuous over several supports, with internal hinges provided to reduce the degree of statical indeterminacy. The design of the cross-section and arrangement of the reinforcement foreshadowed concepts used several decades later in long-span prestressed concrete bridges.

The most significant reinforced concrete bridge built in England during this period was the Waterloo Bridge over the Thames in London, completed in 1939 (fig. 1.34). Influenced by the form of a multiple masonry arch bridge built in 1817, it was designed as a continuous haunched girder. Its longest span was 77 m. The cross-section consisted of two 7.6 m wide box sections, linked by a 10 m wide ribbed slab. The girder depth varied from 7.5 m at the piers to 2.4 m at midspan.

Figure 1.34
Waterloo Bridge, London

In Switzerland, apart from the masterpieces of Maillart, the 1932 Gueroz Bridge by Alexandre Sarrasin and the 1940 Federal Railways Bridge over the Aare River in Berne stand out for their technical and aesthetic merit.

The 90 m arch span of the Gueroz Bridge crosses a 180 m deep canyon (fig. 1.35). The arch consists of two 0.6 m wide ribs. The thin parapets were designed to stiffen the deck. The bridge can be considered as a masterpiece of the technical aspects of bridge design for its minimum use of materials. As with the bridges of Maillart, however, its use of the thinnest possible dimensions would ultimately lead to to difficult maintenance problems.

An existing iron bridge of the Swiss Federal Railways over the Aare in Berne was in need of replacement by the end of the 1930s. A new alignment and an increase to four tracks required a total bridge length of 1150 m. A design competition jury selected for construction a reinforced concrete bridge with a 150 m main arch span

Figure 1.35
Gueroz Bridge, Switzerland

(fig. 1.36). A three-cell box section was used for the arch. A 17.4 m wide, four-web T-girder was used for the deck over entire length of the bridge. Spans of 27 m were used for the approaches. Continuity of the deck girder was interrupted over the arch abutments by internal hinges, thus reducing the contribution of the girder to the overall stiffness of the system. This was standard practise during this period and was probably done to make bridges easier to analyse. Due to the design live load of 30 kN/m^2, roughly five time greater than highway live loads, the proportions of the members are large. In spite of this, however, the bridge presents an overall impression of harmony and elegance. The careful attention to details and construction was repaid by a service life of over 40 years with no major maintenance.

The longest concrete arch span of this period was the Sandö Bridge, built in central Sweden in the early 1940s (fig. 1.37). Swedish engineers had already acquired experience with long-span concrete bridges with the 181 m arch of the Traneberg-sund Bridge, built in 1934 in Stockholm. The 267 m main span of the Sandö Bridge brought concrete arches into what was previously considered the exclusive domain of steel. Construction of the arch began using a wooden tied arch as falsework, similar to the system used for the Plougastel Bridge in France a decade earlier. This falsework collapsed during concreting and the arch was rebuilt using conventional falsework supported on temporary piers. The Sandö Bridge held the world span length record for twenty years, until longer spans were made possible by developments in falsework and construction technology.

Figure 1.36
Swiss Federal Railways Bridge over the Aare, Berne, Switzerland

Figure 1.37
Sandö Bridge, Sweden

During the first century of reinforced concrete, numerous attempts were made to improve the quality and economy of construction. Most of the reinforcing systems that were developed and patented, however, were short lived. The most important of these were the Melan System, by which iron falsework was cast into the structure as reinforcement, von Emperger's system for strengthening arches using cast iron cores, and Visintini's precast truss system. Specially adapted erection equipment was the major impediment to the further development and use of precast technology at that time.

Several of the innovations introduced during this period, however, are now standard practise. Internal vibration, first used in the construction of the Gueroz Bridge in 1932, substantially improved the strength and impermeability of concrete. Just as significant was the introduction of reinforcing steel with surface deformations. The improved bond behaviour of deformed reinforcing steel resulted in substantial increases in allowable steel stresses; the Swiss design standard of 1935, for example, increased allowable stresses from 120 N/mm^2 to 170 N/mm^2 for deformed bars.

The single most important innovation in the history of reinforced concrete technology was the development of prestressed concrete. The principle of prestressing, the creation of an artificial state of stress to improve structural behaviour, had been used since ancient times in shipbuilding and barrelmaking. It had also been used to a limited extent in masonry construction. Its application to reinforced concrete revolutionized bridge construction.

The earliest investigations of prestressed concrete beams were conducted in the nineteenth century. In 1888, the German engineer W. Döring patented a system for the construction of slabs, planks, and beams, by which cracking was reduced through the use of prestressed wires. Prestressing saw no further practical application in construction until its significance and potential were recognized by Freyssinet. From 1928 to 1936, he worked on the development of stressing systems and patented prestressing jacks and anchors. He was then ready to begin applying his ideas to the construction of bridges and buildings. From the very beginning, Freyssinet insisted on the use of high-strength wires bonded to concrete. His success with prestressing is largely due to this practical implementation of the concept.

Although prestressing was well received in Germany, it was not initially implemented according to Freyssinet's system. Dischinger, the country's most experienced reinforced concrete designer, chose to use large-diameter bars of low-strength steel, unbonded to the concrete. His method had several major shortcomings. Due to its low tensile strength, the prestressing steel could only be prestrained to barely twice the expected plastic strain in the concrete due to creep and shrinkage. Large prestressing losses, and hence cracking and large deformations, were therefore unavoidable.

Dischinger used his system to construct the world's first prestressed concrete bridge in Aue, Germany, completed in 1937. The prestressed portion of the bridge (foreground of figure 1.38) consisted of three spans of 25.2 m, 69 m, and 23.4 m. The central part of the main span was suspended at internal hinges, making the system statically determinate. Cantilever tendons were draped over the two intermediate piers; the suspended segment was prestressed as a simply supported beam. Dischinger used 70 mm diameter bars for the tendons, prestressed to

Figure 1.38
Prestressed concrete bridge in Aue, Germany

220 N/mm² or roughly 42 percent of tensile strength. The thick bars, which were bent into polygonal shapes, could not be jacked from the ends. Instead, they were prestressed by pushing them away from the structure at deviation points. In spite of careful design and construction, the bridge was unsuccessful. After 25 years, almost 75 percent of the initial prestress had been lost through creep and shrinkage, leaving the structure essentially unprestressed. This resulted in large cracks and vertical deflections of 200 mm.

(Unbonded prestressing remained without further development for many years, only to be rediscovered some 40 years later. Much better results were obtained when tendons composed of high-strength wires or strands were used. Prestressing with unbonded tendons is now widely used in building and bridge construction.)

Dischinger's negative experience with his first prestressed bridge in Aue led him to undertake a thorough clarification of the long-term deformations of concrete due to creep and shrinkage. His differential equation for the time-varying concrete strain made reliable calculations of prestressing losses possible for the first time (see Section 4.7). Unfortunately, the significance of creep and shrinkage to the design of prestressed concrete structures was overemphasized by many engineers since then. Although the large volume of research undertaken on the subject is of pure scientific value, it had little effect on improving the quality of bridge design and construction. Even challenging problems such as the calculation of camber and redistribution of sectional forces in cantilever construction can be solved with sufficient accuracy using simple methods based on a few well-defined material parameters.

Freyssinet's ideas guided the further development of prestressed concrete for several decades. Although the original goal of prestressing was to eliminate cracks and undesirable deformations through the creation of a beneficial state of stress, the increase in load-carrying capacity gained from the use of high-strength reinforcement was an important positive side effect. Freyssinet preferred to use extremely thin-walled precast components whenever possible. Because he developed schemes for both pre-tensioned and post-tensioned concrete, is ideas could be easily adapted to cast-in-place concrete. Freyssinet considered prestressed concrete as a completely new material, and accepted only *full prestressing*, that is, the complete elimination of tensile stresses in the concrete under service loads. His rigid philosophy of full prestressing was adopted throughout the world and survived for many years.

Figure 1.39
Bridge over the Marne at Luzancy, France

After building several small bridges from precast simply supported beams, Freyssinet began the construction of the 55 m long Luzancy Bridge over the Marne in 1941 (fig. 1.39). The construction, which was interrupted by the war, was completed in 1945. The series of famous Marne bridges at Esbly, Annet, Trilbardou, Ussy and Changis-St. Jean was built following the same principles between 1945 and 1949. All of these bridges were two-hinged frames with spans of 74 m. The cross-section consisted of three box girders with a very thin cast-in-place deck slab. The girder webs in the Marne bridges, which were only 0.1 m thick, were reinforced with prestressed stirrups. Precast segments of roughly 2.4 m in length were transported to the site on barges and prestressed together into three longer erection units. The units were lifted into place from overhead cables, thus eliminating the need for falsework. After final adjustment, the remaining prestressing wires were installed and the 20 mm wide joints between units were grouted.

Post-war reconstruction, followed by rapid economic growth, produced a boom in bridge construction in Europe beginning in the late 1940s. The early years after the war were an important development period for prestressing concrete, in which new design and construction techniques were tested and improved. The analysis and design of prestressed concrete structures was one of the most important research goals at schools of engineering throughout the world. Unfortunately, it took a relatively long time for a clear picture of the behaviour of prestressed concrete structures to emerge from the countless scientific papers published during this period. One of the first important contributions in this regard was the book *Spannbeton für die Praxis*, first published in 1954 and later translated into English as *Prestressed Concrete – Design and Construction*, by Fritz Leonhardt (b. 1909).

By the end of the 1940s, many new prestressing systems had been developed. All were based to some extent on Freyssinet's system. The Dywidag system found wide use in Germany. Its inventor, Ulrich Finsterwalder (b. 1897), had participated in the construction of the first prestressed concrete bridge at Aue as Dischinger's student. Little remained, however, from Dischinger's original system. Although bars were still used for the tendons, they were smaller (26 mm diameter), stronger (tensile strength 900 N/mm^2), and bonded to the concrete after stressing. Leonhardt and his partner W. Baur developed the Leoba prestressing system. He used strands of high strength steel wires (tensile strength 1800 N/mm^2), placed into ducts and wrapped around concrete stressing blocks at the ends of the bridge. The success of the Freyssinet, Dywidag, and Leoba systems in bridge construction was instrumental in removing any remaining doubt about the merits of prestressing.

Belgium's Gustave Magnel (1889–1955) was a great disseminator of prestressing technology. In addition to conducting fundamental theoretical research, he developed his own prestressing system. His Maas River Bridge, built in 1949, had two main spans of 68 m. It was the world's first prestressed continuous girder bridge.

In Switzerland, Max Ritter and Pierre Lardy conducted their first investigations with pretensioned girders in 1941. The country's first prestressed concrete bridge, a small railway grade separation designed by A. Panchaud, was built in 1943 in Fribourg. The structure used precast prestressed concrete girders. The Stahlton firm, founded by engineers M. Birkenmaier, A. Brandestini and M.R. Roš, had a decisive influence on the rapid dissemination and further development of prestressed concrete. Their BBRV prestressing system, developed in 1948, was carefully detailed and highly reliable. It used high-strength wires (tensile strength 1600 N/mm^2) which were individually anchored by means of cold-formed buttonheads. The system quickly found worldwide acceptance. The first small bridges using this system were built in 1950 and 1951. Switzerland's first major prestressed concrete bridge was the 360 m long Weinland Bridge, completed in 1958, which had a main span of 88 m.

The Germans Finsterwalder and Leonhardt acquired leading roles in prestressed concrete bridge construction in the early 1950s. Finsterwalder was the first to recognize that the balanced cantilever method could be used to advantage in the construction of prestressed concrete bridges. His first cantilever constructed bridge was the Lahn Bridge at Balduistein, built in 1951, which had a main span of 62 m. His excellent experience with this structure lead, only two years later, to the construction of the Nibelungen Bridge at Worms, the first prestressed concrete bridge over the navigable portion of the Rhine. The structure consists of three spans of 101.65 m, 114.2 m, and 104.2 m (fig. 1.40). This bridge attracted worldwide attention and cleared the way for the first longspan bridges in prestressed concrete.

Figure 1.40
Nibelungen Bridge, Worms, Germany

Leonhardt's 1954 Untermarchtal Bridge over the Danube, a 334 m-long five-span continuous girder, was the forerunner of another innovation in construction technology (fig. 1.41). To enable falsework and formwork to be used twice, the girder was cast in two sections. This idea formed the foundation of the span-by-span construction method for continuous girders. Prestressing companies responded to this innovation by developing devices to couple tendons at construction joints between segments. New prestressing systems also appeared at this time, for example the German Polensky and Zoellner system and the Swiss VSL system. The latter was originally derived from the Freyssinet system in 1954, but was later completely modified for seven-wire strand.

Figure 1.41
Untermarchtal Bridge over the Danube, Germany

It became increasingly clear that Freyssinet's philosophy of full prestressing was overly restrictive and uneconomical. In 1946, P.W. Abeles (1897–1977) discussed the advantages of reducing the prestress in the tendons and of combining prestressed and nonprestressed tendons. His ideas were forerunners of the modern concept of *partial prestressing*. His investigations showed that full and partially prestressed beams had approximately equal load-carrying capacity, provided the yield force of the steel remained constant. In the 1950s, several countries relaxed their design standards for prestressed concrete. Germany, for example, introduced "limited prestressing", by which concrete tensile stresses up to roughly $4 \, \text{N/mm}^2$, depending on the concrete strength, were permitted. Mild reinforcement was to be provided to resist the tensile stresses. This steel could then be considered in the calculation of ultimate resistance. The modern concept of partial prestressing was introduced in Switzerland in 1968. Calculation of concrete tensile stresses under service loads was no longer required; resistance at ultimate limit state was calculated considering both prestressing steel and mild reinforcing steel.

Full prestressing is often impractical and costly. Supplementing the longitudinal prestressing with mild reinforcement, for example, was a preferable alternative to full prestressing for the four-span skew bridge shown in figure 1.42, built in Maienfeld, Switzerland, in 1961. The 1962 Reichenau Bridge over the Rhine, also in Switzerland, was the first arch bridge with a partially prestressed stiffening girder (fig. 1.43). Here, the combination of mild and prestressed reinforcement also proved preferable to full prestressing. The stiffening girder was given a relatively weak concentric prestress; mild reinforcement was added as required to resist the live load moment peaks, which varied considerably from span to span.

Increases in labour costs in the 1960s led to simplification and rationalization in bridge construction technology. Arches and other systems with complicated, labour-intensive falsework and formwork became less competitive and were only built in exceptional cases. Precast girders made possible the rapid and economical

Figure 1.42
Skew railway bridge, Maienfeld, Switzerland

Figure 1.43
Bridge over the Rhine at Reichenau, Switzerland

construction of short-span bridges. This construction method was used for many of bridges, especially in Italy (fig. 1.44). The large number of joints and bearings required, however, reduced durability and user comfort. Bridges built from precast girders were later improved by connecting the individual spans into continuous girders.

Innovations in falsework technology resulted in significant cost savings for long cast-in-place bridges. A two-phase form girder was used for the first time in the construction of the Kettiger Hang Highway Bridge in Germany, completed in 1960. In 1961, Hans Wittfoht developed a single-phase form girder for the 1100 m long Krahnenberg Bridge (fig. 1.45). Wittfoht's concept served as the basis for many subsequent mechanized falsework systems. Leonhardt proposed the incremental launching technique for the construction of the Rio Caroni Bridge in Venezuela, completed in 1964 (fig. 1.46). All casting of concrete was done behind

Figure 1.44
Multiple-span viaduct with precast beams, Italy

Figure 1.45
Krahnenberg Bridge, Germany

Figure 1.46
Rio Caroni Bridge, Venezuela

the abutments; the completed segments were prestressed and pushed into position using hydraulic jacks, thus eliminating falsework.

For long spans, cantilever construction had no serious competition. Finsterwalder's cantilever-constructed Bridge over the Rhine at Bendorf, Germany was the first prestressed concrete span greater than 200 m (fig. 1.47). Wittfoht developed a variation of this method for the construction of the Siegtal Bridge in Germany (fig. 1.48), by which a launching girder was used to hang the formwork for the segments, to transport materials, and to stabilize the girder.

The first cantilever constructed bridges using precast segments were built in the early 1960s in the Soviet Union and France. The detailing of the joints between the segments proved to be a particularly difficult problem. The 148 m span of the bridge at the Lichačer Factory in Moskow was designed with 0.2 m thick cast-in-place concrete joints between the segments, at which longitudinal reinforcing steel was lap spliced. Due to considerable delays in construction, however, continuous mild reinforcing steel was abandoned in subsequent bridges. Another Russian structure, the Irtysch Bridge (110 m main span), used 20 mm thick mortar joints; dry joints were used for the Ojat Bridge (64 m main span), erected in winter.

In France, the erection of precast segments was known since the construction of Freyssinet's Marne bridges. French engineers developed new solutions to the problems of mortar joints, which slowed construction, and dry joints, which

Figure 1.47
Bridge over the Rhine at Bendorf, Germany

Figure 1.48
Siegtal Bridge, Germany

Figure 1.49
Oléron Bridge, France

complicated the grouting of tendons. The Choisy-le-Roi Bridge over the Seine, completed in 1965 (span lengths 37.5 m, 55.0 m, and 37.5 m) used 1 mm thick epoxy-coated segment joint faces. Similar joint details were also used in 1964 for the 3 km bridge linking the mainland of western France to Oléron Island (fig. 1.49). Precast segments were transported over the already completed portion of the bridge and erected with a launching truss. This system, which enabled the erection of eight 3 m long segments per day, was successfully used on several other bridges, including the Chillon Viaducts in Switzerland and the Rio-Niteroi Bridge in Brazil. The former, completed in 1967, was 2.1 km long, sharply curved in plan, and had spans varying from 96 m to 106 m (fig. 1.50). The latter, completed in 1973, was a 6 km long prismatic girder with 80 m spans (fig. 1.51).

By the beginning of the 1960s, the Dutch were leaders in the erection of heavy elements with ship-mounted cranes, based on their experience with large sea-works. The 5 km long Oosterschelde Bridge was built in record time between 1962 and 1965 (fig. 1.52). The entire bridge, including foundations and piers, consisted of precast components weighing up to 600 tonnes. The segments used in the cantilever construction of the 95 m long spans were 12.5 m long and were connected with 0.4 m thick concrete joints. Construction of the superstructure reached an average rate of 3500 m² per month. The Dutch again demonstrated their leadership in the erection of large precast segments in the construction of the Saudi Arabia-Bahrain Causeway (fig. 1.53), for which complete spans weighing up to 1 400 tonnes were placed from a floating crane. The causeway, completed in 1985, included five bridges with lengths 934 m, 2034 m, 5194 m, 3334 m, and 934 m.

Figure 1.50
Chillon Viaduct, Switzerland

Figure 1.51
Rio-Niteroi Bridge, Brazil

In England and in Australia, segmental bridges were constructed by erecting precast segments onto falsework and prestressing them together. The 100 mm wide joints between segments were filled before stressing with all-fines concrete. This method was used, for example, for the 600 m long Hammersmith Flyover in London, completed in 1961.

A different method of reducing falsework and formwork costs was used in the construction of the Birs Bridge in Basel, Switzerland. The 25 m wide, single-cell

Figure 1.52
Oosterschelde Bridge, Netherlands

Figure 1.53
Saudi Arabia-Bahrain Causeway

Figure 1.54
Felsenau Bridge, Berne, Switzerland

box cross-section was built by casting first the box portion on falsework and later the wide deck slab cantilevers with a special traveller. The same method was used in the 1972–1974 construction of the approach spans of the Felsenau Bridge in Berne (fig. 1.54) and in the construction of the Lake Gruyère Viaduct, completed in 1978 (fig. 1.55). The longitudinal prestressing of both bridges was designed to permit removal of falsework after the box portion of the section had been cast and stressed.

The 8678 m long Lake Maracaibo Bridge in Venezuela, completed in 1962, was the first concrete cable-stayed bridge (fig. 1.56). The designer, Riccardo Morandi, provided cables at the third points of the girder to support the 235 m-long spans over the navigation channel. The portion of the girders extending from the towers to the cable anchors was cast in place on falsework, prestressed, and then attached to cables. The central portion, designed as a suspended span, consisted of four slender precast T-girders. The stay cables were encased in concrete and post-tensioned. Morandi sought to achieve protection for the cables and greater stiffness in this way.

The Waal Bridge in Tiel, Holland (1972) has a 267 m main span (fig. 1.57). The stays for the central span consist of two pairs of two cables, which were also encased in concrete. The construction of the twin box girder was relatively complicated and costly. It was built using cantilever construction out to the

Figure 1.55
Lake Gruyère Viaduct, Switzerland

second stay cable anchor point (95 m from the towers) using temporary stays. The 77 m long central portion was precast, floated to the bridge, and lifted into place from the girder.

The Lake Maracaibo Bridge and the Waal Bridge are typical examples of the first generation of concrete cable-stayed bridges, which used a small number of cables and a stiff girder. The Brotonne Bridge over the Seine, completed in 1976, was the first long-span example of the second generation, characterized by a large number

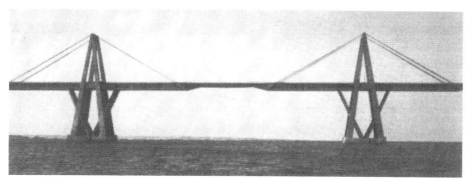

Figure 1.56
Lake Maracaibo Bridge, Venezuela

Figure 1.57
Waal Bridge, Tiel, Netherlands

Figure 1.58
Brotonne Bridge, France

of closely spaced cables and a slender girder (fig. 1.58). The main span is 320 m long. The cables were enclosed in ducts and grouted. The single plane of cables resulted in savings in the towers, yet required a torsionally stiff box section for the deck girder. Maintaining a constant angle of inclination for the stays simplified detailing and construction, but increased steel consumption in the stays and bending stresses in the slender tower. The close spacing of the stay cables permitted the economical construction of the girder using the balanced cantilever

Figure 1.59
Pasco-Kennewick Bridge, U.S.A.

method; a typical segment consisted of precast webs and cast-in-place top and bottom slabs.

Another second-generation structure is the Pasco-Kennewick Bridge over the Columbia River in the U.S.A., completed in 1978 (fig. 1.59). Several important differences can be observed between this bridge and the Brotonne Bridge. The stay cables have been arranged in two planes. Concentrating the anchors at the tops of the towers minimizes tower bending stresses, but created difficulties for detailing and construction. The girder cross-section, which consists of a deck slab and small triangular boxes at either edge, behaves structurally as a double-T girder. The girder is continuously supported by the cables, with no fixed connection to the towers. It was built using the balanced cantilever method. Match-cast precast segments, 8.2 m in length, were floated into position and lifted from the completed portion of the superstructure. The joints between segments were coated with epoxy; small recesses were provided in the concrete section to weld the longitudinal mild reinforcement at the joints.

The design, details, and construction methods used for the Brotonne Bridge and the Pasco-Kennewick Bridge guided the development of concrete cable-stayed bridges into the 1980s. Although most economical for long spans, cable-stayed bridges have been recently been built economically for spans of less than 150 m.

At the present time, the longest-spanning concrete arch, girder, and cable-stayed bridges are as follows:

The Tito Bridge, which links the island of Krk to the Yugoslavian mainland, has an arch span of 390 m (fig. 1.60). It was built by erecting precast box segments

Figure 1.60
Tito Bridge, Island of Krk, Yugoslavia

using the cantilever method. The segments were transported from land to the bridge by an aerial cable. The structural system during construction was a cantilever truss formed by the arch, columns, deck girder, temporary diagonal tension members, and temporary tension ties anchored into rock at either end of the bridge. After closure at the crown, the diagonals and ties were removed. The bridge was opened to traffic in 1980. An unlikely combination of favourable geological, topographical, and economic conditions led to the choice of an arch for this crossing. It is therefore improbable that the record span of this technologically and aesthetically impressive bridge will ever be broken.

Figure 1.61
Gateway Bridge, Brisbane, Australia

Figure 1.62
Barrios de Luna Bridge, Spain

The Gateway Bridge in Brisbane, Australia is the longest spanning cantilever-constructed girder bridge, with a 260 m span (fig. 1.61). It was opened to traffic in 1986. The total height of the structure was limited to 24.7 m, due to its proximity to an airport. The 22 m wide single cell box girder increases in depth from 5.2 m at midspan to 15.0 m at the piers. The choice of 750 mm thick webs, reinforced with vertical stirrups, is unfortunate; their thickness could have been reduced by almost 50 percent if inclined web reinforcement, which can be easily accommodated in cantilever construction, had been used. If not for the severe height restriction, a cantilever-constructed girder bridge would probably not have been built at this site; cable-stayed bridges are more economical and more elegant for spans of this range. The record of the Gateway Bridge is also, therefore, unlikely to be exceeded.

The Barrios de Luna Bridge in Spain (fig. 1.62) is currently the longest spanning concrete cable stayed bridge, with a main span of 440 m. The stays are arranged in two planes. Due to the short side spans, backstays anchor the tower into the rock; an expansion joint is provided at the midpoint of the main span. The three-cell box girder was cast in place using cantilever construction. The bridge was opened to traffic in 1984. Since the 440 m span of this bridge is still far from the limit of economical feasibility for concrete cable-stayed bridges, it is likely that the record span of the Barrios de Luna Bridge will be exceeded in the future.

2 Economy and Aesthetics

2.1 Design Objectives

The fundamental objectives of bridge design are *safety*, *serviceability*, *economy* and *elegance*. A design can be considered successful only when all four of these goals have been achieved. The relative importance of the objectives is defined by the consequences arising when they are not achieved. These vary from the unpleasant feelings evoked by ugly bridges to the loss of life and property caused by unsafe bridges. The order in which the objectives have been listed above can thus be regarded as hierarchical, beginning with safety as most important.

Safety and serviceability are achieved through the systematic application of scientific principles. They thus depend on the analytical skill of the engineer. The criteria used to determine whether these objectives have been met are codified in design specifications and standards. Proficiency in designing safe and serviceable bridges can be acquired through an understanding of the underlying scientific principles, which are presented in detail in chapters 3 through 8.

Economy and elegance, on the other hand, are achieved through nonscientific means. They depend almost entirely on the creativity of the engineer. Economic and aesthetic criteria have not been codified and are largely subjective. Useful guidelines are available to help in improving the cost-effectiveness and visual form of bridges. Proficiency in designing economic and aesthetically pleasing bridges can nevertheless be acquired only through direct design experience, critical observation of completed structures, and full utilization of the engineer's creative talents.

Visual elegance and economy are to some extent interdependent. Aesthetically pleasing bridges are distinguished by transparency, slenderness, and the lack of unnecessary ornamentation, all of which result in an efficient use of materials and hence low construction cost. It is incorrect to infer, however, that the most economical design is necessarily the most elegant.

In addition to meeting the above objectives, designs must also conform to constraints that arise in special situations, such as:

1. Restrictions on available construction time
2. Restrictions on the location of piers in watercourses
3. Environmental protection regulations
4. Noise level restrictions
5. Traffic safety along existing roads

2.2 Economy

2.2.1 Life-Cycle Costs

The cost-effectiveness of bridges cannot be judged on the basis of construction cost alone. Bridge costs are best compared on the basis of *life-cycle cost*, defined as the total cost of construction, operation, amortization, and demolition, including the costs and benefits arising from changes in existing traffic patterns. Attempts to reduce the consumption of construction materials through optimization of span lengths and cross-section dimensions will have little effect on the total life-cycle cost. Cost-effectiveness is rather a function of overall concept, characterized by a properly chosen structural system, cross-section, foundation system, and construction sequence.

Operating expenses are incurred as a result of annual inspection, annual maintenance, and periodic rehabilitation. Yearly operating costs can be calculated by expressing the cost of rehabilitation as an equivalent annual expenditure. Total yearly operating costs of highway bridges are given as a percentage of the construction cost in table 2.1. These figures are valid for bridges that have been designed and constructed to minimize life-cycle costs; annual operating expenditures will be higher for bridges that have been designed to minimize construction cost.

Bridges are removed from service as a result of changes in the transportation system, increases in legal live loads beyond the capacity of the bridge, or excessive maintenance and rehabilitation costs. For planning purposes, the lifetime of a bridge is normally assumed equal to 100 years.

Table 2.1
Annual Operating Costs of Highway Bridges

Item	Annual Cost (% of Construction Cost)
Inspection	0.1
Maintenance	0.5
Rehabilitation (averaged over the lifetime of the bridge)	0.4 to 0.6
Total	1.0 to 1.2

The effect of interest and inflation rates on life-cycle cost is investigated in the following example:

Example 2.1:
Life-cycle costs

Table 2.2 gives the construction costs and annual operating expenditures for two similar bridges, denoted Bridge I and Bridge II. Bridge I has a slightly higher construction cost, lower annual costs, and a longer service life. It is assumed the costs of demolition and the costs and benefits due to use are identical for both bridges. The life-cycle costs of both bridges are compared 80 years after construction. The salvage value of Bridge I after 80 years of service is assumed equal to 20 percent of its construction cost.

Table 2.2
Basic Costs for Bridges of Example 3.1

Bridge	Construction Cost (units)	Annual Costs (units)	Lifetime (years)
I	1000	$(1.0\% \times 1000 =)$ 10.0	100
II	900	$(1.2\% \times 900 =)$ 10.8	80

It is first assumed that the interest rate is equal to 5 percent and the inflation rate is equal to 4 percent. The present value (at time of construction) of all costs is compared for both bridges in table 2.3. On this basis, Bridge II is 2 percent more expensive than Bridge I.

Table 2.3
Present Value of Life-Cycle Costs: 5% Interest, 4% Inflation

Item	Bridge I	Bridge II
Construction cost	1000	900
Salvage value	−90	0
Operating costs	549	592
Total	1459	1492

Different life-cycle costs are obtained when the inflation rate is reduced to 1 percent for construction costs and 2 percent for operating costs, and interest is maintained at 5 percent. The present value of all costs is compared in table 2.4. Under these assumptions, Bridge I is 5 percent more expensive than Bridge II.

Table 2.4
Present Value of Life-Cycle Costs: 5% Interest, 1% Inflation for
Construction Cost, 2% Inflation for Operating Costs

Item	Bridge I	Bridge II
Construction cost	1000	900
Salvage value	−9	0
Operating costs	302	326
Total	1293	1226

2.2.2 Construction Costs

The total construction cost of a concrete bridge can be broken down into the
following structure:

1. Mobilization

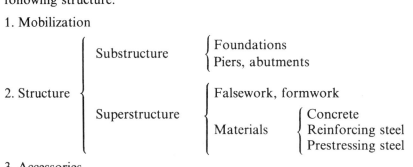

3. Accessories
4. Design and construction management

Mobilization is defined as the work required before construction can begin, for
example providing access to the construction site, preparation of site facilities, and
procurement of equipment. Accessories include bearings, expansion joints,
drainage system, guardrails, deck waterproofing system, and wearing surface.

Estimates of construction costs should be based on cost records of completed
structures. Average costs of mobilization, structure, and accessories have been
calculated from a sample of 19 concrete highway bridges built in Switzerland
between 1958 and 1985 (Menn 1986, 77). Bridges 1 through 4 are elevated
highways in urban areas, bridges 5 through 11 are viaducts in mountainous
terrain, and bridges 12 through 19 are valley crossings. For the remainder of this
section, total construction cost is defined as the sum of the costs of mobilization,
structure, and accessories.

Figure 2.1 shows the total construction cost of each bridge broken down into
components due to mobilization, structure, and accessories. The average costs,
denoted $\bar{X}$, are as follows:

1. Mobilization: 8%
2. Structure: 78%
3. Accessories: 14%

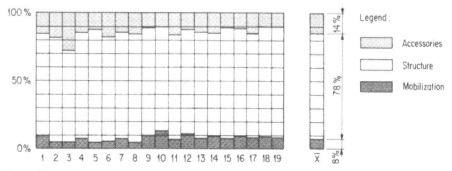

Figure 2.1
Costs of mobilization, structure, and accessories as percentages of total construction cost

Structure costs can be subdivided into the costs of superstructure and substructure. Due to uncertainties in geotechnical and hydraulic data, foundations are usually designed more conservatively than other structural components. Even under very unfavourable foundation conditions, however, substructure costs are usually considerably less than superstructure costs. Figure 2.2 shows the contribution of superstructure and substructure costs to the cost of the structure. The average costs are:

1. Substructure: 30%
2. Superstructure: 70%

The substructure consists of foundations, piers, and abutments. Abutment costs, which are relatively small, are normally included with pier costs. Foundation costs are a primarily function of geotechnical and hydraulic conditions; pier costs are primarily a function of bridge height. Since these factors vary considerably from bridge to bridge, so will the ratio of foundation cost to pier cost. The average costs of foundations and piers, expressed as percentages of substructure cost, structure cost, and total bridge cost are given in table 2.5.

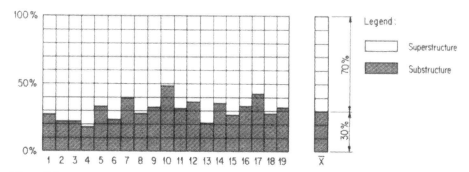

Figure 2.2
Costs of superstructure and substructure as percentages of structure cost

Table 2.5
Substructure Costs

	Piers and Abutments (%)	Foundations (%)	Total Substructure (%)
Substructure cost	24.0	76.0	100.0
Structure cost	7.0	23.0	30.0
Construction cost	5.5	18.0	23.5

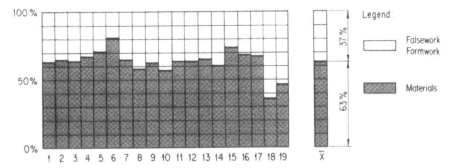

Figure 2.3
Costs of falsework, formwork, and materials as percentages of superstructure cost

Superstructure costs are divided into the cost of falsework and formwork and the cost of materials. The installation and removal of falsework and formwork are labour-intensive operations. During the past 25 years, increases in wages have kept pace with cost savings achieved through mechanization. As a result, the cost of falsework and formwork has changed little relative to the cost of materials. Falsework and formwork costs can be minimized by the proper choice of structural system, cross-section, and construction sequence. Figure 2.3 shows falsework and formwork costs and material costs for each bridge in the sample, expressed as percentages of superstructure cost. Table 2.6 summarizes the average costs as percentages of superstructure cost, structure cost, and total construction cost.

Table 2.6
Superstructure Costs

	Falsework and Formwork (%)	Materials (%)	Total Superstructure (%)
Superstructure cost	37.0	63.0	100.0
Structure cost	26.0	44.0	70.0
Construction cost	20.0	34.5	54.5

Figure 2.4
Costs of concrete, reinforcing steel, and prestressing steel as percentages of material cost

The relative costs of concrete, reinforcing steel, and prestressing steel are functions of span length and cross-section dimensions. Increasing the thickness of the slabs and webs will increase the consumption of concrete and longitudinal reinforcement and reduce the consumption of transverse reinforcement. It is therefore cost-effective to use relatively thick slabs and webs for short span bridges. Slender, heavily reinforced cross-sections are preferable for longer spans, for which a reduction in dead load is desirable. Figure 2.4 shows concrete, reinforcing steel, and prestressing steel costs for each bridge in the sample, expressed as percentages of material cost. Table 2.7 summarizes the average costs as percentages of material cost, superstructure cost, structure cost, and total construction cost.

The average costs of concrete, reinforcing steel, and prestressing steel are roughly equal. Detailing considerations normally govern the selection of cross-section dimensions, and hence the consumption of concrete. Between 60 and 65 percent of the reinforcing steel is normally required for stability of the assembled cage of bars during concreting and for crack control in the finished structure. Only the remaining 35 to 40 percent of reinforcing steel and the prestressing steel can therefore be reduced through calculation. Reducing this steel by 10 percent would decrease the total construction cost by less than 2 percent. Substantial savings, therefore, cannot be achieved by refined calculations alone. The time and effort required for refined calculations is preferably invested in the detailing and layout of the reinforcing and prestressing steel.

Table 2.7
Material Costs

	Concrete (%)	Reinforcing Steel (%)	Prestressing Steel (%)	Total Materials (%)
Material cost	29.0	39.0	32.0	100.0
Superstructure cost	18.0	25.0	20.0	63.0
Structure cost	12.5	17.0	14.5	44.0
Construction cost	10.0	13.3	11.2	34.5

Table 2.8
Breakdown of Average Construction Costs

Item	Cost (% of Total Construction Cost)		
Mobilization			8.0
Structure			
Substructure			
Foundations	18.0		
Piers and abutments	5.5		
Total substructure	23.5	23.5	
Superstructure			
Falsework, formwork	20.0		
Concrete	10.0		
Reinforcing steel	13.3		
Prestressing steel	11.2		
Total superstructure	54.5	54.5	
Total structure		78.0	78.0
Accessories			14.0
Total construction cost			100.0

The average costs obtained from the sample of 19 bridges are summarized in table 2.8. Although the sample was chosen to represent a wide variety of conditions, the most important cost factors do not vary significantly from one bridge to another. Table 2.8 can therefore be useful in preparing preliminary cost estimates.

2.2.3 Preliminary Estimates of Superstructure Costs

Superstructure costs can be reliably estimated with the help of the *geometrical average span length* l_m, defined by the following equation:

$$l_m = \frac{\sum l_i^2}{\sum l_i}$$

The summation is over the total number of spans; the length of span i is denoted l_i. Empirical equations giving the quantities of concrete, reinforcing steel, and prestressing steel as functions of l_m have been derived from a sample of recently constructed bridges. These approximate expressions are presented in this section.

The volume of concrete in the superstructure is obtained by multiplying the total deck surface by the effective girder depth h_m, defined by the following expression:

$$h_m = 0.35 + 0.0045\, l_m$$

(The parameters h_m and l_m are in metres.) This equation is valid provided the actual girder depth, h, satisfies the following inequality:

$$\frac{1}{20} \leq \frac{h}{l_m} \leq \frac{1}{16}$$

The quantity of reinforcing steel is obtained by multiplying the total volume of concrete by the mass of steel per unit volume of concrete, m_s. The parameter m_s is estimated using the equation

$$m_s = 90 + 0.35\, l_m$$

where l_m is in metres and m_s is in kilograms per cubic metre of concrete (kg/m^3). This expression is valid provided the deck slab is not transversely prestressed. Between 65 and $70\,kg/m^3$ of reinforcement is required for stability during construction and crack control; this quantity is independent of span length. The transverse reinforcement required to resist loads is primarily a function of cross-section dimensions. An additional 20 to 25 kg/m^3 is required for commonly used cross-sections, regardless of span length. Most of the steel required above the minimum 65 to 70 kg/m^3 is located in the deck slab. The deck slab should therefore be the focus of attention in the design and arrangement of the superstructure reinforcement.

The mass of prestressing steel per unit volume of concrete, m_P, is a function of span length and construction method. For girders that are cast on conventional falsework, m_P is estimated using the equation

$$m_P = 0.4\, l_m \tag{a}$$

where l_m is in metres and m_P is in kilograms per cubic metre of concrete. This expression is valid for girders that are not transversely prestressed. The quantity of prestressing steel is obtained by multiplying m_P by the total volume of concrete.

Equation (a) can also be used to compute m_P for incrementally launched bridges. The launching procedure induces high stresses at all locations along the length of the girder. Incrementally launched structures consequently require a heavier cross-section ($l_m/h \cong 14$) and more prestressing than a conventionally constructed bridge with identical span lengths. The ratio of steel mass to concrete volume, however, remains approximately equal for both construction methods.

Prestressing steel in cantilever-constructed bridges is arranged very closely to the moment diagram due to loads. The parameter m_P will therefore be somewhat less than for conventional girders:

$$m_P = 0.35\, l_m$$

where l_m is in metres and m_P is in kilograms per cubic metre of concrete.

The estimated costs of concrete, reinforcing steel, and prestressing steel in the superstructure are obtained by multiplying the estimated quantities by unit material costs. The cost of falsework and formwork should be estimated taking into account the proposed construction sequence; if it is greater than 65 percent of the superstructure material costs, another construction method should be considered. Adding the material, falsework, and formwork costs yields the total superstructure cost. The remaining costs can be estimated using table 2.8.

2.3 Aesthetics

A bridge can be perceived as an independent entity or as an element of a larger landscape. Elegance in bridge design can thus be considered as a function of both abstract structural form and the relationship between structural form and environment. These two aspects of bridge aesthetics are often independent of each other. Structures that are aesthetically pleasing as independent objects are not always suited to their surroundings. Conversely, integration into the environment may be achieved in spite of shortcomings in structural form. It is therefore important that neither aspect be neglected by the designer.

The appearance of a proposed design must be evaluated from all possible viewpoints. The use of a large-scale model or three-dimensional computer graphics is strongly recommended; two-dimensional orthographic views alone are insufficient. The relative importance of the viewpoints should be considered. Frequently occurring views are normally more important than those that occur infrequently. The appearance of the bridge as seen from the most important viewpoints should be designed and evaluated with special care.

a) Abstract Structural Form

Aesthetically pleasing structural forms can be characterized in terms of efficiency, harmony, and artistic shaping.

Efficiency. The technical and aesthetic aspects of bridge design are closely related through the concept of efficiency. Our perception of elegance in bridges has been conditioned by familiarity with structures in the natural world, where beauty and efficient use of materials are inseparable. From Roman times to the present day, bridges that have achieved renown for their elegance have almost without exception been remarkable for their efficient use of materials. The visual expression of efficient structural function is thus a fundamental criterion of elegance in bridge design. It is one of the primary distinguishing factors between structural engineering art and architecture.

The role of efficiency in bridge aesthetics is illustrated in figure 2.5. It is apparent that the bridge of figure 2.5a, a slender inclined-leg frame, requires considerably less concrete than the girder and retaining walls of figure 2.5b. The former

a

b

Figure 2.5
Visual expression of efficient structural function: a, efficient structure; b, inefficient structure

structure thus accomplishes the same function as the latter with a more efficient use of materials. The massive, heavy appearance of the bridge of figure 2.5b is directly related to its lack of efficiency. Its counterpart appears much lighter in comparison.

Two of the most important ways of expressing efficiency are transparency and slenderness.

Bridges that lack a suitable degree of transparency appear as solid walls from a wide range of viewing angles. For exceptionally low bridges, transparency is a function of girder depth. For most bridges, however, transparency is a function of the number and width of the columns (fig. 2.6). It is most effectively enhanced by reducing column width; even long bridges with many short spans can be given an adequate degree of transparency provided the columns are suitably narrow.

Maximum transparency is achieved using one column per support axis. Round columns, which are most effective in this regard, often appear to lack lateral stability and thus create a disturbing impression. Rectangular columns are a reasonable compromise between transparency and perceived stability. Single columns can be used regardless of bridge height provided the width of the superstructure, B, is less than 12 m. Column width, b, should be chosen to ensure that the ratio B/b is between 3:1 and 3.5:1. Single columns can also be used for wider bridges ($B > 12$ m), provided the bridge is sufficiently tall and the superstructure consists of a single cross-section. For these bridges, the ratio B/b should be chosen between 3.5:1 and 4.0:1.

Two columns per support axis must normally be used for low, wide bridges ($B > 12$ m) and for all twin bridges. The columns should be slender and have a compact cross-section, for example a circle or a flattened hexagon. The lateral spacing between columns should yield a balanced moment diagram in the superstructure diaphragm. The use of three or more columns per support axis severely reduces transparency and is therefore not recommended.

The transparency of long bridges is drastically reduced by hammerhead columns and multiple-column bents. Their use should therefore normally be avoided. Two-

a

b

Figure 2.6
Transparency and slenderness: a, transparency is enhanced by single columns; slenderness is a
function of bridge length and girder depth; b, transparency is somewhat reduced by twin, wide
columns; slenderness is a function of bridge length and girder width

column frames have been successfully used, however, for bridges with a small number of relatively long spans. An acceptable degree of transparency is obtained provided the individual spans can be recognized.

Slenderness is primarily a function of the superstructure arrangement. It is normally expressed quantitatively in terms of an *effective slenderness* parameter λ, defined as the ratio of span length to girder depth. The parameter λ is most useful as a rough measure of the relative economy of projects. The most economical superstructure is obtained when λ is chosen between 15 and 17. Although girders with effective slenderness ratios as high as 30 are possible, they are considerably more expensive.

The parameter λ is not, however, a reliable measure of visual slenderness, which is a function of the *visually apparent* superstructure dimensions. Visual slenderness can be defined as the ratio of the perceived uninterrupted length of the superstructure to the perceived superstructure dimension perpendicular to length. Depending on the location of the observer relative to the bridge, this dimension can be either girder depth or girder width (fig. 2.6). Span length therefore plays a subordinate role in the visual perception of slenderness, provided the continuity of the superstructure across the intermediate supports is not interrupted.

Apparent superstructure depth is of greatest significance in connection with low bridges. It is also important for short high bridges viewed from afar. Long high bridges always appear slender when observed from a distance, regardless of depth. Apparent depth can be effectively reduced through the use of wide deck slab cantilevers. The shadows they cast onto the girder webs create contrasting parallel strips of light and dark that accentuate the long dimension of the superstructure. Varying the depth of the girder can also be beneficial. The visual slenderness of a long three-span bridge, for example, can be substantially improved by haunching the main span and tapering the side spans. Haunched rigid-frames are preferable to low, single-span simple-beam bridges, which appear heavy for values of λ as high as 25. The interaction of superstructure depth and bridge height must also be considered; an adequate degree of transparency and slenderness cannot be achieved unless the ratio of bridge height to girder depth is greater than 4.

Apparent superstructure width is primarily of importance for the visual slenderness of high bridges. The critical viewpoint is relatively close to the structure; the critical direction of view is upward at a slightly oblique angle. Superstructure width is normally fixed by traffic requirements. The visually apparent width can, however, be reduced by an appropriately chosen cross-section. Single-cell box girders with wide deck slab cantilevers are particularly well-suited in this regard.

Harmony. All of the components of a bridge should be harmoniously integrated into one coherent, organic entity. This is accomplished by both visual means (providing symmetry, order, and regularity) and technical means (properly defining the structural function of each component).

Our perception of harmony in bridges, like our perception of efficiency, has originated from familiarity with naturally occurring structural forms which have grown using minimum energy and materials. These forms invariably possess symmetry; the visual impression of balance and stability they create is always matched by a state of stable equilibrium and low stress. A symmetrical structural system forms the basis of almost all of the bridges that have been acclaimed for their elegance, from Roman times to the present day.

The concept of order is related to the orientation and arrangement of bridge components, many of which can be considered one-dimensional. The number of different inclination angles of similar structural members should be as small as possible, unless the members form a regular, gently curved envelope. Otherwise, the structure may appear ambiguous or unstable from certain viewing angles. Twin bridges that are parallel in plan should have identical longitudinal grades. If possible, the roadways should be arranged so that only one of the two bridges is visible from the most important viewing location.

Both the span arrangement and the cross-section must possess a high degree of regularity. Spans of equal length and a constant cross-section are thus desirable from a visual point of view. They have the added advantage of requiring the least amount of material and producing the most favourable construction conditions. Bridges of varying height should, however, be given spans of varying length. The ratio of span length to bridge height can thus be maintained constant, which results in a more balanced appearance than would be obtained using spans of equal length.

Artistic shaping. The raw structural form required for safety, serviceability and economy is rarely the most elegant. It can normally be refined into an elegant form, however, through artistic shaping of the structural members. The associated additional cost is insignificant. Shaping that follows the flow of internal forces is recommended in most cases (fig. 2.7a). Forms that disregard the flow of forces normally produce a chaotic effect and should thus be avoided. In the hands of a gifted designer, however, member shaping based on purely aesthetic considerations and ornamentation can produce particularly charming results (fig. 2.7b).

b) Structure and Environment

Bridge aesthetics should be considered in the design of highway and railway alignments. Topographical features that enhance the appearance of bridges should be identified and, if possible, incorporated into the alignment. Viaducts along mountain slopes should be gently curved to follow the contours of the landscape (see, for example, figure 1.50).

The character of the landscape should be reflected in the structural form. A subdued form is preferable in flat or gently rolling terrain that lacks conspicuous topographical features. Prominent obstacles such as wide rivers or deep canyons

Figure 2.7
Artistic shaping of structural members: a, derived from the flow of internal forces; b, derived from purely aesthetic considerations

a

b

are best crossed by structures in which one span has been given particular visual emphasis. Bridges in built-up areas should normally be unobtrusive. In certain cases, however, a positive visual effect can be achieved by a structure that stands out from an urban environment.

c) Aesthetics and Economy

Any modification of the structural form made to improve appearance will be reflected in the total construction cost. Due to the relation between aesthetics and efficient use of materials, some modifications may result in cost reductions. Others, such as the artistic shaping of members, may increase the total cost. The combination of all structural modifications made for aesthetic reasons, excluding increases in span length, will normally result in a net change in construction cost of no more than about 2 percent.

Cost increases can be substantial, however, if span lengths are increased for aesthetic reasons. Any savings in substructure costs that may result are usually outweighed by increases in superstructure costs. The most economical span length is relatively short, and appears mediocre and overly cautious in most cases. Longer spans, on the other hand, substantially enhance transparency and convey an impression of efficiency and boldness of conception; the overall visual impression is greatly improved provided the spans remain in proportion to the surrounding landscape. It is therefore recommended that spans slightly longer than the economical minimum be provided, especially for prominently exposed bridges. A cost increase of up to about 7 percent of the cost of the most economical solution should be allowed for this purpose.

Reference

Menn, C. 1986. *Stahlbetonbrücken* (Reinforced concrete bridges). Vienna and New York: Springer-Verlag

3 Materials and Actions

3.1 Materials

3.1.1 Concrete

a) Quality

Concrete must have sufficient compressive strength to make possible a state of equilibrium between stresses in the concrete and reinforcement on the one hand and sectional forces due to external loads on the other. It must also protect the embedded reinforcing and prestressing steel against damage due to corrosion, fire, and vehicle impact, which can severely impair structural capacity. Even when additional protection is provided, the covering layer of concrete is still the most important barrier between the steel and the exterior environment. The degree of protection provided to the steel must therefore be considered as equally important to compressive strength as a measure of concrete quality in bridge construction.

Corrosion is by far the greatest threat to the reinforcement in concrete bridges. It can occur when a sufficiently high concentration of chloride ions is present in the covering layer or when the alkalinity of the covering layer has been reduced through carbonation. The two principal sources of chloride ions are deicing chemicals, regularly used on bridge decks in Europe and North America, and the ocean. The latter affects concrete in bridges through direct contact or through salt contained in the atmosphere of coastal regions. Carbonation is a chemical reaction of concrete and atmospheric carbon dioxide; although it is a less aggressive process than chloride attack, it occurs in all geographical locations. Corrosion can be prevented by blocking the penetration of chlorides and carbon dioxide into the covering layer. The effectiveness of the covering layer in preventing damage to the reinforcement is therefore primarily a function of its impermeability.

Both strength and impermeability are the result of a low water-cement ratio, high quality aggregates, and proper workmanship in mixing, placement, and curing.

The weight of water required for complete hydration, the process by which water and cement are transformed into hardened cement paste, is approximately equal to 35 percent of the cement weight. Roughly 70 percent of this water is chemically bonded to the cement particles; 30 percent is physically bonded to the cement in *gel pores* measuring roughly 2 nm in diameter. Because the water they contain evaporates only at temperatures above 105 °C, the gel pores can be considered as part of the hardened cement paste. The physically bonded water freezes at −78 °C. The danger of spalling due to a buildup of ice pressure in the gel pores can therefore be neglected in concrete bridges.

The chemical reaction between water and cement results in a gross decrease in volume of approximately 10 percent. This shrinkage induced by hydration results in the formation of *capillary pores* in the hardened cement paste, measuring roughly 1 μm in diameter, or 500 times the diameter of the gel pores (Neville 1981, 26). The capillary pores produced in this way are empty and are not inter-connected to each other. They thus do not pose a problem with regard to durability.

The amount of water added to the concrete mixture must include a sacrificial fraction to account for losses due to the evaporation that occurs between mixing and casting. Water-cement ratios varying between 36 and 42 percent by weight are therefore required for complete hydration in actual construction practise, depending on atmospheric humidity, temperature, transportation, and workman-ship. Assuming complete hydration, one cubic metre of concrete will contain approximately 0.18 m³ of hardened cement paste (which includes 0.04 m³ of gel pores), 0.02 m³ of capillary pores, and 0.02 m³ of air pores, entrapped in the fresh concrete during mixing, casting, and consolidation (fig. 3.1). Entrapped air pores are relatively large, often surpassing 1 mm in diameter.

Water in excess of the amount required for complete hydration can be neither physically nor chemically bonded to the cement particles. It remains in the hardened cement paste as *capillary water* contained in capillary pores of roughly

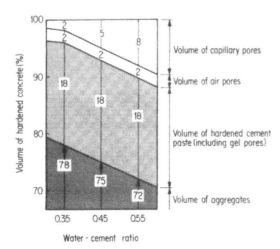

Figure 3.1
Composition of hardened concrete as a function of water-cement ratio for a cement content of 300 kg/m² (Krenkler (1980))

the same size as those produced by shrinkage due to hydration. Assuming a water-cement ratio of 55 percent, one cubic metre of concrete will contain approximately 0.18 m^3 of hardened cement paste (which includes 0.04 m^3 of gel pores), 0.08 m^3 of capillary pores, and 0.02 m^3 of entrapped air pores (fig. 3.1).

The capillary pores created by excess water are interconnected with themselves and the empty capillary pores caused by volume loss due to hydration. They thus increase the permeability of concrete, facilitating the penetration of harmful substances, liquid or gaseous, into the covering layer. All of the chemicals related to the corrosion process – carbon dioxide, chloride ions, and oxygen – readily enter the concrete through the network of pores created by capillary water.

Capillary water freezes at $0 \,°C$. The resulting ice pressure enlarges the capillary network and can be sufficient to fracture the concrete. Damage of this type can largely be prevented by using high-strength, low-porosity concrete. The cube compressive strength should be at least 30 N/mm^2; capillary water should occupy no more than 6 percent of the concrete volume.

When deicing salts are used on concrete bridge decks, most of the heat required to melt ice and snow is removed from a thin layer of concrete at the surface. This produces high differential thermal stresses in the deck. The combined action of thermal shock induced by the melting process and ice pressure in the capillary system cannot be effectively resisted by conventional concrete composed only of aggregates, water, and cement. Artificially increasing the number of air pores by means of air-entraining admixtures has proven helpful in increasing the resistance of concrete to this type of damage. Entrained air pores, which are up to 1000 times larger than capillary pores, restrict the penetration of water by blocking capillary action and provide room for the freezing water to expand freely. Increasing the total air content to 5 percent of the total concrete volume can significantly reduce damage due to the combined action of freezing and deicing salts. The protection provided by air entrainment alone, however, may not be entirely adequate. Due to higher water-cement ratios in the upper layer of the deck slab, the proportion of capillary pores will always be higher at the top of the slab than in the interior. Deck slabs must therefore be given the additional protection of a waterproofing membrane to block the penetration of water and a thick wearing surface to reduce thermal shock.

Only hard, clean aggregates with a favourable ratio of volume to surface area should be selected. The particle size distribution should conform to standardized grading requirements. Aggregates that are not in accordance with an accepted sieve curve must be thoroughly investigated before they are approved for use.

Concrete must be mixed for at least one minute after all ingredients have been combined. Strength and impermeability are substantially reduced when shorter mixing times are used. Longer mixing times may be required when admixtures are used.

Difficulties in concreting often lead to deficiencies in strength and impermeability. Satisfactory placement of concrete with a water-cement ratio of less than 40 percent is possible only with the addition of a super-plasticiser. Conversely, concrete without a super-plasticiser must be placed using a water-cement ratio greater than 40 percent, which increases the content of capillary pores. The resulting increase in permeability must be offset by increasing the thickness of the covering layer. Both mixing and placing are made easier by the use of spherical aggregates with smooth surfaces. Concrete placement is facilitated by a properly detailed arrangement of reinforcing and prestressing steel. The height of drop should be limited to avoid separation of the concrete components. Access openings in the formwork for pouring and vibrating should be provided where necessary.

Proper curing is of utmost importance. Impermeility of the upper surface is easily destroyed by finely distributed cracks due to evaporation at the surface or due to strains induced by the thermal gradient between the interior, warmed by the heat of hydration, and the exterior, cooled by the atmosphere. Uncured concrete surfaces must therefore be covered with mats as soon as possible to prevent drying out and provide thermal insulation. Spraying with water has a detrimental effect, since this further increases the thermal gradient.

Although strength and impermeability are to some extent related, it is incorrect to assume that high strength concrete is all that is required to protect the reinforcement. Defects such as gravel pockets or cracks due to shrinkage and water loss have little effect on the overall strength of structures. They are, however, disastrous with regard to the durability of the embedded steel. Impermeability should therefore always be given proper consideration in the design and construction of concrete structures.

b) Specifications

Specifications for concrete must prescribe minimum permissible compressive strength, $f_{c,\,min}$, and define procedures for testing compressive strength of the concrete that has been cast.

Compressive strength is usually tested using specimens cast at the construction site or at the concrete plant. Cores extracted from hardened concrete in the structure may also be used. The most commonly used shapes for cast specimens are cubes, square prisms, and cylinders. Prism strength, f_{cp}, and cylinder strength, f_c, will be understood to refer to the compressive strength of specimens with a ratio of height to width of 2. Cube strength f_{cw} is approximately 18 percent greater than f_{cp} and f_c for specimens made of identical concrete.

Swiss standard SIA 162 specifies that concrete strength be tested using cube specimens and prescribes a minimum number of specimens, n. The set $\{f_{cw,1}, f_{cw,2}, \ldots, f_{cw,n}\}$ of compressive strengths must satisfy the following condition:

$$f_{cw,\,m} - \alpha(n)s \geqq f_{cw,\,min} \tag{a}$$

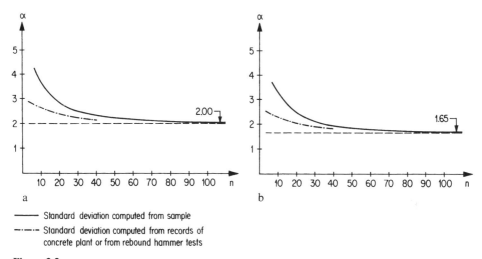

— Standard deviation computed from sample

—·—·— Standard deviation computed from records of
concrete plant or from rebound hammer tests

Figure 3.2
Values of the coefficient α: a, for the 2% fractile of compressive strength (adapted from SIA 162);
b, for the 5% fractile of compressive strength

where $f_{cw, m}$ is the sample mean, s is the sample standard deviation, and $f_{cw, min}$ is the specified minimum strength. This inequality is based on the assumption that the set $\{f_{cw, 1}, f_{cw, 2}, \ldots, f_{cw, n}\}$ is a random sample from a Gaussian normal distribution. The quantity α is a function of n and of a given fractile value of the normal distribution. The 2 percent fractile is specified in SIA 162.

SIA 162 also allows the standard deviation to be determined independently of the sample. The value of s can be computed from the records of the concrete plant or from calibrated rebound hammer tests of previously constructed structural components. Since the standard deviations obtained using either method are more reliable than those obtained from the set of n cube specimens, the value of α in inequality (a) can be reduced accordingly. The parameter α is plotted as a function of n in figure 3.2.

Cement content per cubic metre of hardened concrete must also be specified. Special properties required to ensure adequate behaviour under service conditions may also be included in the concrete specification, for example impermeability and resistance to the effects of freezing, freezing and deicing salts, abrasion, and specific chemicals.

c) Constituent Materials

Cement should conform to an accepted standard specification. Non-standard cements should only be used in exceptional cases and only after their suitability has been confirmed by extensive testing. In particular, cement should contain

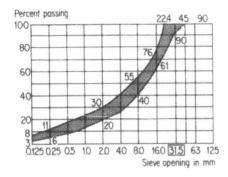

Figure 3.3
Recommended aggregate particle size distribution from SIA 162

no more than 4.5 percent sulfate and 0.1 percent chloride by weight. High-early-strength cement, which generates more heat of hydration than normal cement, should only be used when the ambient temperature is low and strict precautions are taken during curing.

Potable water is suitable for use in the concrete mix without special testing. Water that does not originate from a public drinking supply must be tested before it can be used.

The suitability of the aggregates must always be verified prior to the production of concrete. Particularly high-quality aggregates are required for concrete that must resist freezing or the combined action of freezing and deicing salts. In such cases, aggregates should be tested for organic impurities and a detailed investigation of relevant petrographic properties (including strength, porosity, shape, and surface area) should be conducted. Aggregates should never include micaceous materials.

The recommended particle size distribution from SIA 162 is shown in figure 3.3. Material corresponding to specified aggregate grading requirements is often not readily available to concrete plants. The grading of the aggregates should therefore be regularly monitored during construction to avoid any excess in a given grain size class. The percentage of particles less than 0.02 mm diameter should be limited to no more than 1 percent of the total aggregate mass. In addition, the quantity of particles under 0.125 mm diameter, including cement, should not exceed 350 kg/m^3.

Super-plasticisers are frequently used in bridge construction as a means of reducing the water-cement ratio. Their suitability must always be verified prior to use, especially when they are used in conjunction with other admixtures.

d) Fresh Concrete

Concrete must always be tested immediately before it is placed into the forms. These tests are especially important since they are the last available means of rejecting substandard concrete without serious economic consequences.

The water-cement ratio must be continuously monitored. Concrete for which the water-cement ratio exceeds a specified threshold value should be rejected. Easily conducted consistency tests, for instance the slump test (SIA 162/1), give some indication of the water-cement ratio and the grading of the aggregates. Only when it is known that one of these factors is constant, however, can the other be estimated with any degree of reliability.

Cement content need not be tested in fresh concrete provided the aggregates have been measured by weight beforehand. The cement content per cubic metre of hardened concrete can be determined with sufficient reliability from the volume and gross density of test samples of hardened concrete.

If air-entraining admixtures are used, air content must be tested in fresh concrete that has been consolidated by vibration.

e) Hardened Concrete

The rate of increase in compressive strength with age is a function of type and quality of cement, properties of the aggregates, admixtures, workmanship, and curing conditions (in particular temperature). Strength gain is therefore subject to considerable variability. Under normal conditions, however, the curve of figure 3.4 can be used for rough estimates. Certain applications, such as the post-tensioning of segmentally built girders, require high strength 2 to 3 days after casting. In such cases, strength must be tested using a specimen of the same age or a calibrated rebound hammer. Consideration must also be given to deformations, moisture loss, and thermal shock in determining when formwork and falsework may be removed; for conventional methods of construction, these factors are normally more important than compressive strength in this regard.

The durability of concrete that will be subjected to severe environmental conditions should be investigated using samples of hardened concrete taken from the structure. SIA 162/1 specifies tests of resistance to freezing and of resistance to

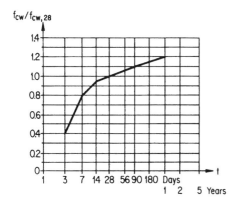

Figure 3.4
Growth of compressive strength with time
(adapted from SIA 162)

the combined action of freezing and deicing chemicals. The former test measures the number of freeze-thaw cycles necessary to produce a 50 percent reduction in modulus of elasticity; the latter measures the mass of loose particles recovered from the surface of a specimen subjected to 30 freeze-thaw cycles while saturated with a 3-percent salt solution.

f) Deformations

Concrete has a nonlinear stress-strain diagram, the shape of which varies with rate of strain, strength, and age. Diagrams are obtained from compressive tests, most commonly using 28-day-old prisms or cylinders with strain applied at a rate of 0.001 per minute. This rate corresponds to static loads of short duration; slower rates can be used when investigating behaviour under long-term loads.

The strain corresponding to compressive strength is denoted ε_0. Its value, which decreases algebraically with decreasing rate of strain, normally lies between -0.0020 and -0.0035 for specimens of unconfined concrete. In displacement-controlled tests, strains algebraically less than ε_0 are accompanied by a decrease in stress. This behaviour is illustrated in figure 3.5 for specimens of unconfined concrete. Lateral confinement of concrete loaded in compression produces an algebraic decrease in ε_0 and a more gradual decrease in stress for strains beyond ε_0 (fig. 3.6).

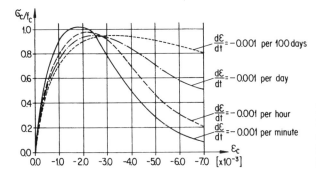

Figure 3.5
Stress-strain diagrams for various rates of strain, displacement-controlled test (adapted from Rasch (1962))

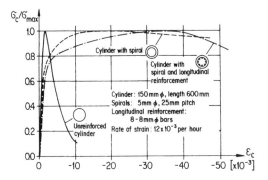

Figure 3.6
Stress-strain diagrams for confined and unconfined concrete, displacement-controlled test (Rüsch and Stöckl (1969))

The stress-strain diagram of concrete is subject to considerable variability. Concrete strains under service conditions should therefore be limited to 0.002 to ensure an adequate margin of safety against crushing.

An idealized stress-strain diagram is normally used to compute the ultimate resistance of reinforced and prestressed concrete cross-sections (see Section 4.3.1).

According to SIA 162/1, the modulus of elasticity, E_c, is determined from the unloading curve of the stress-strain diagram of concrete prisms or cylinders loaded in compression. It is defined as the secant modulus, $(\sigma_2 - \sigma_1)/\Delta\varepsilon$, where σ_2 is approximately one third of compressive strength and σ_1 is 0.5 N/mm^2. Stress is increased at a rate of 0.5 N/mm^2 per second. The value of E_c thus obtained can be used to compute deformations due to static loads of short duration. For dynamic analyses, the modulus of elasticity thus obtained should be increased by roughly 10 percent to account for the higher rate of loading. The increase of E_c with age follows a curve which is geometrically similar to the curve of figure 3.4.

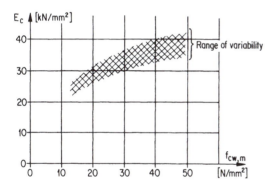

Figure 3.7
Typical relation between modulus of elasticity E_c and average cube strength $f_{cw,m}$ (adapted from SIA 162)

For design calculations, empirical relations can be used to compute E_c as a function of average cube compressive strength, $f_{cw,m}$ (fig. 3.7). (The value of $f_{cw,m}$ can be taken as the 2 percent fractile cube strength plus 10 N/mm^2.) The variability of E_c calculated in this way is significant.

Poisson's ratio, v_c, varies between 0.15 and 0.20. The shear modulus, G_c, is computed from E_c and v_c using the familiar equation

$$G_c = \frac{E_c}{2(1 + v_c)}$$

The coefficient of thermal expansion, α_T, is approximately equal to $10^{-5}/°C$.

The compressive stress-strain behaviour of concrete can be formulated analytically using Hooke's law:

$$\sigma_c = E_c \varepsilon_c \tag{b}$$

provided σ_c is of short duration and is less than or equal to $0.33 f_{cp}$. Concrete undergoes time-dependent elasto-plastic deformations under sustained stress due to *creep*. Equation (b) is therefore not valid for long-term stresses and strains.

The total strain due to a sustained stress σ_c can be expressed as the sum of an elastic strain $\varepsilon_{c,el}$ and a creep strain ε_{cc}:

$$\varepsilon_c = \varepsilon_{c,el} + \varepsilon_{cc}$$

where $\varepsilon_{c,el} = \sigma_c / E_c$. Provided $\sigma_c \leqq 0.33 f_{cp}$, ε_{cc} can be rewritten as

$$\varepsilon_{cc} = \phi(t, \tau) \varepsilon_{c,el}$$

where t is the age of the concrete and τ is the age when the stress σ_c is applied. The slope of the *creep function* $\phi(t, \tau)$ is steep immediately after load is applied and gradually decreases to zero at time infinity.

For constant t, ϕ increases with:

1. Decreasing atmospheric humidity
2. Increasing fineness of cement
3. Increasing water-cement ratio
4. Decreasing concrete density
5. Decreasing cross-section dimensions
6. Decreasing age of concrete at time of loading, τ

These influences can be formulated analytically in the following equation:

$$\phi(t, \tau) = \phi_n k(\bar{\tau}) f(t - \tau) \tag{3.1}$$

The creep coefficient ϕ_n is a function of material properties and environmental conditions. It is plotted in figure 3.8 as a function of relative humidity and water-cement ratio. This graph is valid provided the proportion of cement and aggregates smaller than 0.125 mm diameter is 350 kg per cubic metre of concrete. Increasing this proportion by 50 kg/m³ will increase ϕ_n by 5 to 10 percent.

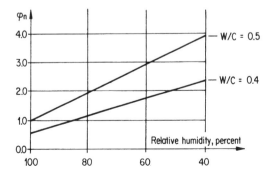

Figure 3.8
Effect of relative humidity and watercement ratio by weight (W/C) on creep coefficient ϕ_n (adapted from SIA 162)

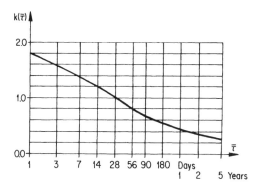

Figure 3.9
Correction factor k as function of fictitious time parameter $\bar{\tau}$ (adapted from SIA 162)

The function $k(\bar{\tau})$ is a correction factor for the age of concrete at time of loading (fig. 3.9). The fictitious time scale $\bar{\tau}$ accounts for temperature during the time elapsed between casting of concrete and application of load:

$$\bar{\tau} = \tau \frac{\sum_{i=1}^{\tau} (T_i + 10\,°C)}{30\,°C}$$

where τ and $\bar{\tau}$ are in days and T_i is the average temperature on day i.

The function $f(t - \tau)$ describes the time-varying behaviour of creep. As shown in figure 3.10, $f(t - \tau)$ depends on an effective thickness parameter h_{ef}:

$$h_{ef} = \frac{2A_c}{U} \tag{c}$$

where A_c denotes cross-sectional area and U denotes perimeter exposed to drying out.

Concrete also undergoes time-varying deformations that are not induced by stress. The progressive drying of concrete after hardening results in a decrease in

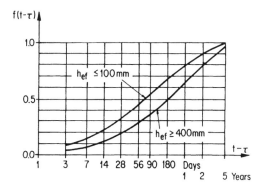

Figure 3.10
Time-varying component of the creep function $f(t - \tau)$ for two typical values of h_{ef} (adapted from SIA 162)

volume. This phenomenon is called *shrinkage*. As with creep, shrinkage strains ε_{cs} increase sharply immediately after hardening and approach a final value at time infinity. The following factors result in increased shrinkage strains:

1. Decreasing atmospheric humidity
2. Increasing proportion of fine aggregates and cement
3. Increasing fineness of cement
4. Increasing water-cement ratio
5. Decreasing concrete density
6. Decreasing cross-section dimensions

The shrinkage strain in concrete cured under normal conditions can be expressed as

$$\varepsilon_{cs}(t, t_0) = \varepsilon_{cs, n}(g(t) - g(t_0)) \tag{3.2}$$

where t_0 is the age at which shrinkage strains begin, i.e., when the concrete is exposed to the atmosphere. The *shrinkage coefficient* $\varepsilon_{cs, n}$ is a function of material properties and relative humidity. Figure 3.11 is valid provided the proportion of cement and aggregates smaller than 0.125 mm diameter is 350 kg per cubic metre of concrete. Increasing this proportion by 50 kg/m^3 will increase $\varepsilon_{cs, n}$ by 5 to 10 percent.

The function $g(t)$, graphed in figure 3.12, describes the time-varying behaviour of shrinkage. It is influenced by the thickness parameter h_{ef}, defined by equation (c). The shrinkage strain ε_{cs} decreases with increasing values of t_0. It is therefore recommended that exposure to the atmosphere be postponed as long as possible by covering the concrete and keeping it moist during curing.

The material parameters that affect both short-term and long-term deformations of concrete are subject to considerable variability. The properties of cement, aggregates, and admixtures can cause significant deviation from the curves of

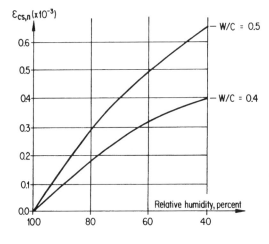

Figure 3.11
Effect of relative humidity and watercement ratio by weight (W/C) on shrinkage coefficient $\varepsilon_{cs, n}$ (adapted from SIA 162)

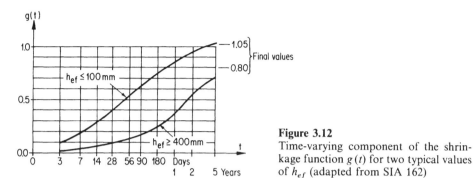

Figure 3.12
Time-varying component of the shrinkage function $g(t)$ for two typical values of h_{ef} (adapted from SIA 162)

figures 3.7 through 3.12. These figures should therefore only be used when it is known from experience or testing that, as a minimum, the parameters E_c, ϕ_n, and $\varepsilon_{cs,n}$ agree with figures 3.7, 3.8, and 3.11, respectively.

3.1.2 Reinforcing Steel

a) Quality Assurance

Reinforcing steel is governed by standards that specify its metallurgical and mechanical properties. For bridge design engineers, the most important properties include yield strength, ductility, bond, resistance to fatigue, and weldability.

Testing of metallurgical and mechanical properties is conducted either at the mill or at an independent laboratory. Bars must also be carefully inspected on the construction site to ensure conformity to reinforcing steel standards and design documents. The inspection should include a verification of identifying marks on the bars and of bend radii, spot checks of bar weight per unit length, and detection of rolling defects, dirt, and other faults. Workmanship must also be carefully monitored, in particular when reinforcing steel is field bent. Bends should not be made without standard radius pins for bars greater than 12 mm diameter; heat must never be used to bend bars.

b) Types of Steel

Reinforcing steel is commonly classified according to strength and method of production. In most industrialized countries, reinforcing bars are manufactured from high-strength steel. Several different processes are used in their production.

Naturally hard steel (designated S 500 a in SIA 162) achieves its strength entirely on the basis of its chemical composition. Its stress-strain diagram is characterized by a distinct yield plateau. Cold-formed steel (SIA 162 – S 500 b) is strengthened

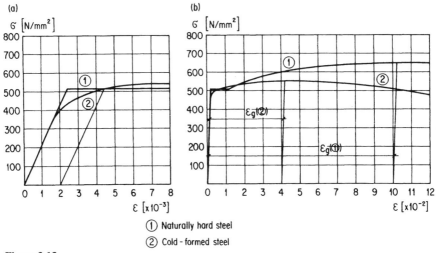

Figure 3.13
Typical stress-strain diagrams for reinforcing steel: a, fine scale; b, coarse scale

after rolling by drawing or twisting. Since its stress-strain diagram has no distinct yield plateau, the yield point of cold-formed steel is defined nominally as the stress that produces a residual permanent strain of 0.2 percent. Tempered steel (SIA 162−S 500c) is strengthened by thermal treatment and quenching. Its stress-strain behaviour is similar to that of cold-formed steel. An alternative procedure for the manufacture of small-diameter bars consists of coiling the bars after rolling (SIA 162−S 500d). The coils, which are considerably longer than straight bars, reduce waste and facilitate transportation. They are straightened immediately prior to fabrication; the cold-working of the straightening operation results in a significant increase in strength. The mechanical properties of bars fabricated from coils are thus identical to those of cold-worked steels. Typical stress-strain diagrams for naturally-hard and cold-formed or tempered steel are shown in figure 3.13. The modulus of elasticity, E_s, is equal to 210 kN/mm^2 for all types of reinforcing steel. The limit of proportionality is defined in SIA 162/1 as the stress producing a permanent strain of 5.0×10^{-5}.

c) Properties

Strength. Tensile strength and yield stress are determined from standardized tensile tests of reinforcing bar specimens (SIA 162/1). SIA 162 specifies limits for the 5 percent fractile, f_{yk}, and the minimum value, $f_{y,\,min}$, of yield stress for the types of steel commonly used in Switzerland; the sample must consist of no fewer than 33 specimens. The specified minimum value, $f_{y,\,min}$, is taken as the design value of yield stress, f_{sy}.

Tensile strength and yield stress must be computed relative to the effective section, which is equal to mass per unit length divided by density ($\varrho_s = 7850$ kg/m^3). It is relatively easy to reduce the amount of steel ($A_s' < A_s$) and to increase the yield stress ($f_{sy}' > f_{sy}$), while maintaining constant yield force ($A_s' f_{sy}' = A_s f_{sy}$). This deviation from standard must never be allowed, however, since the reduction in area results in higher steel strains and larger crack widths under service conditions. Mass per unit length must therefore be carefully verified for samples of each of the standardized bar sizes.

Bond. Reinforcing steel must be capable of developing sufficient bond with the concrete for efficient, economical anchorage and lap splices. Good bond is also required to ensure the proper distribution of cracks under service conditions. The transfer of force between steel and concrete is accomplished primarily through raised deformations on the surface of the reinforcement. When poorly maintained rolling equipment is used, these deformations may be insufficiently pronounced, impairing the force transfer mechanism. The bond strength of bar samples should be periodically verified using a pull-out test (SIA 162/1).

Ductility. Reinforcing steel must be sufficiently ductile to enable fabrication and to ensure that structures can deform plastically at ultimate limit state. The ductility required for fabrication can be tested with an aging-rebend test (SIA 162/1). The maximum plastic deformation in structures at ultimate limit state is a function of the maximum plastic strain in the reinforcing steel measured between yield point and tensile strength. This strain is approximately equal to ε_g, which can be conveniently computed after completion of the tensile test using the equation

$$\varepsilon_g = 2\varepsilon_{10} - \varepsilon_5$$

where ε_5 and ε_{10} are final average strains, measured over intervals of original length 5 and 10 bar diameters, respectively, centered on the location of the rupture. The steel can be considered sufficiently ductile when ε_g is greater than 0.02.

Resistance to Fatigue. The fatigue resistance of reinforcing steel can sometimes be of controlling significance in the design of railway bridges. The maximum allowable stress range in the reinforcement is a function of the predicted number of live load cycles during the lifetime of the bridge, commonly specified as 2 million. Fatigue resistance of reinforcing steel is normally measured in the laboratory using naked test specimens. The allowable stress range of straight bars cast into concrete is approximately 80 percent of the allowable stress range of naked specimens for the same number of load cycles. The allowable stress range of bars bent to a radius of 5 bar diameters is approximately 60 percent of the stress range of naked specimens; this figure reduces to 30 percent for bars bent to a radius of 2.5 bar diameters.

Weldability. The heat produced by welding causes the embrittlement of high-carbon, naturally hard steel and a significant loss of strength in cold-formed and tempered steel. Welds that must develop the full tensile strength of the bar must therefore be carefully inspected. Welds should be avoided where reinforcement is subject to fatigue and at all bar bend locations.

3.1.3 Prestressing Steel

a) Quality Assurance

The following properties of prestressing steel must be carefully specified and tested: strength, ductility and toughness, bond, resistance to fatigue, and relaxation behaviour. The metallurgical and mechanical properties of prestressing steel are tested at the mill or in independent laboratories. Inspection on the construction site is often more difficult than inspection of reinforcing steel and is usually limited to a superficial visual examination of the completed tendons in their ducts. Because it is easily damaged, prestressing steel must be stored and handled with special care.

Proper grouting of bonded tendons is of critical importance to the durability of prestressing steel. Grout, which is normally made of portland cement, water, and additives, must not contain chlorides nor any other substance that promotes corrosion. The water-cement ratio should be less than 40 percent by weight. Grouting operations must always be carefully monitored and must not be undertaken when the temperature inside the ducts is less than $5\,°C$.

Prestressing steel is particularly susceptible to corrosion. It is therefore essential that improperly grouted tendons, gravel pockets, and defects in the covering layer of concrete be prevented and, when necessary, located and repaired.

b) Types of Steel

Prestressing steel is supplied from the mill in the form of wires, thin rods, 7-wire strands, and bars. Typical stress-strain diagrams for the most common types of prestressing steel are given in figure 3.14.

Prestressing wires are produced in diameters ranging from 3 mm to 8 mm. They are normally thermally treated after rolling and then cold drawn in 5 or 6 stages. This process gives the wires a crystal structure that is aligned parallel to the longitudinal axis, resulting in high strength, ductility, and toughness. The wires are then stretched and heat-treated to reduce relaxation. Tempered wires obtain their high strength through intense heating and subsequent quenching. Their crystal structure is not aligned parallel to the longitudinal axis, but is rather directed radially. Ductility and toughness are thus significantly reduced; small defects or rust scars in tempered wires can cause brittle fracture. Tempered wires are therefore not recommended for use in post-tensioned structures, but can be used for pre-tensioning when properly handled.

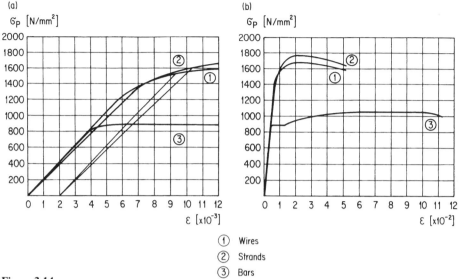

Figure 3.14
Typical stress-strain diagrams for prestressing steel: a, fine scale; b, coarse scale

Wires are available with a smooth or slightly deformed surface. Wires 6 mm and 7 mm in diameter have a modulus of elasticity of 205 kN/mm². They are used primarily for parallel-wire tendons. Thinner wires have a higher tensile strength. Because they are more difficult to handle, wires 4 mm in diameter are manufactured into 7-wire strands in the mill. The modulus of elasticity of 7-wire strands is 195 kN/mm².

Steel can also be cold-drawn into prestressing rods in the 9 mm to 12 mm diameter range. Rods are used as individual prestressing units rather than grouped together into tendons. Due to their larger diameter, rods have a lower tensile strength than wires.

Prestressing bars made of naturally hard steel are manufactured in diameters varying from 20 mm to 36 mm. After rolling, they are stretched and heat treated. Prestressing bars normally have cold or hot rolled threads and can thus be easily anchored and coupled using threaded fittings. Their bond strength is comparable to that of reinforcing steel. The stress-strain diagram of prestressing bars is characterized by a distinct yield plateau; the modulus of elasticity is 210 kN/mm². Prestressing bars are relatively brittle. Special care must be taken when they are used in falsework construction, where brittle failure can occur as a result of repeated imposed deformations or welding scars.

c) Properties

Stength. The stress-strain diagrams of wires, strands, and thin rods are characterized by gradual softening without a distinct yield plateau. The design value of yield stress, f_{Py}, is normally defined as the product of the specified minimum 5 percent fractile of tensile strength f_{Pt} and a factor, typically equal to 0.9. Table 3.1 gives values of f_{Py} obtained from SIA 162 for commonly used wire and rod diameters. The value of f_{Py} is based on *specified* tensile strength and may not be increased even though the *actual* tensile strength may be greater. Effective wire and rod diameters used in calculations are computed from density and mass per unit length.

The value of f_{Py} for naturally hard prestressing bars can be determined as for reinforcing steel, and is approximately 80 percent of tensile strength. Table 3.2 gives values of f_{Py} obtained from SIA 162.

Ductility and Toughness. Prestressing steel must be sufficiently ductile to allow the structure to deform plastically at ultimate limit state. It must also be sufficiently tough to resist brittle fracture, which can occur as a result of small surface defects such as rust scars. Prestressing steel is particularly vulnerable to brittle fracture during stressing operations.

Bond. Bond is of particular importance in pre-tensioned elements, where it is the primary mechanism for transferring the prestressing force to the concrete. Good bond is also necessary in post-tensioned structures, to ensure equality of strain in the prestressing and reinforcing steel after cracking occurs.

Resistance to Fatigue. Prestressing steel is normally required to withstand 2 million cycles of design fatigue live load. The fatigue resistance of prestressing steel is determined from tests of naked specimens (SIA 162/1). The maximum stress range of smooth, 140 mm-long wire specimens is approximately 270 N/mm² at 2 million cycles. The stress range for deformed wire specimens of identical length is roughly 80 percent of the stress range of the smooth wire. The fatigue resistence of wire tendons that are cast into concrete, stressed, and grouted is substantially lower than the resistance of naked specimens. The allowable stress range in bonded post-tensioning steel should be limited to 5 percent of tensile strength.

Relaxation. Relaxation is defined as the time-dependent increase in steel strain at constant stress. It is thus fundamentally different from creep in concrete, which is a

Table 3.1
Design Yield Stress f_{Py} of Prestressing Wires and Rods (from SIA Standard 162)

Diameter (mm)	f_{Py} (N/mm²)
3	1670
4	1640
5 and 6	1590
7 and 8	1530
9 to 12	1410

Table 3.2
Design Yield Stress f_{Py} of Naturally Hard Prestressing Bars (from SIA Standard 162)

Diameter (mm)	f_{Py} (N/mm²)
20	1000
26	1000 [a]
26 to 36	830

[a] The 26 mm diameter bars are available in two different strengths in Switzerland

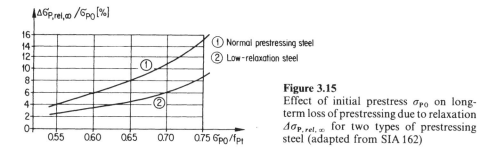

Figure 3.15
Effect of initial prestress σ_{P0} on long-term loss of prestressing due to relaxation $\Delta\sigma_{P,rel,\infty}$ for two types of prestressing steel (adapted from SIA 162)

decrease in stress at constant strain. Relaxation progresses more rapidly than creep. Loss of prestress can be reduced by using low-relaxation steel, which has been stretched and heat treated. The behaviour of conventional and low-relaxation prestressing steel is compared in figure 3.15.

3.2 Actions

3.2.1 Loads

Loads are actions that affect the internal and external *equilibrium* of structures. For structural systems composed of beam elements, loads can be characterized as *nonzero* functions q in the following differential equation

$$\frac{d^2}{dx^2}\left(EI(x)\frac{d^2 w}{dx^2}\right) = q(x)$$

where $EI(x)$ is flexural stiffness, w is deflection, and q is to be regarded as a generalized function, including concentrated loads and moments. Loads must always be equilibrated by reactions and sectional forces, regardless of the compatibility conditions assumed or the degree of statical indeterminancy.

The following loads must be considered in bridge design:

1. Dead Load. The total dead load, g, is the sum of the self-weight, g_0, and the superimposed dead load, Δg. Self-weight is defined as the weight of all structural components and is equal to the product of volume and density. An average density of 2550 kg/m^3 can be assumed for reinforced and prestressed concrete structures. This figure is based on an average concrete density of 2450 kg/m^3 and a steel consumption of 100 kg/m^3. Superimposed dead load is defined as the weight of all non-structural components, including wearing surface, guardrails, and conduits.

2. Live Load. Live load is defined as the static weight of vehicles carried by the bridge and any dynamic effects produced by their movement. Live loads are specified in codes and standards (see, for example, SIA 160). They normally consist of load models, which are simplified representations of the actual vehicles.

Railway loads are typically modeled as an idealized train, consisting of several concentrated loads for the locomotive and a uniform load for freight and passenger cars. Highway loads are derived either from legal axle and truck loads or from measurements of actual vehicles in service. They are normally modeled as at least one concentrated load and a uniform load.

Experience has shown that the magnitude and pattern of highway loads can change significantly during the lifetime of a bridge. Complicated models are therefore of dubious value since whatever accuracy they bring to caclulations is lost as soon as traffic patterns change. A live load model consisting of a single concentrated load and a distributed load is preferable in all cases. The magnitude of the distributed load should vary according to the span length of the structural component under consideration.

The static live loads described above must be increased to account for the dynamic response of bridges to the movement of vehicles. This dynamic increment of live load, often called *impact*, is a function of the natural frequency of the structural component under consideration. Resonant vibrations are possible when the natural frequency of a component is between 2 Hz and 4 Hz or between 8 Hz and 12 Hz. (The former range of frequencies correponds to truck body vibrations, the latter to truck axle vibrations.) In such cases, the dynamic increment for a single truck can be as high as 70 percent for a smooth wearing surface and even higher for rougher surfaces (Cantieni 1983).

The envelope of figure 3.16 shows a relation between dynamic increment and natural frequency proposed by Cantieni (1983) based on measurements taken on Swiss highway bridges. The relation is valid for a single truck on a smooth wearing surface. Resonance at higher frequencies becomes more significant with increasing roughness.

Impact for full live load is much lower than impact for a single truck. It varies typically between 10 and 20 percent, depending on the region of influence of the load for the particular structural component under consideration.

A separate live load model must normally be used to verify fatigue resistance. Fatigue loads for railway bridges are defined in appropriate design standards. For

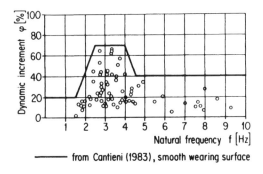

Figure 3.16
Dynamic increment due to a single truck, smooth pavement

highway bridges, fatigue is normally of importance only in the design of stay cables. The fatigue load for highway bridges can be derived directly from legal loads and actual traffic conditions. It can, for example, be taken as a single truck of the highest weight class, with impact, and a distributed load calibrated to the patterns and frequency of bridge traffic. In normal cases, the distributed load varies between $0.8 \, kN/m^2$ and $1.0 \, kN/m^2$.

3. Wind Load. Wind loads have a very small influence on the construction cost of most bridges. A detailed investigation of their effects is therefore not warranted in most cases. Sectional forces due to wind can be calculated using extreme values for wind pressure and drag coefficients obtained directly from codes and standards.

A thorough investigation of wind loadings may, however, result in safer and more economical structures in special cases, such as very high bridges, long-span bridges, and cantilever-constructed bridges. The object of such an investigation is a detailed quantitative description of the wind regime at the bridge site, which normally includes the following information:

1. Design hourly average wind speed at a reference height of 10 m above ground, $\bar{v}_{10}$
2. The functional relation between average wind speed and elevation
3. Design instantaneous maximum wind speed v_{max}
4. Spatial correlation of wind speed at neighbouring points along the structure

The design hourly average wind speed at reference height, $\bar{v}_{10}$, is chosen to correspond to a specified *return period*, T, in years. The probability that the yearly maximum hourly average wind speed v is greater than $\bar{v}_{10}$ must be less than $1/T$. (Speed v is also defined at reference height.) The value of $\bar{v}_{10}$ is calculated from wind records collected as close as possible to the bridge site. Since T is often greater than the number of annual wind records, statistical methods must be used to estimate $\bar{v}_{10}$.

Maximum hourly average speeds and maximum gust speeds have been found to obey a Gumbel distribution (Gumbel 1958). (The following discussion is valid for either.) The probability that the yearly maximum wind speed is less than v is given by the following expression:

$$p_v = \exp\{-\exp[-(Av + B)]\} \tag{a}$$

where A and B are parameters to be estimated from the available wind records.

A set of annual maximum wind speeds $\{v_1, v_2, \ldots v_n\}$, recorded over a period of n years, is required to estimate A and B. Each speed v_i is assigned a rank m_i in ascending order, such that $v_{m_i} \leqq v_{m_i + 1}$. The probability that the yearly maximum wind speed is less than v_i is estimated by p_i, defined as follows:

$$p_i = \frac{m_i}{n + 1}$$

The points (p_i, v_i) are transformed to points (x_i, v_i) using the relation $x_i = -\ln(-\ln p_i)$. A relation between x and v can be estimated using linear regression on the n data points (x_i, v_i):

$$x = av + b$$

The quantities a and b are estimates of the parameters A and B, respectively, in equation (a). Given a specified return period T, therefore, the design wind speed can be calculated using the following equation:

$$v = -\frac{1}{a}\ln\left[-\ln\left(1 - \frac{1}{T}\right)\right] - \frac{b}{a}$$

These concepts are illustrated by the following example:

Example 3.1:
Calculation of design wind speed using Gumbel's distribution

Annual maximum wind speeds v_i are given in Table 3.3. Linear regression yields $a = 0.0791$ and $b = -7.135$. The regression line is shown in figure 3.17. Design wind speeds are 118 km/h and 139 km/h for return periods of 10 years and 50 years, respectively.

Table 3.3
Annual Maximum Wind Speeds for Example 3.1 (Reber and Menn 1982)

i	Year	v_i (km/h)	Rank m_i	p_i $(= m_i/14)$	x_i $(= -\ln(-\ln p_i))$
1	1967	121	13	0.93	2.60
2	1968	113	12	0.86	1.87
3	1969	85	3	0.21	−0.43
4	1970	92	7	0.50	0.37
5	1971	111	10	0.71	1.09
6	1972	86	4	0.29	−0.23
7	1973	98	9	0.64	0.82
8	1974	84	2	0.14	−0.67
9	1975	89	5	0.36	−0.03
10	1976	112	11	0.79	1.42
11	1977	81	1	0.07	−0.97
12	1978	91	6	0.43	0.17
13	1979	98	8	0.57	0.58

Given a design wind speed, v, and the corresponding return period, T_v, the probability that the yearly maximum wind speed exceeds v in any given interval of N years is given by the following expression:

$$P_v(T_v, N) = 1 - \left(1 - \frac{1}{T_v}\right)^N \tag{b}$$

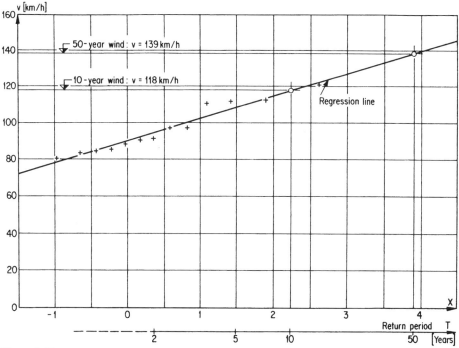

Figure 3.17
Linear regression on the data of Table 3.3

This equation can be useful in developing design wind loads that are valid for a relatively short construction period, in which the probability of extreme winds is reduced. Assuming T_v is 50 years, the following probabilities are obtained from equation (b):

$$P_v(50,50) = 0.64$$
$$P_v(50,2) \ = 0.04$$

The probability that v is exceeded at least once during an assumed construction period of 2 years is thus only 6 percent of the probability that it will be exceeded at least once in the 50-year design lifetime.

The variation of hourly average wind speed, $\bar{v}$, with height, z, is normally expressed as follows:

$$\bar{v}(z) = \bar{v}_{10} \left(\frac{z}{10}\right)^{\alpha} \tag{c}$$

where z is in metres and $\bar{v}_{10}$ is the design hourly average speed at a reference height of 10 m. The speed $\bar{v}_{10}$ is obtained using the statistical method described above or from an applicable standard. The parameter α is a function of the roughness of the surrounding terrain and varies from 0.10 to 0.40. Its value is obtained from measurements taken at the site of hourly mean wind speeds at several elevations. Alternatively, it can be taken from published tables that give α for several general terrain types (Simiu and Scanlan 1986, 43).

The instantaneous maximum wind speed, v_{max}, is normally defined for a 3-second interval corresponding to the response time of fast-response anemometers. The ratio $v_{max}(z)/\bar{v}(z)$ decreases with increasing z; its value can be estimated from site measurements or using empirical formulas (Scruton 1981, 11).

Design wind pressures p are calculated from the design wind speeds v using the equation

$$p = \tfrac{1}{2}\varrho v^2 \, C_p$$

where ϱ is the density of air and C_p is a pressure coefficient which depends on characteristics of the structural member and the flow. Values of C_p are tabulated in most design standards and codes. For unusual structures, C_p can be measured using a scale model in a wind tunnel.

Wind load is obtained by multiplying p by the exposed area. For obvious reasons, load factors for design at ultimate limit state must not be applied to the design wind speed or to the return period, but rather to the load itself.

Although this quasi-static approach does not explicitly consider the time-varying nature of wind loadings or structural response, it is valid for the vast majority of bridges. Structures with natural frequencies less than 1 Hz may nevertheless be prone to wind-induced oscillations that may impair safety and serviceability. The dynamic behaviour of these structures should therefore be carefully investigated. The theoretical foundations of such an investigation are described by Simiu and Scanlan (1986).

4. Horizontal Loads Due to Traffic. Loads due to acceleration, deceleration, and centrifugal force are dynamic actions. They have been idealized in most design standards, however, as equivalent static loads. Although relatively large for railway bridges, these loads are normally of secondary importance for highway bridges.

5. Collision. Collision loads must be considered in the design of guardrails and occasionally in the design of piers. Equivalent static loads on guardrails are given in most design standards. Collision loads on piers must sometimes be estimated; the principle of conservation of energy can be used to equate the kinetic energy of the vehicle or ship before impact and the work done by the equivalent static load Q as it displaces through a stopping distance s. Assuming constant deceleration, it follows that

$$Q = \frac{mv^2}{2s}$$

where m is the mass of the moving body and v its speed before collision.

The accuracy of the value of Q computed above depends on the accuracy of the stopping distance s, for which only rough estimates are normally available, and on the validity of the assumption of constant deceleration.

6. Earth Pressure. Loads due to earth pressure are of primary importance in the design of abutments, piers, shaft foundations, and piles. These loads can be reliably determined only as a result of thorough geotechnical investigations. The effects of active earth pressure and earth pressure at rest are usually restricted to soil located in the immediate vicinity of the structure. Passive earth pressure can, however, affect a mass of earth which is substantially wider than the direct contact area between soil and structure. A row of piers located in inclined terrain, for example, may be required to resist passive pressure resulting from displacements of the soil down the slope. The mass of earth to be retained by a given pier will normally be wider than its exposed frontal area, due to arching of the soil between piers. The load on the pier will therefore be substantially increased.

7. Earthquake. Seismic loads are modelled with equivalent static loads in areas of the world where earthquake risk is minor. These loads normally control the design of the bearings or piers that transfer horizontal load from superstructure to the foundations. In regions of high seismic risk, the dynamic response of bridges to seismic action must be considered. In all cases, however, an appropriate choice of structural system and details can minimize sectional forces due to seismic action, eliminating the need for costly increases in member resistance.

3.2.2 Prestressing

Prestressing with stressed steel tension elements produces a self-equilibrating state of stress, in which the tensile force in the steel is balanced by a compressive force of equal magnitude in the concrete. The two forces act at the same location and in opposite directions in statically determine structures. The stress in the prestressing steel σ_P is plotted as a function of moment in figure 3.18. The decompression moment, M_D, is defined as the moment producing a concrete stress of zero at the previously compressed tension face. For moments less than M_D, σ_P is larger than

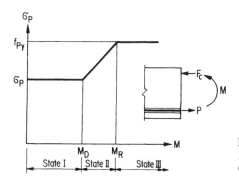

Figure 3.18
Effect of moment on stress in prestressing steel (schematic)

necessary to resist the applied sectional force. As the moment is increased beyond M_D to the ultimate moment M_R, the stress in the prestressing steel is approximately equal to the stress that would be required for internal equilibrium without prestressing.

Deformations before cracking (Stage I) can be calculated by the superposition of deformations due to prestressing and external actions. After cracking (Stage II), the total deformation is equal to the deformation in the uncracked system at decompression plus the deformation due to the additional strains in the cracked system. The redundant forces in a cracked, statically indeterminate system can be calculated as usual from the deformations due to prestressing and external actions in the uncracked state, provided the cracks remain small.

The designer is completely free in the layout of prestressing tendons and the choice of prestressing force to achieve safety, serviceability, economy, and elegance. The *degree of prestress*, however it is defined, is in itself no measure of the quality of the structure. It is important to define, at the preliminary stages of design, the role of prestressing in terms of the desired performance of the structure, in particular with regard to serviceability and economy. The role of prestressing defined in this way will be referred to as the *prestressing concept*.

3.2.3 Restrained Deformations

Restrained deformations are produced by support movement, temperature change, shrinkage, and creep in statically indeterminate structures. The response of the structure to restrained deformations is characterized analytically by solutions to the homogeneous differential equation

$$\frac{d^2}{dx^2}\left(EI(x)\,\frac{d^2 w}{dx^2}\right) = 0$$

The corresponding internal forces are directly proportional to member stiffness and are in no way proportional to the magnitude of the actions themselves. The stiffness of a system is sharply reduced by cracking under service conditions and is reduced to zero with the formation of a plastic hinge mechanism at ultimate limit state. The effect of restrained deformations decreases correspondingly. Theoretically, no sectional forces due to restrained deformations are present at ultimate limit state in ductile systems. In general, therefore, restrained deformations are only significant for the behaviour of the structure under service conditions, particularly with regard to cracking and deformations. The reinforcement required for crack control is practically independent of the magnitude of the restrained deformation.

References

Cantieni, R. 1983. *Dynamische Belastungsversuche an Strassenbrücken in der Schweiz* (Dynamic load tests of highway bridges in Switzerland). Forschungs- und Arbeitsberichte Abteilung 116, Massivbau, Bericht Nr. 116/1. Dübendorf: Eidgenössische Materialprüfungs- und Versuchsanstalt.

Gumbel, E.J. 1958. *Statistics of Extremes*. New York: Columbia University Press.

Krenkler, K. 1980. *Chemie des Bauwesens* (Chemistry of construction). Berlin, Heidelberg, and New York: Springer-Verlag.

Neville, A.M. 1981. *Properties of Concrete*. 3rd. ed. London: Pitman.

Rasch, C. 1962. *Spannungs-Dehnungslinien des Betons und Spannungsverteilung in der Biegedruckzone bei konstanter Verformungsgeschwindigkeit* (Stress-strain curves of concrete and stress distribution in the flexural zone for constant rate of deformation). Deutscher Ausschuß für Stahlbeton, vol. 120. Berlin: Wilhelm Ernst & Sohn.

Reber, U. and C. Menn. 1982. Windeinwirkung auf Brücken (Wind effects on bridges). *Schweizer Ingenieur und Architekt* 100: 773–782.

Rüsch, H. and S. Stöckl. 1969. *Versuche an wendelbewehrten Stahlbetonsäulen unter Kurz- und Langzeit wirkenden zentrischen Lasten* (Investigations of spirally reinforced concrete columns under short and long term concentric loads). Deutscher Ausschuß für Stahlbeton, vol. 205. Berlin: Wilhelm Ernst & Sohn.

Scruton, C. 1981. *An Introduction to Wind Effects on Structures*. Engineering Design Guides, no. 40. Oxford University Press.

SIA 160. 1989. *Norm 160: Einwirkungen auf Tragwerke* (Standard 160: Actions on structures). Zurich: Schweizerischer Ingenieur- und Architekten-Verein.

SIA 162. 1989. *Norm 162: Betonbauten* (Standard 162: Concrete structures). Zurich: Schweizerischer Ingenieur- und Architekten-Verein.

SIA 162/1. 1989. *Norm 162/1: Betonbauten, Materialprüfung* (Standard 162/1: Concrete structures, testing of materials). Zurich: Schweizerischer Ingenieur- und Architekten-Verein.

Simiu, E. and R.H. Scanlan. 1986. *Wind Effects on Structures*. 2d. ed. New York: Wiley Interscience.

4 Fundamentals of Analysis and Design

4.1 Design for Safety and Serviceability

Before any structure may be built, its safety and serviceability must be proven during the design phase. This is accomplished by comparing quantitative measures of the safety and serviceability of each structural component and system to minimum acceptable values defined by codes and standards. Since they are often independent of each other, safety and serviceability must be verified separately.

4.1.1 Safety

A structure that is in external and internal equilibrium under a given loading is said to be at *ultimate limit state* when any increase in load, however small, results in loss of equilibrium and hence in collapse of the structure. The load corresponding to ultimate limit state is called the *ultimate load*, denoted q_u. *External* equilibrium refers to the relation between loads, reactions, and sectional forces, and is formulated analytically by the following equations:

$$\frac{dV}{dx} = -q \qquad \frac{dM}{dx} = V \qquad \frac{dT}{dx} = -m_t \tag{a}$$

(The sectional forces must also satisfy any given statical boundary conditions.) *Internal* equilibrium refers to the equilibrium of the sectional forces and the internal forces, and is formulated as follows:

$$S \leqq R \tag{b}$$

which must be true at all points in the structure. The quantity S is a sectional force due to the given loads and R is the maximum internal force that can be developed by the section, called the *ultimate resistance*.

According to the theory of plasticity, any load that is equilibrated externally and internally and satisfies the statical boundary conditions will be less than or equal to the ultimate load, and hence will not induce collapse (Thürlimann et al. 1983).

Structures can therefore be considered safe when the statical boundary conditions are satisfied and when equations (a) and inequality (b) are true at all locations.

This definition of safety, however, is merely of theoretical interest since it is based on the assumption that loads and resistance are known exactly. In fact, loads and resistance are functions of many different factors, such as traffic patterns and workmanship, which cannot be quantified reliably. It is therefore preferable to define safety in probabilistic terms, based on the statistical distributions of loads and resistance. Structures can then be considered safe when the probability of loss of equilibrium is less than an acceptably low target value.

Design calculations do not explicitly consider probabilities of collapse and the statistical distributions of loads and resistance. Rather, this information is used to define design values of material resistance, specified loads, and safety factors, which are given in codes and standards. In its simplest form, the probabilistic definition of safety consists of satisfying the statical boundary conditions and equations (a) with the specified loads and satisfying the following inequality at all points in the structure:

$$\gamma S \left(\sum_i q_i \right) \leq R \tag{c}$$

where S is the sectional force due to the set of specified loads $\{q_i\}$, R is the ultimate resistance of the section computed from the design values of material resistance, and γ is a safety factor.

The factor γ is calibrated to ensure that the probability of collapse is less the specified target value. The design values of material resistance are commonly defined as extreme fractile values of their respective statistical distribution. Specified loads, however, may either be defined as actual service loads (e.g. dead and live loads) or as extreme fractile values (e.g. wind and earthquake loads). It is therefore practical to define partial safety factors for resistance and each type of load. Swiss standards SIA 160 and SIA 162 specify *resistance factors*, γ_R, and *load factors*, $\gamma_{S,i}$. Factors γ_R account for sources of statistical variability in resistance not attributable to the materials themselves, for example dimensional tolerances and workmanship. Factors $\gamma_{S,i}$ must be calibrated to the actual statistical models used for each specified load. In addition, safety factors should make allowance for small errors in engineering calculations and for small changes in loads and resistance over time. The use of partial safety factors transforms inequality (c) into

$$S \left(\sum_i \gamma_{S,i} q_i \right) \leq \frac{1}{\gamma_R} R \tag{d}$$

The left-hand side of this inequality will be referred to as the *design sectional force*, and will be denoted S_d:

$$S_d = S \left(\sum_i \gamma_{S,i} q_i \right)$$

Structures are therefore safe when the statical boundary conditions, equations (a), and inequality (d) are satisfied with prescribed design values of material resistance, specified loads, and safety factors. This definition of safety is based on the *statical method* of the theory of plasticity; the load $\gamma_R \sum_i \gamma_{S,i} q_i$ is a lower bound for the ultimate load. The *kinematic method* of plasticity, which yields an upper bound for the ultimate load, should never be used for design. It may, however, be useful as a means of checking designs based on the statical method.

Inequality (d) is normally used for *checking* concrete cross-section dimensions that have been previously selected and for the *direct design* of the reinforcement. Proceeding in this way at all points in the structure yields the greatest economy for the assumed design sectional forces.

A prerequisite to the practical application of the concepts presented in this section is *ductile* structural behaviour. The designer must ensure that the structure is capable of deforming plastically after the ultimate resistance of the cross-section is achieved.

4.1.2 Serviceability

Limit states at which bridges are structurally safe but otherwise unfit for service are called *serviceability limit states*. The most important aspects of serviceability are durability, function, and appearance. *Appearance* refers here to the prevention of unsightly defects such as water stains and visible cracks, rather than to the aesthetic aspects of design discussed in Chapter 2. The *function* of the bridge as a whole is defined in terms of traffic safety and comfort. The function of individual components, such as expansion joints and bearings, must also be considered. *Durability* is closely related to function and appearance, both of which can be severely impaired by deterioration of concrete and reinforcement. It is also related to structural safety, since deterioration of reinforcement can result in a serious loss of resistance.

Satisfactory behaviour under service conditions cannot be verified incontestably during design in the same way as safety, since the direct relationship between safety and equilibrium has no analog in the context of serviceability. The following strategies are therefore used to ensure that bridges will perform well during service:

1. Behaviour that can be readily quantified and calculated can be considered acceptable when it is within an acceptable range of values. Acceptable values may be specified in codes and standards or may be determined by the owner, in consultation with the engineer, for a specific project. They are often based on non-technical considerations and are usually not absolute. Vibration criteria, for example, are normally more severe for bridges that are used by pedestrians as compared to those used by vehicles alone. This category includes most aspects of structural behaviour under service conditions, including deflections, vibrations, and cracking. Many aspects of non-structural behaviour, for example the run-off of rain water on the deck, are also included.

2. Behaviour that is not easily quantified and calculated cannot be verified against acceptable values. An example from this category is the corrosion of reinforcing steel. A quantitative description of the corrosion process, although possible, is not well suited to the needs of bridge designers. Undesirable behaviour of this type is best prevented by properly specifying materials, careful workmanship and inspection during construction, and good detailing practise. Details that facilitate inspection, maintenance, and replacement of defective components are of particular importance in this regard.

Serviceability evolves over time, due to changes in the actions to which bridges are subjected and the ability of bridges to withstand them. Regular inspection is therefore of utmost importance for the early detection of potential problems.

Serviceability is discussed further in Section 4.8.

4.2 Calculation of Sectional Forces

4.2.1 Fundamentals

Sectional forces under service conditions and at ultimate limit state are required for design. Sectional force diagrams are based on the equilibrium conditions of the structure, often expressed in the form of a differential equation. The bending moments in frame structures, for example, are calculated as follows:

$$\frac{d^2 M}{dx^2} = -q(x)$$

The sectional forces at the supports of statically indeterminate systems are obtained either from compatibility conditions or from plasticity conditions, depending on the state of stress in the structure.

Cracking under service conditions must be restricted to small, well-distributed cracks; neither large cracks caused by yielding of reinforcement nor plastic hinges are allowed. The calculation of sectional forces under service conditions is therefore based on the compatibility conditions of the elastic system and on the stiffness of the uncracked structure. Elastic theory can therefore be used; the resulting sectional forces are referred to as the *elastic solution*. The redistribution of sectional forces is nevertheless possible under service conditions due changes in stiffness caused by cracking. The greater the effect of the change in stiffness on the compatibility conditions, the more pronounced will be the redistribution.

Since the requirements of equilibrium place no restriction on cracking, the formation of plastic hinges may be considered at ultimate limit state. Boundary values of sectional forces can then be calculated from plasticity conditions, rather than from compatibility conditions. The assumption of plastic hinges is permissible under the following conditions:

1. The system must be sufficiently ductile
2. The influence of system deformations on the equilibrium conditions must be considered
3. The formation of plastic hinges must not critically impair the ultimate resistance of the section in shear

The ultimate load, q_u, was defined in Section 4.1.1 as the maximum load satisfying the statical boundary conditions, external equilibrium, and internal equilibrium. In general, it is calculated only when a supplementary check of safety is desired. In normal cases, design at ultimate limit state is loosely based on the elastic solution. This results in an economical design and acceptable crack distribution and steel stresses under service conditions. It is unreasonable, however, to adhere blindly to elastic theory and the "exact" elastic solution. The effects of the additional accuracy on safety and economy are usually insignificant. Simplified load distributions, for example, can substantially reduce computational effort without significant loss of accuracy. The exact elastic solution, moreover, is unsuitable when significant redistribution of sectional forces actually does occur at ultimate limit state.

4.2.2 Sectional Forces Due to Loads, Prestressing, and Restrained Deformations

Sectional forces due to loads, prestressing, and restrained deformations are fundamentally different from each other in the structural behaviour they induce. Sectional forces should therefore be calculated using a procedure appropriate for the type of action that causes them.

a) Loads

Sectional forces due to service loads are calculated according to elastic theory. The actual compatibility conditions can be approximated by those of the homogeneous, uncracked system. The simplification of member stiffness is acceptable provided the deformations of the simplified model do not deviate substantially from the deformations of the actual system. Otherwise, the stiffness of the cracked section should be considered.

At ultimate limit state, the compatibility conditions of the elastic system need no longer be satisfied. The redistribution of sectional forces is therefore permissible for ductile systems. Redistributions can be regarded as the superposition of self-equilibrating sectional force diagrams, which have no effect on overall equilibrium, onto the elastic solution. The differential equations for self-equilibrating bending moments are

$$\frac{d^2 M}{dx^2} = 0 \quad \text{for frame structures}$$

and

$$\frac{\partial^2 m_x}{\partial x^2} + 2\frac{\partial^2 m_{xy}}{\partial x \partial y} + \frac{\partial^2 m_y}{\partial y^2} = 0 \qquad \text{for slabs}$$

The number of linearly independent self-equilibrating sectional force diagrams is equal to the degree of statical indeterminacy of the system. For continuous beams, these diagrams are readily calculated from imposed displacements of the supports. For slabs, they are preferably computed from the superposition of stresses due to an arbitrary load, q, applied to the original system, and stresses due to $-q$ applied to a modified system. This procedure is discussed further in Section 7.9.

The design sectional forces in frame structures are normally based on the elastic solution. For continuous girders, the envelope of sectional forces is computed from the elastic solutions for dead load plus the controlling live load cases. As shown in figure 4.1, the width of the envelope can be reduced by redistributing the sectional forces due to live load. The sectional forces due to dead load, however, are usually not redistributed. The effect of large redistributions at ultimate limit state on cracking and steel stresses under service conditions must always be investigated.

It is usually not necessary to calculate a complete stress history corresponding the actual construction sequence. Sectional forces due to dead load can be computed for an idealized continuous girder, built on conventional falsework with all concrete cast simultaneously. Regardless of construction sequence, the action of creep gradually transforms the dead load stress distribution into the stresses of this simplified model. Small differences between the assumed and the actual distributions of sectional forces are not significant for design, especially at ultimate limit state, where one can be transformed into the other by a small redistribution. Although stress histories are not required for the design of completed structures, any critical stress conditions that may occur during construction must nevertheless be checked.

The inherent ductility of slabs is higher than that of frames, enabling the redistribution of sectional forces not only due to live load, but also due to dead load. The redistribution of sectional forces due to dead load can help to relieve local stress concentrations in the elastic solution and to simplify the arrangement of reinforcement which, if laid out according to the elastic solution, would be overly complicated.

The internal forces in panels subjected to in-plane loading can be calculated using simplified truss models. Although considerable freedom is possible in the arrangement of the model, the actual flow of forces in the elastic solution should be respected. In the neighbourhood of a concentrated load, for example, the elastic stress trajectories are as shown in figure 4.2a. These correspond to the truss model of figure 4.2b, in which the compression force spreads out at an angle of roughly 60 degrees.

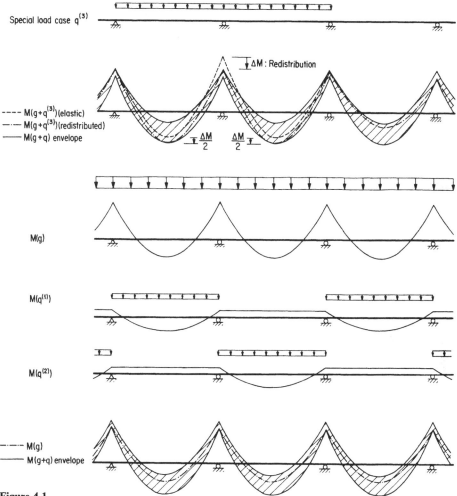

Figure 4.1
Reduction of moment envelopes by redistribution

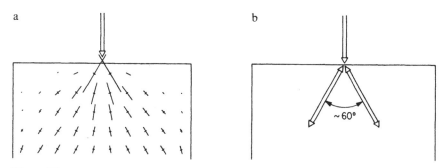

Figure 4.2
Flow of forces in panels: a, elastic stress trajectories; b, truss model

b) Prestressing

Prestressing, which is a self-equilibrating state of stress, induces no net sectional forces. It is nevertheless worthwhile to associate the concrete stresses induced by prestressing to forces, using the following integrals:

$$N_P = \int \sigma_c(P)\,dA\;(=F_c) \qquad M_P = \int \sigma_c(P)\,z\,dA$$

(The coordinate z is measured from the centroid of the section.) The sectional forces thus defined are equilibrated by the force in the prestressing steel, P, and any redundant sectional forces required for compatibility.

In the absence of externally applied loads, the resultant compressive force in the concrete, F_c, is equal and opposite to the tensile force in the steel, P. In statically determinate systems, these two forces act at the centroid of the prestressing steel. The moment due to prestressing is thus P times the eccentricity of the prestressing force, e. The sectional forces in the statically determinate system, denoted S_{0P}, induce deformations in the structure. Redundant sectional forces are therefore required for compatibility in statically indeterminate systems. The total sectional force due to prestressing, S_P, can be formulated as follows:

$$S_P = S_{0P} + S_{sP}$$

where S_{0P} is the sectional force in a statically determinate primary system, and S_{sP} is the redundant force. Since S_{sP} is a function only of the prestressing force P, it remains essentially unchanged when the stiffness of the system is reduced by cracking under service conditions. A significant decrease in S_{sP} occurs only after substantial plastic rotation at ultimate limit state has taken place.

Under service conditions, both components of S_P must be considered. For frame structures, it is usual to calculate S_{0P} for a statically determinate primary system and then to solve for S_{sP} from compatibility conditions using elastic theory. For slabs, S_P is normally calculated directly from equivalent loads due to the deviation and anchor forces of the tendons. In general, the axial force due to prestressing need not be calculated.

At ultimate limit state, the difference between prestressed and unprestressed cross-sections no longer exists; the sectional forces S_{0P} have no significance at ultimate limit state. Regardless of prestress under service conditions, flexural resistance is equal to the yield force in the steel, F_{Py}, times the internal lever arm, z. For statically determinate structures, therefore, design at ultimate limit state is based on the following expression:

$$M_d \leq \frac{1}{\gamma_R} F_{Py}\, z$$

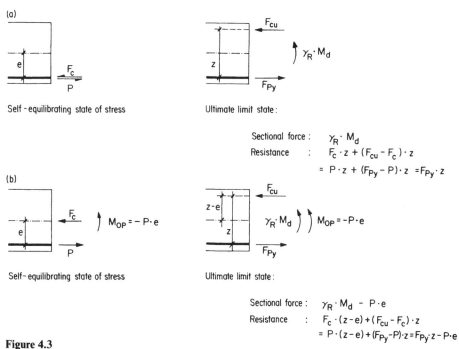

Figure 4.3
Analytical formulation of the ultimate limit state of prestressed concrete sections: a, without considering Pe; b, considering Pe

This inequality can, however, be artificially reformulated to include $M_{0P} = -Pe$ as a sectional force:

$$\gamma_R \cdot M_d - Pe \leqq P\,(z-e) + (F_{Py} - P)\,z$$

As shown in figure 4.3, these two expressions are identical.

The contribution of the redundant sectional forces due to prestressing at ultimate limit state is a function of the amount of plastic deformation expected. The forces S_{sP} are not considered when the design is based on the ultimate load of the system, since the compatibility conditions need not satisfied after full redistribution of the sectional forces has occurred. As discussed in Part (a) of this section, designs are normally loosely based on the elastic solution, with limited plastic rotation and redistribution of sectional forces. In such cases, a substantial portion of S_{sP} must be considered. The contribution of S_{sP} can be incorporated into the design using the factor γ_P:

$$S_d + \gamma_P\, S_{sP} \leqq \frac{1}{\gamma_R}\, R$$

The value of γ_P can be freely chosen, but should normally be between 0.8 and 1.4. Different values of γ_P can be selected for each redundant force of the primary system. For obvious reasons, however, γ_P must be constant for all of the sectional forces induced by a given redundant force.

c) Restrained Deformations

Restrained deformations due to temperature change, shrinkage, creep, and support displacements induce sectional forces that are directly proportional to stiffness, and hence highly sensitive to any change in stiffness. Since creep strains and cracking readily occur under service conditions, reductions in stiffness are likely even at relatively low stress levels. A linear relationship between restrained deformations and the sectional forces they induce thus does not exist.

Under service conditions, sectional forces are of interest only for the calculation of crack widths and deformations. The contribution of restrained deformations to cracking and deformations, however, can be calculated more easily and reliably using geometrical, as opposed to statical, methods (see Sections 4.8.4 and 4.8.5). The sectional forces due to restrained deformations under service conditions, therefore, need not be calculated.

At ultimate limit state, the sectional forces due to restrained deformations disappear completely with the formation of plastic hinges. Restrained deformations need therefore only be considered in systems of limited ductility, where the assumption of plastic deformations is not valid.

d) Summary

The principal differences between loads, prestressing, and restrained deformations are summarized in table 4.1.

Table 4.1
Comparison of Loads L, Prestressing P, and Restrained Deformations RD

	Sectional Forces in Primary System	Redundant Forces	Effect of Cracking
L	Function of primary system	Function of primary system	Small redistribution of sectional forces
P	Independent of primary system	Independent of primary system	Small redistribution of sectional forces
RD	No sectional forces	Independent of primary system	Strong reduction of sectional forces

4.3 Calculation of the Resistance of the Cross-Section

4.3.1 Fundamentals

Simple and rational models of cross-section resistance are always preferable to more complicated and less straightforward models that may promise an additional degree of precision. The desired accuracy of the resistance model must always correspond to the available accuracy of the load model. Load models for bridges are subject to considerable statistical variability. Live load, in particular, is prone to increases over the life of the bridge which are especially difficult to predict. The additional accuracy of a complicated resistance model normally has no significant impact on overall economy. Economy in structures, especially bridges, is not the result of refinements in analysis, but depends primarily on the overall concept of the structure itself and the construction procedure.

The interaction of different types of sectional forces must be considered in calculating the resistance of the cross-section. It is normally sufficient to combine the maximum value of one type of sectional force with the simultaneously occurring values of the remaining types. A detailed investigation of all possible combinations of sectional forces is usually not necessary, since the controlling combination will usually be apparent from inspection.

The calculation of cross-section resistance is based on the stress-strain diagrams of the materials used. Although stress-strain behaviour is influenced by many factors, for example rate of strain, their effect on the resistance of the cross-section can either be neglected or can be accounted for in a simplified manner. Simplified stress-strain diagrams should therefore be used for each of the constituent materials of prestressed concrete: concrete, reinforcing steel, and prestressing steel.

The concrete stress-strain diagram used for the design of structural components is based on the stress-strain diagram of unreinforced cube, prism, or cylinder specimens. The characteristic shape of the stress-strain curve of laboratory specimens (see figure 3.5) is normally simplified. Swiss standard SIA 162 prescribes the solid curve shown in figure 4.4, composed of a parabolic and a linear segment. It also permits the use of the further simplified rectangular diagram shown dashed, and defined as follows:

$$\sigma_c = 0 \quad \text{for} \quad |\varepsilon_c| \leq |0.2\varepsilon_{cu}|$$
$$\sigma_c = f_c \quad \text{for} \quad |\varepsilon_c| > |0.2\varepsilon_{cu}|$$

The *design ultimate compressive strain*, ε_{cu}, is specified as 0.0035.

The parameter f_c denotes the *design value* of concrete compressive strength, used in calculations of cross-section resistance. It is defined by the following equation in SIA 162:

$$f_c = 0.65 f_{cw,\,min} \tag{a}$$

where $f_{cw,\,min}$ is the specified minimum cube strength. The factor 0.65 accounts for:

1. The decrease in compressive strength with decreasing rate of strain. (The specified value of $f_{cw,\,min}$ corresponds to a rate of 0.001 mm/mm per minute (SIA 162/1). Since dead load is the most important component of total load, the design value of compressive strength should correspond to a lower rate of strain. Rüsch (1960) has reported that compressive strength for sustained loads is roughly 80 percent of $f_{cw,\,min}$.)
2. The difference between the strength of cube specimens and prism specimens ($f_{cp,\,min} \cong 0.85 f_{cw,\,min}$)
3. The difference in strength between prism specimens and flexural compression zones

Design shear strength, τ_c, is expressed as a function of $f_{cw,\,min}$ in SIA 162:

$$\tau_c = 0.4 + 0.02 f_{cw,\,min} \tag{4.1}$$

where τ_c and $f_{cw,\,min}$ are in N/mm². Equations (a) and (4.1) are valid provided $f_{cw,\,min}$ is between 10 N/mm² and 40 N/mm².

The diagrams of figure 4.4 give only compressive stress-strain behaviour; the tensile strength of concrete is neglected. The strength of concrete in tension should not, in general, be relied on. Temperature, shrinkage, creep, and settlement of supports induce self-equilibrating states of stress which cannot be reliably estimated or controlled. The tensile stresses due to these actions can easily exceed the tensile strength of the concrete, destroying its capacity to resist loads in tension. Tensile strength must, however, be considered in thin slabs without shear reinforcement, since tensile stresses transverse to the plane are necessary for shear resistance. High self-equilibrating tensile stresses transverse to the plane are effectively prevented provided the slab is sufficiently thin.

SIA 162 prescribes one bilinear stress-strain diagram for all types of reinforcing steel (fig. 4.5). The simplified diagram corresponds well to the actual behaviour of naturally hard steel. For cold-formed steel, which does not exhibit a distinct yield plateau, the approximation is is conservative. Stresses in the simplified diagram are less than the actual stresses in cold-formed steel for strains greater than 0.002.

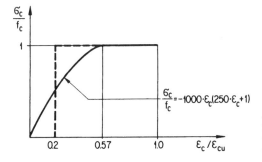

Figure 4.4
Design stress-strain diagrams for concrete adapted from SIA 162 ($\varepsilon_{cu} = -0.0035$)

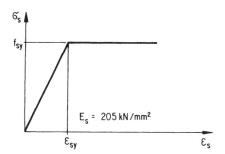

Figure 4.5
Design stress-strain diagram for reinforcing steel
(SIA 162)

Strain hardening is not considered in the design stress-strain curve. The design yield stress, f_{sy}, is defined as the specified minimum yield stress $f_{sy, min}$.

SIA 162 also specifies a bilinear design stress-strain diagram for prestressing steel (fig. 4.6). The design yield stress, f_{Py}, is defined in terms of the specified minimum tensile strength $f_{Pt, min}$. For cold-drawn wires and strands, f_{Py} is specified as $0.9 f_{Pt, min}$. For bars, f_{Py} is specified as $0.8 f_{Pt, min}$.

Bond stresses between post-tensioning steel and concrete must be transferred through the grout and duct material. The transfer of forces is thus not as effective as between mild reinforcing steel and concrete, which is direct. The increase in steel stress in the vicinity of cracks will thus be greater for mild reinforcing steel than for post-tensioning steel. This phenomenon need not be considered, however, in calculating the resistance of cross-sections with mild and prestressed reinforcement, since both types of steel reach their respective yield stress at ultimate limit state.

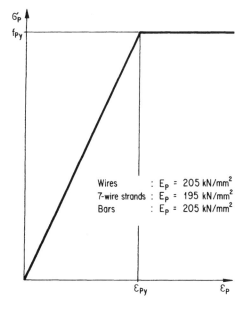

Figure 4.6
Design stress-strain diagram for prestressing
steel (SIA 162)

The stress in prestressing steel used for the calculation of ultimate resistance must be consistent with the actual deformations of the structure. The increase in stress in unbonded tendons, for example, is a function of the total tendon elongation between anchors. A special investigation is therefore required to calculate stresses in unbonded prestressing steel at ultimate limit state (see Section 4.6.7). The contribution of longitudinal prestressing to shear resistance must normally be computed using the effective prestress, σ_P, and not f_{Py}. Since these tendons are practically horizontal, any significant increase in steel stress requires a corresponding increase in strain in the longitudinal direction. Increases in steel stress due to shear deformations may only be considered for tendons inclined at angles greater than 45° with respect to the longitudinal axis of the member.

It is normally assumed that reinforcing and prestressing steel are effective only in axial tension and compression. Doweling action in the reinforcement is therefore neglected in the calculation of cross-section resistance.

4.3.2 Flexure and Axial Force

The calculation of combined flexural and axial resistance is based on the assumption that plane sections remain plane during deformation. Strain distributions at ultimate limit state are thus linear. Stresses in concrete and steel are readily determined from the design stress-strain diagrams for an arbitrary ultimate state of strain. The ultimate resistance (M_R, N_R) is then obtained by integrating these stresses.

For simple flexure, the strain distribution at ultimate limit state is based on an extreme compression fibre strain equal to the design ultimate compressive strain, $\varepsilon_{cu} = -0.0035$, and on the equilibrium condition $N_R = 0$, which implies that

$$F_{cu} = F_{Py} + F_{sy}$$

This state of equilibrium is shown in figure 4.7.

To ensure that the cross-section is sufficiently ductile to deform plastically, the steel strain at ultimate limit state should be at least 2.5 times the yield strain, ε_{sy}. Assuming $\varepsilon_{sy} = 0.0022$, this condition is equivalent to the following inequality:

$$x \leqq 0.4\,d \tag{4.2}$$

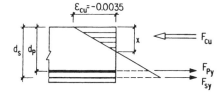

Figure 4.7
F_{Py} Ultimate state of strain in flexure (extreme fibre
F_{sy} compressive strain $= -0.0035$)

where x is the depth of the neutral axis and d is the distance from the extreme compression fibre to the centroid of tension reinforcement. When the concrete in the compression zone is properly confined with transverse ties, the extreme fibre compressive strain can be increased beyond -0.0035. In such cases, however, the depth of the neutral axis, x, should normally not exceed $0.5\,d$.

The resistance model shown in figure 4.7 is valid provided all reinforcement is bonded to the concrete. The stress in unbonded prestressing tendons at ultimate limit state must be specially calculated, taking into account the overall deformations of the structural system (see Section 4.6.7). Otherwise, it can be conservatively assumed that only the effective prestress, σ_P, is available.

The resistance of the cross-section to the combined action of flexure and axial force is normally obtained from $M_R - N_R$ interaction diagrams, which give all possible combinations of ultimate moments, M_R, and ultimate axial forces, N_R. The use of interaction diagrams is preferable to the calculation of one component of resistance (M_R or N_R) for a given value of the other component, since an iterative computation is required. The calculation of the complete diagram is normally based on the ultimate states of strain shown in figure 4.8 b. Strain at the extreme compressive fibre is limited to -0.0035; maximum strain in the outer layer of tension reinforcement is normally chosen between 0.005 and 0.007. A typical interaction diagram is shown in figure 4.9.

The design sectional forces in slender compression members depend on the deformations of the system at ultimate limit state, which are a function of the ultimate state of strain and hence of ultimate resistance. For slender columns, therefore, sectional forces and cross-section resistance are coupled. States of strain should thus be chosen not to maximize cross-section resistance, but rather to result in a reasonable combination of sectional force and resistance that corresponds to a realistic value for the ultimate load.

(a)

(b)

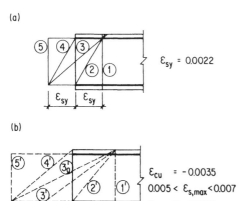

$\varepsilon_{sy} = 0.0022$

$\varepsilon_{cu} = -0.0035$

$0.005 < \varepsilon_{s,max} < 0.007$

$\varepsilon_{sy} = 0.0022$

Figure 4.8
Ultimate states of strain for flexure plus axial force: a, reduced resistance, used for slender members; b, effective resistance.

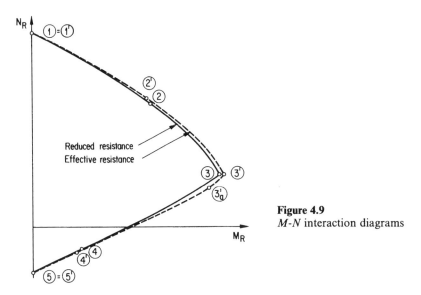

Figure 4.9
M-N interaction diagrams

The ultimate states of strain shown in figure 4.8b correspond to large curvatures and, for slender members, large system deformations and second-order moments. Although the resistance of the cross-section is maximized, the associated sectional forces are disproportionately large. The ultimate load is actually reached at a state of strain with smaller curvatures. The *reduced* states of strain shown in figure 4.8a are defined by extreme fibre strains equal to the yield strain of the steel, ε_{sy}. They result in a cross-section resistance that is only slightly lower (fig. 4.9), yet correspond to substantially smaller sectional forces due to reduced curvature. These states of strain yield a much improved estimate of the ultimate load of slender members.

The resistance of slender members subjected to flexure and axial force is discussed further in Section 8.1.3.

4.3.3 Shear

The resistance of cross-sections to shear is calculated using a truss model. The web members of the truss consist of concrete compression struts and steel tension ties. The shear resistance of the cross-section is equal to $V_{R,s}$, the resistance of the tension ties (stirrups). The resistance of the concrete compression struts, $V_{R,c}$, must be greater than $V_{R,s}$ to ensure ductile behaviour at ultimate limit state.

A simple Warren truss can be used (fig. 4.10). The angle of inclination of the compression diagonals, α, should not deviate more than 15° from α_0, the inclination of principal compressive stresses in the webs at the controid of the uncracked section. For members that are not axially compressed by load or

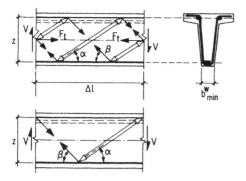

Figure 4.10
Truss model for shear resistance

prestressing, therefore, α can be chosen between 30° and 60°. Since prestressing reduces the value of α_0, correspondingly smaller values of α can be used. The panel length of the truss, Δl, is given by the following equation:

$$\Delta l = z\,(\cot\alpha + \cot\beta)$$

where z is the distance between centroids of the top and bottom longitudinal reinforcement and β is the angle of inclination of the stirrups.

A typical element of length Δl is loaded by shear forces V as shown in figure 4.10. The component of V resisted by the stirrups is $V/\sin\beta$. Assuming the horizontal spacing between stirrups is s, the following equation can be used to obtain the required area of steel, A_s^w:

$$\frac{V}{\sin\beta} = A_s\,f_{sy} = \frac{A_s^w\,\Delta l}{s}\,f_{sy}$$

The shear resistance of the stirrups is thus

$$V_{R,s} = \frac{A_s^W\,f_{sy}}{s}\,z\,(\cot\alpha + \cot\beta)\sin\beta \tag{4.3}$$

A longitudinal tensile force, $F_t = V_{R,s}\,(\cot\alpha - \cot\beta)$, is required for equilibrium in the element. This force acts at the centroidal axis of the element and is the sum of the horizontal components of the forces in the diagonals. Tension F_t can be distributed equally to the top and bottom chords of the girder and combined with the flexural tensile and compressive forces for the design of the reinforcement.

The total compressive force resisted by the concrete diagonals is $V/\sin\alpha$. This force is applied over an area $A_{c,\,\min}$, defined as follows (fig. 4.11):

$$A_{c,\,\min} = b_{\min}^w\,\Delta l\sin\alpha = b_{\min}^w\,z\,(\cot\alpha + \cot\beta\,\sin\alpha)$$

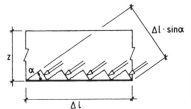

Figure 4.11
Compression strut of figure 4.10 distributed over panel length Δl

where $b^w_{\min}$ is defined in figure 4.10. Cracks induced by shear reduce the compressive strength of the concrete in the direction of the compression diagonals. The design compressive strength of the diagonals, $f_{c,\,red}$, can be calculated using the following equation, adapted from SIA 162:

$$f_{c,\,red} = f_c \left(1 - 0.33 \frac{\alpha_0}{45°} \right) \sin \beta \qquad (4.4)$$

(This expression has general applicability and can also be used, for example, in calculating the transfer of compressive force from the bottom slab of a box section to the webs (see Section 5.3.3).) The concrete compression component of shear resistance is thus equal to

$$V_{R,c} = f_{c,\,red}\, A_{c,\,min}\, \sin \alpha = f_{c,\,red}\, b^w_{\min}\, z (\cos \alpha + \cot \beta\, \sin \alpha) \sin \alpha \qquad (4.5)$$

The vertical component of tension in inclined prestressing tendons also contributes to the shear resistance of the section. The total shear resistance, V_R, can be expressed as $V_{R,s} + V_{R,P}$, where the contribution of prestressing, $V_{R,P}$, is defined as follows:

$$V_{R,P} = P \sin \beta_P$$

(The parameter P denotes force in the prestressing steel; β_P is the angle of inclination of the tendon.) The value of P must be consistent with the actual deformations of the structure. Unless a detailed investigation shows otherwise, P should be taken as the effective prestressing force after all losses, P_∞, when β_P is less than 45°.

The shear resistance of non-prismatic beams must consider the contribution of the inclined tensile and compressive forces in the chords of the truss model. In addition, the element length, Δl, used to calculate $V_{R,s}$ in equation (4.3) is changed. Figure 4.12 shows an idealized case in which the upper and lower chords have the same angle of inclination, $\delta/2$, with respect to the axis of the member. For vertical stirrups, the element length Δl is given by

$$\Delta l = z \frac{\tan \alpha}{\tan^2 \alpha - \tan^2 \dfrac{\delta}{2}}$$

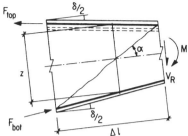

Figure 4.12
Model for the shear resistance of non-prismatic beams

This equation reduces to

$$\Delta l = z \cot \alpha$$

when $\delta/2$ is small relative to α. Shear force is considered perpendicular to the axis of the member. For the model of figure 4.12, therefore, the contribution of the chord forces, $F = \pm M/z$, is given by

$$V_{R,M} = \pm 2F \sin \frac{\delta}{2} \cong \pm \frac{M}{z} \sin \delta$$

Resistance can be either increased or decreased by $V_{R,M}$, depending on the sign of M.

4.3.4 Torsion

Torsion in hollow box and solid sections is resisted by a closed shear flow. The torsional resistance of both can thus be modeled using an idealized hollow section, characterized by its *effective area*, A_{ef}, and effective wall thickness, t_{ef}. The effective area is defined in SIA 162 as the area of the polygon enclosing the longitudinal reinforcement for torsion. For hollow sections, t_{ef} is the actual wall thickness; for solid sections, t_{ef} is equal to $d_0/8$, where d_0 is the diameter of the largest circle that can be inscribed in A_{ef}.

The constant shear flow, v, induced by torsional moment T is

$$v = \frac{T}{2 A_{ef}}$$

Integrating v over element i of the cross-section yields the torsional shear force, $V_i = z_i^T v$, where z_i^T is the effective depth of element i (fig. 4.13).

The resistance of element i to V_i is calculated using a truss model, as described in Section 4.3.3. To ensure a state of equilibrium between the compressive force in

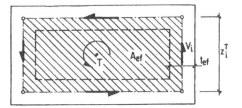

Figure 4.13
Model for torsional resistance

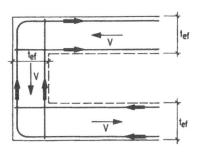

Figure 4.14
Transverse reinforcement on both faces of the walls of hollow sections required for equilibrium

the concrete struts and the tensile force in the steel, reinforcement must be provided at the interior and exterior faces of the walls of the effective section (fig. 4.14). The torsional resistance of the entire section is governed by the individual element with the least shear resistance.

4.3.5 Bending Resistance of Slabs and Tensile Resistance of Panels

The bending resistance of slabs can be calculated as described in Section 4.3.2 for beams. The bending resistance per unit width of slab, m_R, is given by the following equation (fig. 4.15):

$$m_R = a_s f_{sy} (d - 0.4 x)$$

where a_s is the area of reinforcement per unit width and the depth of the neutral axis, x, is given approximately by

$$x = 1.25 \frac{a_s f_{sy}}{f_c}$$

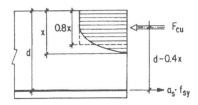

Figure 4.15
Model for the flexural resistance of slabs

Alternatively, m_R can be expressed in terms of the mechanical reinforcement ratio, ω:

$$m_R = \omega d^2 f_c \left(1 - \frac{\omega}{2}\right) \tag{a}$$

where

$$\omega = \frac{a_s}{d} \frac{f_{sy}}{f_c}$$

Slabs have a greater capacity for moment redistribution than beams, especially when load is carried in two directions. Significant redistributions occur under service conditions at the formation of the first cracks. Inequality (4.2) can therefore be relaxed somewhat, even for unconfined concrete; adequate ductility is ensured then the following inequality is satisfied:

$$x \leqq 0.5 d$$

Reinforcement that is not orthogonal or that is arranged in more than two directions occurs frequently in bridge construction, particularly in slab bridges. In such cases, flexural resistance can be calculated from an equivalent distribution of orthogonal reinforcement. Standard formulas for the transformation of coordinates are used. Given n arbitrarily oriented groups of parallel bars with corresponding mechanical reinforcement ratios ω_1 through ω_n, equivalent reinforcement ratios ω_x, ω_y, and ω_{xy} are obtained for an arbitrarily chosen set of $x - y$ axes using the following equations:

$$\omega_x = \sum_i \omega_i \cos^2 \gamma_i$$

$$\omega_y = \sum_i \omega_i \sin^2 \gamma_i$$

$$\omega_{xy} = \sum_i \omega_i \sin \gamma_i \cos \gamma_i$$

where the angle γ_i is shown in figure 4.16. These values are then transformed into the equivalent orthogonal reinforcement ratios, ω_1 and ω_2, in the principal directions ξ and η:

$$\omega_1 = \omega_\xi = \omega_x \cos^2 \phi_1 + \omega_y \sin^2 \phi_1 + \omega_{xy} \sin 2\phi_1$$

$$\omega_2 = \omega_\eta = \omega_x \sin^2 \phi_1 + \omega_y \cos^2 \phi_1 - \omega_{xy} \sin 2\phi_1$$

where

$$\tan 2\phi_1 = \frac{2\omega_{xy}}{\omega_x - \omega_y}$$

The corresponding flexural resistances $m_{R\xi}$ and $m_{R\eta}$ are calculated using equation (a):

$$m_{R\xi} = \omega_\xi d^2 f_c \left(1 - \frac{\omega_\xi}{2}\right) \qquad m_{R\eta} = \omega_\eta d^2 f_c \left(1 - \frac{\omega_\eta}{2}\right)$$

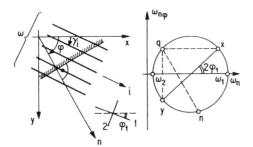

Figure 4.16
Transformation of the mechanical
reinforcement ratio, ω

These transformation equations can also be applied to the tensile resistance of
arbitrarily reinforced panels subjected to in-plane loading.

4.4 Safety of Beams, Slabs, and Panels

4.4.1 Beams

The reinforcement is designed to resist the tensile forces due to the combined
effects of all simultaneously occurring sectional forces (flexure, axial force, shear,
and torsion). These sectional forces must also be used to check the assumed
concrete cross-section dimensions.

4.4.2 Slabs

The flexural resistance of slabs is adequate provided the following inequalities are
satisfied at all locations on the slab and for all directions n:

$$-\frac{m'_{Rn}}{\gamma_R} \leq m_{dn} \leq \frac{m_{Rn}}{\gamma_R} \tag{a}$$

where m_{dn} is the design bending moment, m_{Rn} is flexural resistance for positive
moments, and m'_{Rn} is flexural resistance for negative moments. (Flexural resistance
is always considered positive.)

Inequality (a) can be rewritten in the following form when the reinforcement is
arranged in orthogonal directions x and y (Wolfensberger 1964):

$$\left(\frac{m'_{Rx}}{\gamma_R} + m_{dx}\right)\left(\frac{m'_{Ry}}{\gamma_R} + m_{dy}\right) + m_{dxy}^2 \geq 0$$

and   (4.6)

$$\left(\frac{m_{Rx}}{\gamma_R} - m_{dx}\right)\left(\frac{m_{Ry}}{\gamma_R} - m_{dy}\right) - m_{dxy}^2 \geq 0$$

where the quantities in parentheses must be non-negative. The moments m_x, m_y, and m_{xy}, are as shown in figure 5.38. Inequalities (4.6) can be linearized, yielding the following expressions:

$$-m_{dx} + |m_{dxy}| \leq \frac{m'_{Rx}}{\gamma_R} \qquad -m_{dy} + |m_{dxy}| \leq \frac{m'_{Ry}}{\gamma_R} \qquad (b)$$

and

$$m_{dx} + |m_{dxy}| \leq \frac{m_{Rx}}{\gamma_R} \qquad m_{dy} + |m_{dxy}| \leq \frac{m_{Ry}}{\gamma_R} \qquad (c)$$

Top reinforcement is designed using inequalities (b). If the left-hand sides of both expressions are positive, reinforcement is provided in both orthogonal directions. If both left-hand sides are negative, no top reinforcement is required in either direction. If the left-hand side of only one of the inequalities is negative, the reinforcement is designed using the following modified expressions:

Case 1: $\quad -m_{dx} + |m_{dxy}| \leq 0$: $\quad m'_{Rx} = 0 \quad$ and $\quad m_{dy} + \dfrac{m^2_{dxy}}{-m_{dx}} \leq \dfrac{m'_{Ry}}{\gamma_R}$

Case 2: $\quad -m_{dy} + |m_{dxy}| \leq 0$: $\quad m'_{Ry} = 0 \quad$ and $\quad m_{dx} + \dfrac{m^2_{dxy}}{-m_{dy}} \leq \dfrac{m'_{Rx}}{\gamma_R}$

If the left-hand side of the modified inequality is negative, no top reinforcement is required in either direction.

An identical procedure is followed for the design of bottom reinforcement, using inequalities (c). Reinforcement is provided in both directions if the left-hand side of both expressions is positive; no reinforcement is provided in either direction if both left-hand sides are negative. If the left-hand side of only one inequality is negative, one of the following expressions must be used:

Case 1: $\quad m_{dx} + |m_{dxy}| \leq 0$: $\quad m_{Rx} = 0 \quad$ and $\quad m_{dy} + \dfrac{m^2_{dxy}}{-m_{dx}} \leq \dfrac{m_{Ry}}{\gamma_R}$

Case 2: $\quad m_{dy} + |m_{dxy}| \leq 0$: $\quad m_{Ry} = 0 \quad$ and $\quad m_{dx} + \dfrac{m^2_{dxy}}{-m_{dy}} \leq \dfrac{m_{Rx}}{\gamma_R}$

No bottom reinforcement is required in either direction if the left-hand side of the modified inequality is negative.

When reinforcement is arranged non-orthogonally or in more than two directions, the design moments m_{dx}, m_{dy}, and m_{dxy} must first be transformed into the principal directions of the reinforcement, ξ and η, obtained as described in Section 4.3.5. The following expressions are used:

$$m_{d\xi} = m_{dx} \cos^2 \phi_1 + m_{dy} \sin^2 \phi_1 + m_{dxy} \sin 2\phi_1$$
$$m_{d\eta} = m_{dx} \sin^2 \phi_1 + m_{dy} \cos^2 \phi_1 - m_{dxy} \sin 2\phi_1$$
$$m_{d\xi\eta} = \tfrac{1}{2}(m_{dy} - m_{dx}) \sin 2\phi_1 + m_{dxy} \cos 2\phi_1$$

The inequalities given above for x-y coordinates are equally applicable to design moments and section resistances transformed into the equivalent orthogonal ξ-η coordinates.

Shear reinforcement is not required when the following condition is satisfied:

$$v_d \leq \frac{1}{\gamma_R} \tau_c z \tag{d}$$

where v_d is design shear force per unit width, z is effective depth, and τ_c is design shear strength (equation (4.1)). If inequality (d) is not satisfied, more detailed calculations are required. Shear reinforcement can be used where appropriate to increase the resistance of the slab. In all cases, however, a sufficient percentage of longitudinal reinforcement must be extended and properly anchored beyond the supports. Slabs must also be capable of resisting punching shear whenever concentrated loads or reactions are transferred to slabs.

Large torsional moments often occur at the free edges of slabs, particularly in skew slab bridges. The additional shear force induced by torsion at this location must be considered. A slab of depth h is assumed. According to the definition of sectional forces in slabs (Section 5.3.1), the torsional moment m_{xy} is equilibrated by an antisymmetrical, linear distribution of shear stresses. These stresses can be integrated into a shear flow, v, satisfying the following condition:

$$m_{xy} = v h_0$$

where $h_0 = 2h/3$. It thus follows that

$$v = \frac{3}{2} \frac{m_{xy}}{h}$$

The effective wall thickness, t_{ef}, is $h/3$. The additional shear force due to torsion, $V_x(m_{xy})$, is thus the product of the shear flow v and the internal lever arm $h_0 = 2h/3$ (fig. 4.17):

$$V_x(m_{xy}) = \left(\frac{3}{2}\frac{m_{xy}}{h}\right)\left(\frac{2h}{3}\right) = m_{xy}$$

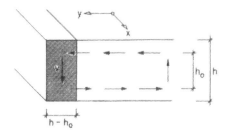

Figure 4.17
Additional shear due to torsion

The total sectional forces over an edge strip of width $h/3$ are thus:

$$M_x = m_x \frac{h}{3} \qquad V_x = v_x \frac{h}{3} + m_{xy}$$

The slab reinforcement transverse to the edge must be well anchored in the edge beam.

4.4.3 Panels

The reinforcement of panels can be designed and arranged on the basis of the internal forces obtained from a truss model. When the in-plane stress resultants from a plane stress model, n_x, n_y, and n_{xy}, are used instead, the following inequality must be satisfied for all directions n:

$$n_{dn} \leq \frac{n_{Rn}}{\gamma_R}$$

where n_{dn} is design tensile force and n_{Rn} is tensile resistance (fig. 4.18).

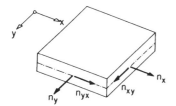

Figure 4.18
Sectional forces in panels

For panels reinforced in the orthogonal directions x and y, this condition can be transformed into the following inequality:

$$\left(\frac{n_{Rx}}{\gamma_R} - n_{dx}\right)\left(\frac{n_{Ry}}{\gamma_R} - n_{dy}\right) - n_{dxy}^2 \geq 0$$

which is analogous to inequalities (4.6) used for the flexural design of slabs in Section 4.4.2. The design of reinforcement arranged non-orthogonally or in more than two directions is accomplished using the transformation formulas given for slabs in Sections 4.3.5 and 4.4.2.

4.5 Detailing of Reinforcement

The flow of forces can be followed through a structure using a truss model composed of tension members, corresponding to the actual reinforcement, and idealized concrete compression members. The internal forces obtained from such a model will satisfy equilibrium. Truss models are particularly useful for developing proper details for anchorage of reinforcement, lap splices, and reinforcement for transverse tension.

4.5.1 Anchorage and Splicing of Reinforcement

Reinforcing bars are anchored by compressive forces in the concrete oriented at roughly 30° to the axis of the bar and acting opposite the tensile force (fig. 4.19). The length required for complete transfer of the yield force in the bar is called the *development length*, and is denoted l_b. Development length is a function of many factors, including:

1. Properties of the reinforcement: diameter, surface conditions, and any mechanical anchorage devices, such as hooks or plates
2. Tensile strength of the concrete
3. Concrete stress in the anchorage zone (transverse compression or transverse tension)
4. Location (top or bottom of the cross section) and orientation (vertical or horizontal) of the reinforcement
5. Distance between reinforcement and concrete surface

Values of l_b are normally specified in codes and standards. If the full yield stress of the bar is not used, development length can be reduced accordingly.

Tensile force can also be transferred from one bar to another by diagonal concrete compressive forces, inclined at roughly 45° to the axis of the bar (fig. 4.20). The tensile forces transverse to the splice, required for equilibrium, can often be resisted by the transverse reinforcement normally provided for crack control. When large diameter bars or prestressing tendons are lap spliced, however, the transverse tension must be resisted by specially designed transverse reinforcement.

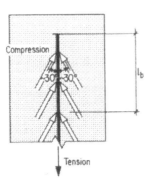

Figure 4.19
Flow of forces in anchorage zones (l_b: development length)

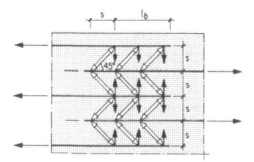

Figure 4.20
Flow of forces in lap splices

As shown in figure 4.20, the required lap splice length is the sum of the development length, l_b, and the spacing of the bars, s. The value of s should therefore be as small as possible. When large forces are to be transferred, however, s should not be less than the maximum diameter of aggregates.

4.5.2 Detailing of Reinforcement at Joints of Rigid Frames

Truss models are also useful for the detailing of reinforcement at corners of rigid frames, including the simple case of two solid rectangular sections and more complex cases involving hollow sections. The following examples describe the flow of forces and corresponding reinforcement arrangement for several typical cases.

Example 4.1:
Corner of frame under closing load, solid rectangular sections for beam and column

Moments that tend to close corners induce tension on the exterior faces and compression on the interior faces. The diagonal compressive force required for equilibrium deviates these forces at the corner (fig. 4.21 a). Since construction is facilitated by locating the construction joint level with the beam soffit, the reinforcement is normally spliced in the beam as shown in figure 4.21 a. Tensile force is transferred from one bar to the other by means of smaller compression diagonals (fig. 4.21 b).

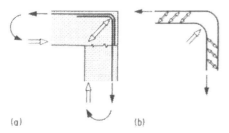

(a)   (b)

Figure 4.21
Frame corner under closing load: a, global flow of forces; b, local flow at lap splice

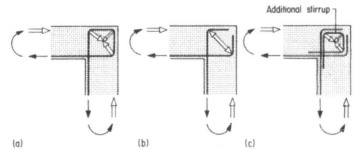

Figure 4.22
Recommended details for frame corner under opening load

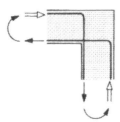

Figure 4.23
Frame corner under opening load: detail to be avoided

Example 4.2:
Corner of frame under opening load, solid rectangular sections for beam and column

Moments that tend to open corners induce tension on the interior faces and compression on the exterior faces. The state of equilibrium shown in figure 4.22a ensures a positive transfer of forces, regardless of the development length of the bars used. The residual tensile forces in the exterior corner are equilibrated by a diagonal compressive force. The detail of figure 4.22b can only be used when the hooks provided are sufficient to anchor the bars. Otherwise, additional closed stirrups must be used (fig. 4.22c).

The 90° hooks provided in the detail of figure 4.23 are ineffective in anchoring the bars, since the compression in the concrete acts in the same direction as the tension in the bar. This detail should therefore be avoided.

Example 4.3:
"T" joint under closing load, solid rectangular sections for beam and column

Figure 4.24a shows the correct reinforcement splice detail at a "T" joint. The tensile force is transferred directly from the horizontal to the vertical reinforcement. Transverse tension will be small provided the horizontal and vertical bars are not far apart transversely. The details shown in figures 4.24b and 4.24c cannot transfer the tensile force directly and should thus be avoided. Although

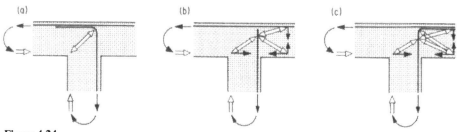

Figure 4.24
"T" joint: a, recommended detail; b and c, details to be avoided

equilibrium is possible in both cases, the additional tensile forces transverse to the main reinforcement may exceed the tensile strength of the concrete. Additional horizontal and vertical reinforcement will then be required. Furthermore, the embedment length provided for the vertical reinforcement in figure 4.24b may not be sufficient.

Example 4.4:
Transfer of moment from a box girder to a solid rectangular column

As shown in figure 4.25, moment ΔM can be transferred from a box girder to a solid column in essentially the same way as for the solid "T" joints of the previous

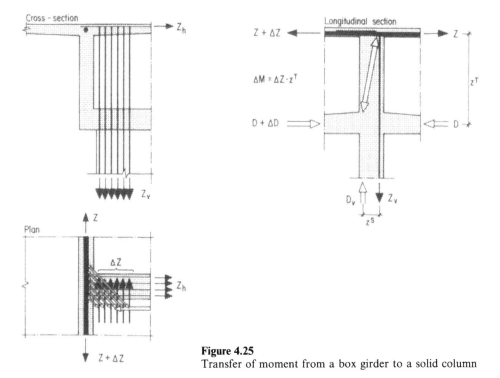

Figure 4.25
Transfer of moment from a box girder to a solid column

example. The girder reinforcement is usually localized at the top of the webs. The column steel that is carried up into the deck, however, is usually distributed uniformly over a greater width in the deck slab. Longer hooks for the column reinforcement and special reinforcement for transverse tension, Z_h, are therefore required. The vertical reaction from the girder is only fully transferred to the column below the girder soffit. The column reinforcement in the upper portion of the diaphragm must therefore resist the full tension, Z_v, without benefit of a relieving compressive force. The force Z_v can be considerably greater than ΔZ since the lever arm of the column is normally much smaller than the lever arm of the girder.

Example 4.5:
Transfer of moment from a box girder to a box column

A transfer of forces from box girder to box column as described in the previous example is only possible when the diaphragms are designed accordingly (fig. 4.26). As shown in figure 4.27, an analogous flow of forces cannot be established when

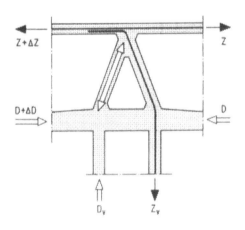

Figure 4.26
Transfer of moment from a box girder to a box column: triangular diaphragms

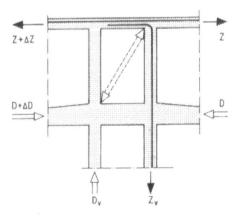

Figure 4.27
Transfer of moment from a box girder to a box column: vertical diaphragms

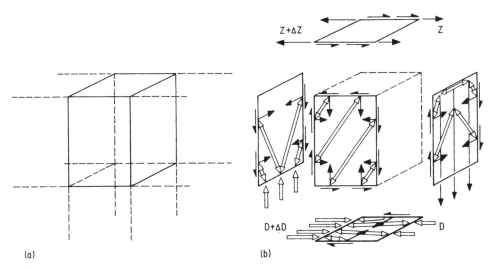

Figure 4.28
Transfer of moment from a box girder to a box column through shear in diaphragms and webs: a, schematic arrangement; b, flow of forces

the flanges of the column are carried into the girder as two vertical diaphragms. It is therefore necessary to transfer the moment in a more roundabout way, through shear in the webs and diaphragms. The corresponding flow of forces, shown in figure 4.28, is then used to design and detail the reinforcement. Transverse reinforcement required for the transfer of shear from one panel to another (web, slab, or diaphragm) will be fully stressed at the edges of the panels and must therefore be properly anchored.

4.5.3 Flow of Forces in Panels

The flow of forces in panels can also be reliably determined using truss models. Although simple models with relatively widely-spaced truss elements can be used, the arrangement of the compression elements should not deviate significantly from the compressive stress trajectories of the elastic solution. The compressive forces induced by a concentrated load should thus spread out at roughly 30° on either side of the line of action of the load. Away from zones of local disturbance, a linear state of strain can be assumed. The flow of forces from load to reaction should be direct and should minimize deformations. Deflections of the nodes of the truss, which result in large isolated cracks, should be prevented by supplementary tension members.

Elastic stress trajectories and corresponding truss models are given in figure 4.29 for several typical cases.

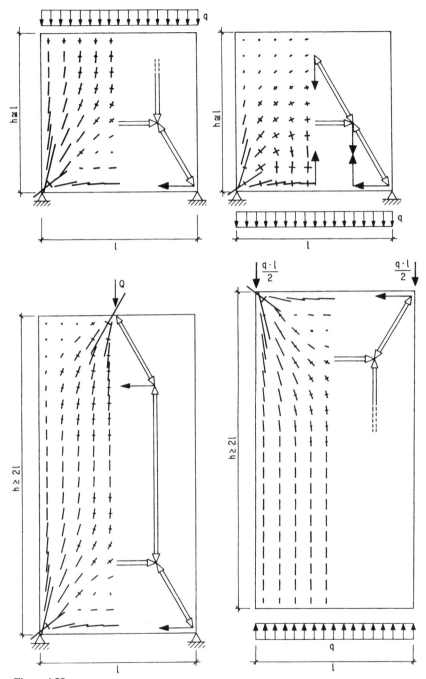

Figure 4.29
Stress trajectories and truss models for panels: a, uniform load applied to upper edge; b, uniform load applied to lower edge; c, tall panel with concentrated load at centreline; d, tall panel with concentrated loads at panel edges; e, square panel with edge shear forces; f, square panel with edge shear forces and axial load ($\Delta n_y b = 2n_{yx} \Delta l$)

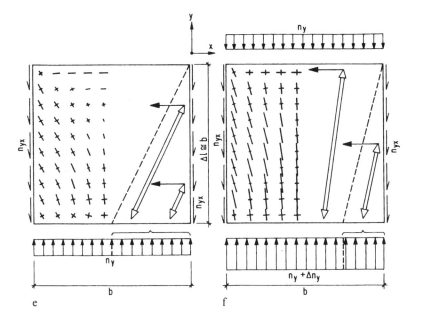

4.5.4 Flow of Forces in Box Girders and T-Girders

Box girders and T-girders can be regarded as prismatic folded plate structures, in which the principal sectional forces are resisted by in-plane membrane forces. The flow of internal forces can therefore be represented by truss models. The introduction of concentrated loads into the structure and the propagation of internal forces are modelled exactly as for panels under in-plane loading, as described in Section 4.5.3.

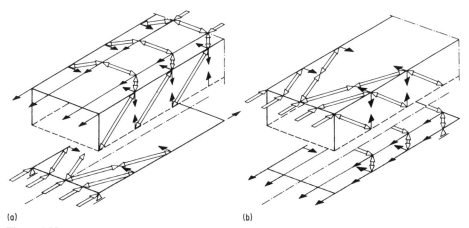

(a)   (b)

Figure 4.30
Flow of forces in box girders: a, at an intermediate support; b, at an end support

Using truss models, it is easy to show that the main longitudinal reinforcement should be anchored in the webs or as close to the webs as possible. This arrangement results in the shortest flow of forces and minimizes transverse reinforcement. When the reinforcement is anchored in the bottom slab at the ends of girders, equilibrium can only be satisfied when the bottom slab is extended beyond the supports (fig. 4.30).

4.6 Prestressing

4.6.1 The Purpose of Prestressing

Prestressing is a special state of stress and deformations which is induced to improve structural behaviour. Structures can be prestressed either by artificial displacements of the supports or by steel reinforcement that has been pre-strained before load is applied. The forces induced by the former method are sharply reduced by creep and shrinkage and are generally ineffective at ultimate limit state. Due to these inherent disadvantages, support displacements are rarely used for prestressing. The forces induced by pre-strained reinforcement can, however, survive the effects of shrinkage and creep, provided the initial steel strain is sufficiently larger than the anticipated shortening in the concrete. The required pre-strains are best achieved using high-strength steel. High-strength steel that has been pre-strained can normally be stressed to its full yield strength at ultimate limit state.

High-strength steel for prestressing is available in the form of wires, strands, and bars. The properties of each of these types of steel are described in Section 3.1.3. The use of prestressing steel generally results in lower construction costs as compared to conventional reinforced concrete. Prestressing steel has a lower ratio of unit cost to yield stress than mild reinforcing steel. Furthermore, it occupies less space in the framework, thus making possible the use of lighter, more slender concrete cross-sections.

There is no substitute for prestressing in the construction of long-span concrete girders, since these structures are neither technically nor economically feasible in conventional reinforced concrete. Without prestressing, the arrangement of reinforcement is difficult and large deformations will be produced, even for relatively short spans. For example, a conventionally reinforced, simply supported bridge with span 40 m and depth 2 m will deflect 50 to 60 mm immediately after the removal of falsework. Due to shrinkage and creep, this deflection will increase to a long-term value of 120 mm to 150 mm.

Prestressing can be full, limited, or partial. *Full prestressing* is designed to eliminate concrete tensile stresses in the direction of the prestressing under the action of design service loads, prestressing, and restrained deformations. In structures with *limited prestressing*, the calculated tensile stresses in the concrete must not exceed a specified permissible value. Behaviour at ultimate limit state must nevertheless be checked in both cases. *Partial prestressing* places no restrictions on concrete tensile stresses under service conditions. Concrete stresses need not, therefore, be calculated. Partial prestressing encompasses the entire range of possibilities from conventionally reinforced to fully prestressed concrete. Designs must ensure adequate behaviour at ultimate limit state and under service conditions, both of which must be verified directly.

Partial prestressing is generally more economical than full or limited prestressing. Although structures that are partially prestressed require a significant portion of mild reinforcement for crack control and distribution, this steel contributes to the ultimate resistance of the section. Whatever mild steel is added to improve behaviour under service conditions thus reduces the amount of prestressing steel necessary for safety at ultimate limit state. The prestressing force must always be carefully monitored during construction, since deviations from the prescribed prestressing force can lead to cracking, deformations, and fatigue.

In structures with limited or full prestress, the area of prestressing steel necessary to control tensile stress under service conditions, after all prestressing losses have occurred, usually exceeds the area required for safety at ultimate limit state. This inefficient use of prestressing cannot be compensated by elimination of mild reinforcement, since cracking due to self-equilibrating stresses and restrained deformations can occur in all structures, even those that are fully prestressed. Mild reinforcement in fully prestressed structures has only a negligible effect on concrete stresses under service conditions and cannot, therefore, reduce the amount of prestressing steel required. In addition, full prestressing often causes large upward deflections in slender sections when the ratio of live load to dead load is high.

4.6.2 Methods of Prestressing

Structures can be prestressed either by *pre-tensioning* or *post-tensioning*.

Pre-tensioning is used primarily for the prefabrication of concrete components. The prestressing steel is stressed between fixed abutments, forms are installed around the steel, and the concrete is cast. After the concrete has hardened, the prestressing steel is detached from the abutments. Anchorage of the steel, and hence the transfer of prestressing force from steel to concrete, is achieved entirely through bond stresses at the ends of the member.

In post-tensioned construction, the prestressing steel is only stressed after the concrete has been cast and hardened. The steel must therefore be enclosed in ducts and anchored using special devices. The ducts are most commonly embedded in the concrete and filled with grout after stressing to bond steel to concrete and to provide protection against corrosion. The ducts can also be located outside of the concrete section and left unbonded for the entire life of the structure. In such cases, grout is only a means of protecting the steel.

Pre-tensioning is usually more economical for large-volume precasting operations, since the costs of anchors and grouting can be eliminated. It is quite common, however, to combine both methods of prestressing in a given structure.

4.6.3 Post-Tensioning Systems

The principal post-tensioning systems differ primarily in the way the prestressing force is applied to the structure, the design of anchors, and the type of steel used.

Prestressing tendons in the *BBRV* system are composed of wires. At either end of the tendon, the wires pass through holes in a threaded anchor head and are individually anchored by small cold-formed knobs, called "buttonheads" (fig. 4.31). A steel collar, threaded onto the anchor head, bears against an anchor plate. The tendon is stressed by pulling the anchor head and tightening the collar up to the plate. The *Polensky and Zoellner* system is no longer produced, but has been widely used in the past. The system used either wires or strands anchored using a cone and sleeve (fig. 4.32). The *VSL* system anchors strands with steel wedges in a perforated steel anchor head (fig. 4.33). Steel wedges are also used for the anchors of the *Freyssinet* and *Dywidag* strand systems (fig. 4.34). Dywidag also has a prestressing system for individual threaded bars, anchored using a nut and an anchor bell (fig. 4.35).

Each prestressing system offers several different types of anchors to suit the requirements of many special conditions. These range from simple fixed ("dead-end") anchors to relatively complicated coupling devices. The symbols shown in figure 4.36 can be used on design drawings to identify the various anchor types.

Post-tensioning ducts must be fixed into place using supports rigidly mounted to the reinforcement cage (fig. 4.37). In the past, it was common practise to assemble

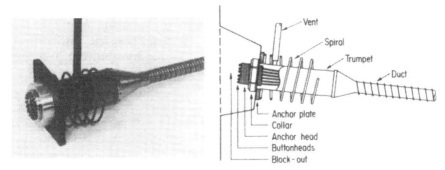

Figure 4.31
BBRV stressing anchor

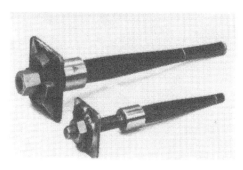
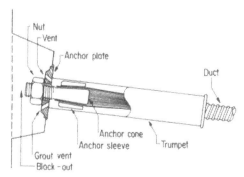

Figure 4.32
Polensky and Zoellner stressing anchors

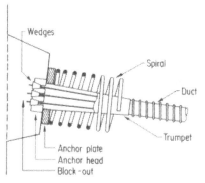

Figure 4.33
VSL stressing anchor

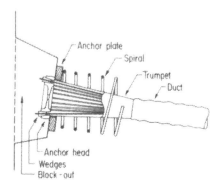

Figure 4.34
Freyssinet stressing anchor

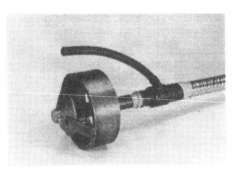

Figure 4.35
Dywidag stressing anchor for single bars

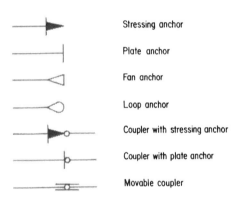

Stressing anchor

Plate anchor

Fan anchor

Loop anchor

Coupler with stressing anchor

Coupler with plate anchor

Movable coupler

Figure 4.36
Anchor symbols

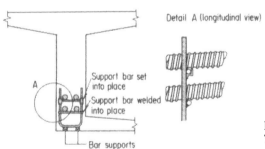

Figure 4.37
Tendon supports

prestressing steel, ducts, and anchors into tendons in the shop. The complete tendons were then placed into the forms with the reinforcing steel and concreted in. In present-day practise, however, it is most common to cast only ducts and anchor plates into the concrete. The prestressing steel is then placed into the duct after the concrete has hardened. Individual strands can be pushed into ducts directly from spools and then cut to length. Dents and the intrusion of cement into the empty ducts inhibit the placement of prestressing steel. Ducts must therefore be carefully inspected for defects.

4.6.4 Detailing

Large concentrated forces are transferred to the concrete at the anchors of post-tensioning tendons. The anchor force is resisted by compressive forces which spread out at 30° to 45° on either side of the tendon. The flow of forces in the region of disturbance must be carefully followed through the structure and checked at ultimate limit state, both when the tendons are stressed and after completion of the bridge. Reinforcement must be provided to resist the transverse tensile forces produced at the deviation points of the compressive forces.

The examples presented in figure 4.38 illustrate the use of simple truss models for the anchor zone. The flow of forces is a function of the following parameters: the angle formed by the compressive struts behind the anchor, tendon force P, support reaction A, and the state of stress (cracked or uncracked) away from the zone of disturbance.

Flexural reinforcement is only effective when it can be brought into equilibrium with an equal and opposite compressive force in the concrete. This is accomplished with diagonal compression struts, as shown in figure 4.39. The tensile and compressive forces are thus not located at the same section; the distance separating them is a function of the dimensions of the cross-section and the location of the reinforcement. This distance is called the *moment development length*, and is denoted l_a.

Figure 4.39 illustrates the concept of moment development length for two related cases: one tendon with yield force F_{Py} in the middle of the bottom slab and two tendons each with yield force $F_{Py}/2$ in either web. Included for each figure are normal stress diagrams computed from elastic theory, a corresponding truss model, and a diagram of the moment, $M(P)$, that can be equilibrated at a given section. The compressive forces in the truss model are assumed to spread out at an angle of roughly 45° without additional shear reinforcement. The moment development lengths obtained from elastic theory and from the truss model are roughly the same and are given by

$$l_a \cong \frac{b}{2} + h + \frac{b}{4} \quad \text{(tendon in bottom slab)}$$

and

$$l_a \cong h + \frac{b}{4} \quad \text{(tendons in webs)}$$

Low values of l_a generally correspond to efficient reinforcement layouts. The savings achieved by minimizing l_a are due not only to reduced tendon length, but also to the elimination of the transverse reinforcement required for equilibrium in

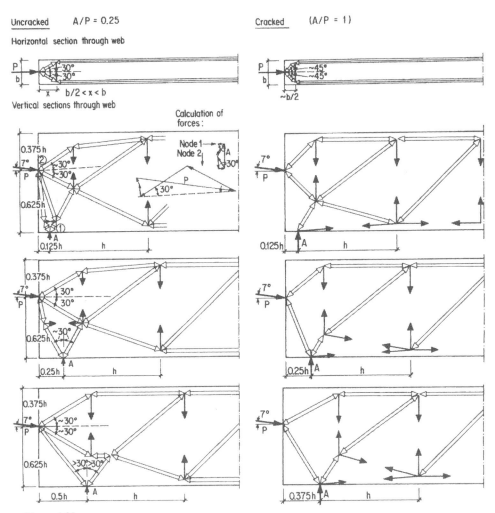

Figure 4.38
Truss models for anchor zones

the lower slab (fig. 4.39a). Whenever possible, therefore, tendons should be anchored in the webs themselves or as close to the webs as possible. These principles are also valid for mild reinforcing steel, provided it is not required in the slab for the control of cracks.

The tensile force in a curved tendon produces deviation forces, q_P, in the plane of curvature (fig. 4.40). These forces are resisted by the concrete; the local stresses thus induced must always be investigated. When only a thin layer of concrete is available to resist q_P, tendons must be anchored into the cross-section with properly designed reinforcement. In such cases, the strength of the concrete alone should never be relied on to restrain the pull-out of tendons, even if the

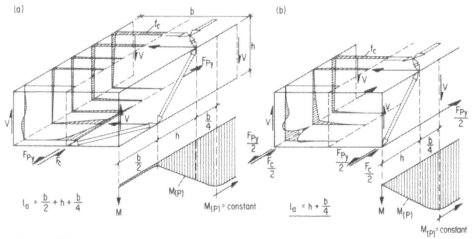

Figure 4.39
Moment development length l_a: a, tendons located in bottom slab; b, tendons located in webs

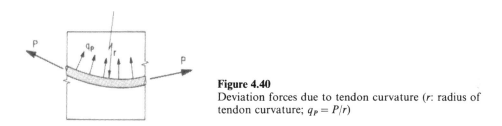

Figure 4.40
Deviation forces due to tendon curvature (r: radius of tendon curvature; $q_P = P/r$)

reinforcement required appears negligible. It is incorrect to assume that the curvature of the tendon is equal to the curvature of the girder; local concentrations of curvature, and hence local peaks in q_P, should be anticipated. Figure 4.41 shows several commonly occurring details in bridge construction that must be properly reinforced to resist the pull-out of tendons.

The state of stress induced by prestressing is self-equilibrating. The inward-directed deviation force due to tendon curvature, q_P, is therefore equal and opposite to the deviation force in the concrete, q_c. The force q_c is the integral of the outward-directed components of concrete compressive stress, σ_{cx}:

$$q_c = \int_{A_c} dq_c = \int_{A_c} \frac{1}{r} \sigma_{cx} dA_c$$

When the cross-section is compact and solid, q_P and q_c induce normal stresses transverse to the axis of the member (fig. 4.42). The nearer the tendon is located to the inside of the section, the greater and more dangerous are the transverse tensile stresses. In open cross-sections, the deviation forces induce normal stresses transverse to the axis of the member, transverse bending moments, and, under

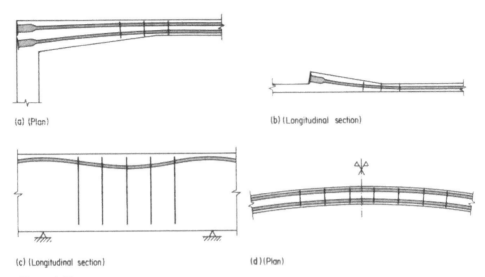

(a) (Plan)

(b) (Longitudinal section)

(c) (Longitudinal section)

(d) (Plan)

Figure 4.41
Reinforcement to resist pull-out of curved tendons: a, girder ends; b, bottom slab; c, twin piers; d, curved webs

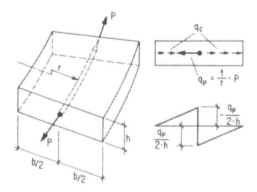

Figure 4.42
Transverse normal stresses in a curved girder with solid cross-section

certain circumstances, torsional moments. Figure 4.43 shows a curved, centrically prestressed I-girder without external load. The deviation forces produce transverse bending moments in the web. At the elevation of the tendon, assuming t is much greater than h, they are given approximately by

$$m = \frac{Ph}{8r(2b+h)}(4b+h)$$

Transverse bending can also be caused by locally curved tendons (figs. 4.41 a, b). In the region of local tendon curvature, the concrete compressive force that balances the tensile force in the tendon will have spread, at least partially, into the entire cross-section. Only a small component of the deviation force q_c will thus

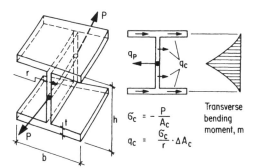

$$G_c = -\frac{P}{A_c}$$

$$q_c = \frac{G_c}{r} \cdot \Delta A_c$$

Transverse bending moment, m

Figure 4.43
Transverse bending moments in a curved, centrically prestressed I-girder

remain in the immediate vicinity of the tendon to resist the full deviation force q_P. This local imbalance of forces will induce transverse bending moments.

4.6.5 Analysis of Prestressed Cross-Sections

Prestressing with pre-strained steel reinforcement induces a self-equilibrating state of stress in the cross-section. The tensile force in the steel and the compressive force in the concrete, obtained from integration of the concrete stresses, are equal and opposite. In statically determinate structures, the two forces act at the same location in the cross-section. The sectional forces in the concrete due to prestressing can thus be easily determined from equilibrium. The following expressions are based on the assumption that the direction of the prestressing force deviates only slightly from a vector normal to the cross-section (fig. 4.44):

$$N_c = -P_x \qquad M_{c,x} = -P_x\left(a_y\frac{da_z}{dx} - a_z\frac{da_y}{dx}\right)$$

$$V_{c,y} = -P_x\frac{da_y}{dx} \qquad M_{c,y} = -P_x a_z$$

$$V_{c,z} = -P_x\frac{da_z}{dx} \qquad M_{c,z} = -P_x a_y$$

$$T_c = -P_x\left[(a_y - c_y)\frac{da_z}{dx} - (a_z - c_z)\frac{da_y}{dx}\right]$$

These sectional forces are denoted collectively as S_{0P}.

In statically indeterminate systems, the deformations induced by the self-equilibrating state of stress result in redundant sectional forces, denoted S_{sP}. The total sectional force due to prestressing can thus be expressed as the following sum:

$$S_P = S_{0P} + S_{sP}$$

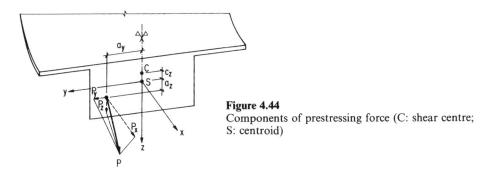

Figure 4.44
Components of prestressing force (C: shear centre; S: centroid)

The moment due to prestressing, for example, is given by

$$M_P = M_{0P} + M_{sP} \tag{a}$$

The redundant forces due to prestressing are primarily a function of the forces in the statically determinate primary system, S_{0P}, and do not depend directly on stiffness. Reductions in stiffness due to cracking are normally balanced by increasing deformations due to the self-equilibrating state of stress; the net change in S_{sP} is thus small.

For uncracked sections, stresses due to prestressing can be calculated based on a homogeneous concrete section. The contribution of mild reinforcing steel is neglected. For approximate analyses and estimates, M_P is obtained simply and quickly from equivalent loads corresponding to the deviation and anchor forces of the tendons. By concentrating losses due to friction at the intermediate supports and assuming that prestressing losses due to other actions are constant within a given span, the equivalent loads can be computed using constant values of prestressing force for each span.

When the girder is loaded with deviation and anchor forces, the total moment due to prestressing M_P is obtained directly. The redundant moment M_{sP} can thus be calculated from the equation

$$M_{sP} = M_P - M_{0P} \tag{b}$$
where
$$M_{0P} = -Pe \tag{c}$$

(The sign convention of figure 4.45 is followed throughout.)

Figure 4.46 shows a girder with constant bending stiffness, fixed at both ends, and prestressed with a parabolic tendon. The prestressing force, P, is constant. The deviation force due to prestressing is given by the following expression:

$$q_p = -\frac{8fP}{l^2} \tag{4.7}$$

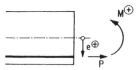

Figure 4.45
Sign convention for moments due to prestressing

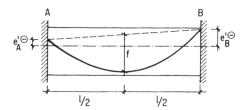

Figure 4.46
Girder with both ends fixed and parabolic tendon

The total moments due to prestressing at either end are thus

$$M_P(A) = -\frac{q_p l^2}{12}$$

$$M_P(B) = -\frac{q_p l^2}{12}$$

The eccentricity of the anchor forces produce the end moments

$$M_{0P}(A) = -Pe'_A$$
$$M_{0P}(B) = -Pe'_B$$

(equation (c))

The redundant moments required to satisfy the compatibility conditions at the girder ends are thus

$$M_{sP}(A) = -\frac{q_p l^2}{12} - M_{0P}(A) = P\left(\frac{2}{3}f + e'_A\right)$$

(equation (b))

$$M_{sP}(B) = -\frac{q_p l^2}{12} - M_{0P}(B) = P\left(\frac{2}{3}f + e'_B\right)$$

Using equation (a), the total moments due to prestressing at ends A and B can thus be rewritten as follows:

$$M_P(A) = \tfrac{2}{3}Pf$$
$$M_P(B) = \tfrac{2}{3}Pf$$

Tendons with reverse curvature near the supports induce local deviation forces, $\bar{q}_P$ (fig. 4.47). These can be calculated using the following equation:

$$\bar{q}_P = \frac{8(\Delta f)P}{4a^2}$$

(4.8)

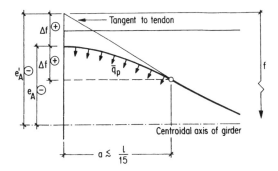

Figure 4.47
Reverse curvature of tendons near the supports

The moments induced by $\bar{q}_P$ at the supports of a fixed-end girder are given approximately by

$$\bar{M}_P \cong -\frac{\bar{q}_P a^2}{2} = -P(\Delta f)$$

The moments at the supports of a fixed-end girder due to prestressing $(q_P + \bar{q}_P)$ can thus be calculated from the following equations:

$$M_{sP}(A) = P\left(\tfrac{2}{3}f - \Delta f_A + e_A\right) \qquad\qquad M_{sP}(B) = P\left(\tfrac{2}{3}f - \Delta f_B + e_B\right)$$

$$M_{0P}(A) = -Pe_A \qquad\qquad\qquad\qquad M_{0P}(B) = -Pe_B$$

$$M_P(A) = P\left(\tfrac{2}{3}f - \Delta f_A\right) \qquad\qquad\quad M_P(B) = P\left(\tfrac{2}{3}f - \Delta f_B\right)$$

Similarly, the redundant moment induced by a parabolic tendon in girders fixed at one end is expressed as follows (fig. 4.48):

$$M_{sP}(B) = -\frac{q_P l^2}{8} - \frac{1}{2}\,M_{0P}(A) - M_{0P}(B) = P\left(f + \frac{1}{2}\,e_A + e_B'\right)$$

Reverse curvature of the tendon at end B is considered approximately using the following expression for the end moment due to $\bar{q}_P$:

$$\bar{M}_P(B) \cong -\frac{\bar{q}_P a^2}{2} = -P(\Delta f)$$

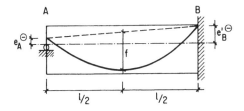

Figure 4.48
Girder with one end fixed and parabolic tendon

For girders fixed at one end, therefore, the moments at the supports due to prestressing $(q_P + \bar{q}_P)$ are obtained from the following equations:

$$M_{sP}(A) = 0 \qquad\qquad M_{sP}(B) = P(f + e_B + \tfrac{1}{2}e_A - \Delta f_B)$$
$$M_{0P}(A) = -Pe_A \qquad\qquad M_{0P}(B) = -Pe_B$$
$$M_P(A) = -Pe_A \qquad\qquad M_P(B) = P(f + \tfrac{1}{2}e_A - \Delta f_B)$$

The redundant moments due to prestressing in continuous girders can be readily computed by distributing the fixed-end moments derived above.

4.6.6 Steel Stresses for Typical Loading States

Partially prestressed structures pass through several characteristic states of stress as load is increased from zero to ultimate load. These states are identified and described in this section for the cross-section of a beam subjected to an increasing moment due to external load, M. Ductile behaviour is assumed. The analysis has been simplified without significant error by neglecting the tensile strength of concrete and by assuming that prestressing steel and mild reinforcing steel have identical bond behaviour.

1. *Pure Self-Equilibrating State of Stress;* $M = 0$ (fig. 4.49). Since $M = 0$, the compressive stresses in the concrete and the tensile stresses in the prestressing steel are in equilibrium with each other. Assuming the forces in the reinforcing steel are small and can be neglected, the resultant compressive force in the concrete, F_{cE}, acts at the same level as the prestressing force, F_{PE}. This state of stress is only of theoretical interest, since at least a portion of the weight of the girder will be effective as soon as the prestressing force is applied. The bending moment due to external load, M, will thus always be greater than zero.

2. *Initial State of Stress;* $M = M_0$ (fig. 4.50). This is the state of stress that actually occurs immediately after the initial prestressing force, F_{P0}, is applied. The falsework is at least partially unloaded by the prestressing; a portion of the self-weight is thus effective as external load. The magnitude of M_0 is a function of the interaction of structure and falsework and is therefore difficult to calculate exactly.

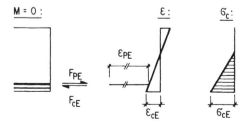

Figure 4.49
Pure self-equilibrating state of stress due to prestressing; $M = 0$

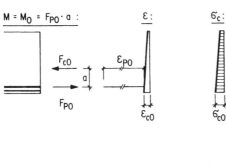

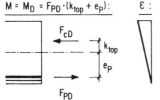

Figure 4.50
Initial state of stress immediately after the tendons have been stressed; $M = M_0$

Figure 4.51
Decompression; $M = M_D$

3. Decompression; $M = M_D$ (fig. 4.51). The moment that reduces the compressive stress at one of the extreme fibres to zero is called the *decompression moment*, denoted M_D. Since the tensile strength of concrete is neglected, moments greater than M_D will cause cracking. Related to M_D is the moment M_{sD}, which produces a strain of zero at the outer layer of reinforcement. Since the layer of concrete covering the reinforcing steel is usually thin compared to the depth of the cross-section, it can be assumed for practical calculations that $M_D = M_{sD}$.

4. Yield of Reinforcing Steel; $M = M_{sy}$ (fig. 4.52). The reinforcing steel normally yields before the prestressing steel. The corresponding moment due to external load is denoted M_{sy}.

5. Yield of Prestressing Steel and Reinforcing Steel; $M = M_y$ (fig. 4.53). The moment M_y that causes yielding in both prestressing and reinforcing steel is called the *plastic moment*. To ensure sufficient ductility of the cross-section, the associated extreme fibre compressive stress in the concrete must be less than the concrete compressive strength f_c.

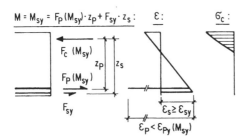

Figure 4.52
Yield of reinforcing steel; $M = M_{sy}$

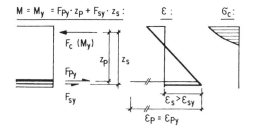

Figure 4.53
Yield of prestressing steel and reinforcing steel; $M = M_y$

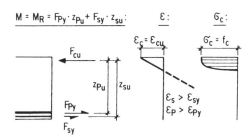

Figure 4.54
Failure of the cross-section; $M = M_R$

6. *Failure of the Cross-Section, $M = M_R$* (fig. 4.54). After the entire reinforcement has yielded, the internal forces F_{Py} and F_{sy} remain constant. Sectional forces slightly higher than M_y are nevertheless possible, since the internal lever arms z_P and z_s can be increased somewhat. Increasing M beyond M_y is accompanied, however, by large deformations. The cross-section fails at the moment, M_R, that causes crushing in the concrete at the ultimate compressive strain ε_{cu}.

The range of possibilities for reinforcement in partially prestressed sections is limited only by the requirements of safety and serviceability. The effective prestressing force and the relative proportions of mild reinforcing steel and prestressing steel can, theoretically speaking, be freely chosen. The remainder of this section examines the influence of these factors on the behaviour of partially prestressed cross-sections.

For each of the two cross-sections shown in figure 4.55, the stress in the prestressing steel, σ_P, has been computed as a function of bending moment, M. Each section is reinforced in eight different ways. Four different values are considered for the ratio of initial prestress to yield stress:

1. $\sigma_{P0}/f_{Py} = 0.00$
2. $\sigma_{P0}/f_{Py} = 0.35$
3. $\sigma_{P0}/f_{Py} = 0.50$
4. $\sigma_{P0}/f_{Py} = 0.70$

Each of these four cases is combined with the following two ratios of reinforcing steel to prestressing steel:

A. $\omega_s/\omega_P = 0$
B. $\omega_s/\omega_P = 1$

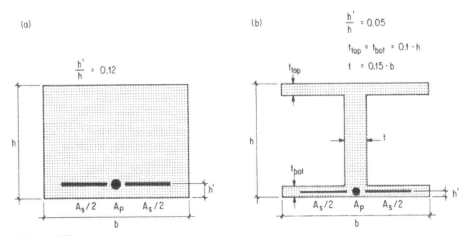

Figure 4.55
Cross-sections used for investigation of partial prestressing: a, rectangular section; b, I-section

where the mechanical reinforcement ratios for reinforcing and prestressing steel are defined by the following equations:

$$\omega_s = \frac{A_s}{bh}\frac{f_{sy}}{f_c} \qquad \omega_P = \frac{A_P}{bh}\frac{f_{Py}}{f_c}$$

Case A thus corresponds to a section with only prestressing steel; case B corresponds to a section in which reinforcing and prestressing steel contribute equally to the ultimate resistance. The total yield force in the reinforcement is assumed constant. For the rectangular section (fig. 4.55a), $\omega_s + \omega_P = 0.15$ in all cases; for the I-section (fig. 4.55b), $\omega_s + \omega_P = 0.1$ in all cases.

The analysis is based on the idealized stress-strain diagrams for concrete, reinforcing steel, and prestressing steel given in figures 4.56, 4.57, and 4.58 respectively. The calculated relations between σ_P and M are presented in dimensionless form in figures 4.59 and 4.60.

Before the decompression moment M_D is reached, the existing tensile force in the steel due to the effective prestress is greater than is necessary to balance M. In this

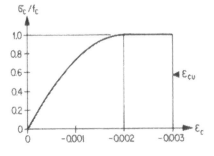

Figure 4.56
Idealized stress-strain diagram for concrete used for investigation

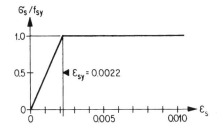

Figure 4.57
Idealized stress-strain diagram for reinforcing steel used for investigation

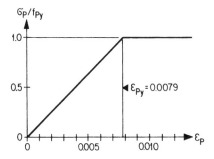

Figure 4.58
Idealized stress-strain diagram for prestressing steel used for investigation

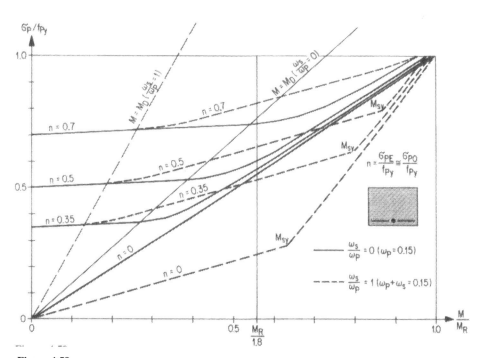

Figure 4.59
Stress in prestressing steel, σ_P, as a function of M (rectangular section)

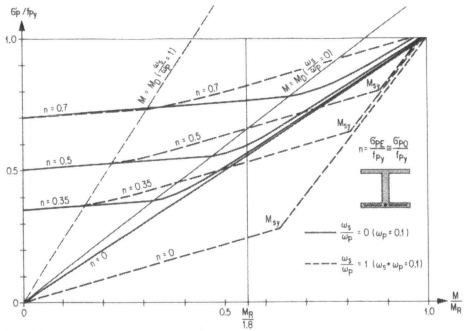

Figure 4.60
Stress in prestressing steel, σ_P, as a function of M (I-section)

state, the prestressed cross-section is uncracked and homogeneous, and thus has a high bending stiffness. The stress in the prestressing steel increases only slightly with increasing moment.

After M_D is reached, the steel stress must increase nearly in direct proportion with the M to maintain equilibrium of the section. The stress in the reinforcing steel, which is practically zero for $M < M_D$, increases at the same rate as σ_P until its yield stress, f_{sy}, is reached.

An exact calculation of σ_P as a function of M is usually not necessary for design. The stress in the steel can be estimated with reasonable accuracy using a simplified trilinear diagram (fig. 4.61). The increase in stress up to decompression is neglected and M_y is assumed equal to M_R. The three points that define the diagram (decompression moment M_D, yield moment of reinforcing steel M_{sy}, and ultimate moment M_R) can thus be simply calculated using the following equations:

$$M_D \cong \sigma_{P0} A_P (e_P + k_{top})$$
$$M_{sy} \cong (\sigma_{P0} + f_{sy}) A_P z_{Pu} + f_{sy} A_s z_{su}$$
$$M_R \cong f_{Py} A_P z_{Pu} + f_{sy} A_s z_{su}$$

where k_{top} is the upper kern point of the section.

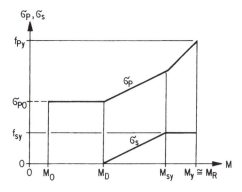

Figure 4.61
Idealized behaviour of stress in prestressing steel, σ_P, and stress in reinforcing steel, σ_s

4.6.7 Prestressing with Unbonded Tendons

Post-tensioning tendons that have been designed as bonded must sometimes contribute to the flexural resistance of the cross-section during construction, before they have been bonded to the concrete by grouting. This is often the case in cantilever construction and in span-by-span construction with form girders. The resistance of the section with unbonded tendons is often considerably less than its resistance after the tendons have been bonded.

Tendons that are left permanently unbonded can also be used in bridges. The most common such application in new construction is the use of external tendons in precast segmental structures. Here, the advantages of unbonded prestressing include speed of construction, elimination of grouting (which can rarely be properly executed for bonded tendons in precast segmental construction), and durability. The prestressing steel, enclosed in plastic ducts and injected with grease or wax, is protected from harmful environmental conditions, yet is accessible for inspection, maintenance, and replacement. Unbonded tendons can also be used for the rehabilitation of existing bridges. For deck slabs with high concentrations of chlorides but otherwise sound concrete and reinforcement, for example, providing an additional layer of concrete reinforced transversely with unbonded tendons may be a less costly alternative to deck replacement.

There is a fundamental difference between the structural behaviour of bonded and unbonded prestressing. For bonded tendons, the increase in stress after decompression, $\Delta\sigma_P$, varies along the length of the member and can be computed at a given section directly from the local state of strain. For unbonded tendons, however, $\Delta\sigma_P$ is a function of the total elongation of the tendon, Δl_P, and can be conservatively assumed constant between anchor points.

The elongation of an unbonded tendon is a function of the global deformations of the girder. Exact computations are often complicated. For tendons that are unbonded only during construction, it may be preferable to calculate ultimate resistance assuming $\Delta\sigma_P = 0$. The stress in the prestressing steel at ultimate limit state is thus limited to the effective prestress minus losses due to friction, σ_{P0}. This simplification is always conservative.

For tendons that remain permanently unbonded, it is normally preferable to compute ultimate resistance with the actual value of $\Delta\sigma_P$. Of greater importance than a rigorous calculation of the increase in stress, however, is a tendon layout that maximizes $\Delta\sigma_P$ for a given deformation of the girder. This is achieved by anchoring the tendons at each support. (The prestressing steel from two adjacent spans should be lapped to maintain continuity.) This tendon arrangement has the additional advantage of shortening tendon length, thus reducing prestressing losses due to friction.

The area of prestressing steel necessary for safety at ultimate limit state can be calculated based on equilibrium and the ultimate deformations of the structure. The design bending moments, M_d, are first computed according to the theory of plasticity; M_d must be in equilibrium with the design loads q_d and satisfy the statical boundary conditions. The force in the unbonded tendon necessary for internal equilibrium, F_d, is obtained from

$$F_d = \frac{M_{d,\,\text{max}}}{z}$$

at the points of maximum design moment. The parameter z is the internal lever arm, typically measured between the centroid of prestressing steel and the middle surface of the compression slab. Since the tendon force F_d is constant between anchors, the lever arm can be readily computed at all other sections of the girder.

The extreme compressive fibre strain at the section of maximum design moment, ε_{cu}, should be conservatively chosen not to exceed 0.0025. The state of strain in the concrete at this location is thus determined from ε_{cu} and the depth of the concrete compressive stress block, x, required to equilibrate F_d.

The state of strain is then computed at the remaining locations along the girder from the design bending moments and material stress-strain diagrams. This calculation is relatively complicated and can be done by computer. Alternatively, it can be assumed that curvature is zero at all sections where M_d is less than the decompression moment, M_D. Linear interpolation can then be used to obtain curvatures at those points for which M_d is between M_D and $M_{d,\,\text{max}}$. This approximation is sufficiently reliable for preliminary calculations.

The elongation of the unbonded tendon is now computed using the following equation:

$$\Delta l = \int \varepsilon_c \, dx$$

where ε_c is the strain in the concrete section at the level of the prestressing steel. The increase in strain in the prestressing steel due to the ultimate deformations of the system is thus given by

$$\Delta\varepsilon_P = \frac{\Delta l}{l}$$

where l is total tendon length between anchors.

The area of prestressing steel, A_P, required for equilibrium is obtained from the relation

$$A_P = \gamma_R \frac{F_P}{\sigma_P}$$

The stress compatible with the ultimate deformations, σ_P, is calculated from the following equations (the quantity ε_{P0} denotes initial strain in the prestressing steel):

$$\begin{array}{lll} & \sigma_P = f_{Py} & \text{for} \quad \varepsilon_{P0} + \Delta\varepsilon_P \geq \varepsilon_{Py} \\ \text{and} & & \\ & \sigma_P = (\varepsilon_{P0} + \Delta\varepsilon_P)\,E_P & \text{for} \quad \varepsilon_{P0} + \Delta\varepsilon_P < \varepsilon_{Py} \end{array}$$

4.6.8 Loss of Prestress

The prestressing force at any point in the structure is always less than the force measured at the jack during stressing. The difference between the effective prestressing force and the jacking force is called the *loss of prestress*. Losses occur primarily due to the following causes:

1. Frictional forces between tendon and duct during stressing
2. Shortening of concrete after stressing due to creep and shrinkage
3. Relaxation of prestressing steel after stressing

The permanent load moment, M_{g+P}, is the sum of two components of opposite sign and of approximately equal magnitude. It is therefore sensitive to small changes in M_P due to loss of prestress. Changes in M_{g+P} will be reflected in the deformations of the structure. Prestressing losses must therefore be investigated to ensure that the residual long-term prestressing force is adequate to control deformations due to permanent load. Special attention must be paid to loss of prestress in partially prestressed structures, which may lead to cracking under permanent load and hence to a substantial reduction in stiffness.

a) Loss of Prestress Due to Friction

The decrease in prestressing force due to friction at a distance x from the stressing location is given by

$$\Delta P(x) = P_0\,(1 - e^{-\mu\alpha(x)})$$

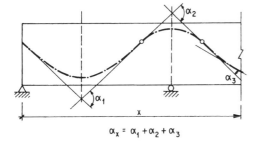

$$\alpha_x = \alpha_1 + \alpha_2 + \alpha_3$$

Figure 4.62
"Intentional" angle change of a prestressing tendon, α_x

where P_0 is the jacking force, μ is the coefficient of friction, and $\alpha(x)$ is the total angle change between the stressing location and point x. When $\mu\alpha(x)$ is less than 0.2, the following linearized formula can be used:

$$\Delta P(x) = P_0\,\mu\alpha(x)$$

The function $\alpha(x)$ can be expressed as the following sum:

$$\alpha(x) = \alpha_x + x\Delta\alpha$$

where α_x and $x\Delta\alpha$ are the "intentional" and "unintentional" angle changes, respectively. The former component is the sum of the deviation angles in the vertical and horizontal planes, α_i, measured from the design drawings (fig. 4.62):

$$\alpha_x = \alpha_1 + \alpha_2 + \alpha_3 + \ldots + \alpha_n$$

The unintentional angle change accounts for any deviation of the actual tendon profile from the design profile, due to construction tolerances and displacements of the tendon during concreting. It is assumed to vary linearly with x. The coefficient $\Delta\alpha$ is a function of the spacing of the cable supports, the way the cable is held in place, and the care taken during concreting. A value of $\Delta\alpha$ roughly equal to 0.005 agrees reasonably well with frictional losses measured in actual field conditions.

The value of the friction coefficient, μ, varies considerably according to the properties of the prestressing steel and the duct. Although a standard value of $\mu = 0.20$ can normally be assumed, it is not uncommon for values of μ of up to 0.40 to occur.

In the absence of data obtained from special investigations, the prediction of losses due to friction is very uncertain. Tendon elongations must therefore be carefully monitored in the field during stressing operations and compared to computed elongations. Field measurements of elongations are relatively unreliable for tendons that are stressed at only one end, since the observed elongation is sensitive to the frictional forces in the immediate vicinity of the stressing location. The prestressing force can be directly checked, however, for tendons stressed at both ends. After jacking at one end, the lift-off force is measured at the other; comparing the jacking force to the lift-off force gives a good indication of the frictional loss along the tendon.

The observed elongations and lift-off forces must not differ from computed values by more than 5 percent. Otherwise, the deformations of the structure must be recalculated using the actual tendon forces. Falsework should not be removed before a careful evaluation of the stressing records has been made and any systematic errors have been identified.

b) Loss of Prestress Due to Creep and Shrinkage

Time-dependent plastic shortening in the concrete due to creep and shrinkage causes shortening in the prestressing steel and thus a loss of prestressing force. The reduction in prestressing force due to creep and shrinkage is a function of the state of stress due to dead load and prestressing. Several different cases must therefore be distinguished:

1. Uncracked Section, Bonded Prestressing Steel. This is the most common case for prestressed concrete bridges. The concrete at the level of the prestressing steel is precompressed under dead load plus prestressing. It is assumed that any change in concrete strain can be transferred directly to the prestressing steel. The loss of prestressing force due to creep and shrinkage can therefore be calculated at a given section based on the compatibility of strains in steel and concrete, assuming an appropriate creep law.

These concepts can be used to derive an expression for the loss of prestress, ΔP, in a member subjected to initial prestressing force, P_0, and external axial compression, N. Both forces act concentrically. The time-varying strain in the concrete, $\varepsilon_c(t)$, is computed from the following creep law (see Section 4.7.1):

$$\Delta \varepsilon_c = \frac{\sigma_{c0}}{E_c} \phi + \frac{\Delta \sigma_c}{E_c} (1 + \mu \phi) + \varepsilon_{cs}$$

where ϕ is the concrete creep coefficient and ε_{cs} shrinkage strain. The parameter μ can be taken as 0.8. The initial stress in the concrete, σ_{c0}, and the change in stress, $\Delta \sigma_c$, are given by the following expressions:

$$\sigma_{c0} = \frac{-P_0 + N}{A_c}$$

$$\Delta \sigma_c = -\frac{\Delta P}{A_c}$$

where A_c is the area of the concrete section. Compatibility requires that $\Delta \varepsilon_P = \Delta \varepsilon_c$. It therefore follows that

$$\Delta P = \Delta \varepsilon_P \, E_P \, A_P = \Delta \varepsilon_c \, E_P \, A_P$$

Substituting for $\Delta\varepsilon_c$ and solving for ΔP, the following expression is obtained:

$$\Delta P = n\varrho \, \frac{\sigma_{c0} A_c \phi + \varepsilon_{cs} E_c A_c}{1 + n\varrho \, (1 + \mu\phi)} \tag{4.9}$$

where $n = E_P/E_c$ and $\varrho = A_P/A_c$.

Equation (4.9) also gives reliable results for prestressing losses in members subjected to both flexure and axial load. In such cases, σ_{c0} and $\Delta\sigma_c$ are computed at the elevation of the prestressing tendon and must include the flexural and axial components of stress. A version of equation (4.9) for this case is derived in Section 4.7.4.

2. Cracked Section, Bonded Prestressing Steel. When cracks penetrate up to or beyond the level of the tendons, the stress in the prestressing steel will be whatever value is necessary to maintain internal equilibrium, regardless of creep and shrinkage. This stress is always greater than the initial prestress. The effect of creep and shrinkage is thus limited to additional shortening and deflections in the girder, without reduction in steel stress. The concrete is stress-free at the level of the tendons and thus does not deform due to creep. The strain in the prestressing steel is likewise not reduced by shrinkage of the concrete at this level. Although the stress in the prestressing steel is reduced somewhat by shortening of the flexural compression zone due to creep and shrinkage, this effect is small and can be neglected.

3. Transition from Uncracked to Cracked State. The decrease in M_P due to loss of prestress may be sufficient to increase M_{g+P} beyond the cracking moment. In such cases, though, the stress in the prestressing steel cannot decrease below the value required for equilibrium of internal forces in the cracked section. Furthermore, the loss of prestress will normally be small, since it is caused only by the shrinkage that occurs before cracking. (Since the concrete strains at the level of the steel are tensile before cracking, no loss of prestress will be incurred due to creep.)

4. Unbonded Prestressing Steel. The prestressing force can be considered constant along the full length of unbonded tendons between anchors, provided the frictional forces between tendon and duct are neglected. The loss of prestress is therefore a function of the total shortening of the concrete between anchors and is constant over the length of the tendon.

c) Loss of Prestress Due to Relaxation of Prestressing Steel

As stated in Section 3.1.3, the loss of prestress due to relaxation, $\Delta\sigma_{P,\,rel}$, is a function of the properties of the steel and the ratio of initial stress to tensile strength, σ_{P0}/f_{Pt}. Relaxation progresses more rapidly than creep and shrinkage in concrete. Whereas 50 percent of the final relaxation strain is reached at roughly 28 days, this same percentage of the final creep and shrinkage strain is achieved only after 90 days. In spite of this fundamental difference, a detailed calculation of the interaction between relaxation, creep, and shrinkage is usually not necessary. Given the range of variability of the coefficient of friction and the material parameters of steel and concrete, such a calculation will be of doubtful accuracy. It is normally sufficient to calculate $\Delta\sigma_{P, rel}$ as a function of σ_{P0} only, using figure 3.15. The interaction between relaxation, creep, and shrinkage can then be considered by calculating the loss of prestress due to creep and shrinkage using the following reduced initial prestress:

$$\sigma_P = \sigma_{P0} - 0.5\,\Delta\sigma_{P, rel, \infty}$$

where $\Delta\sigma_{P, rel, \infty}$ denotes the long-term loss due to relaxation.

4.6.9 Prestressing Concepts

As discussed in Section 4.6.6, the ultimate resistance of partially prestressed structures is a function of the combined yield force of prestressing and reinforcing steel. The appropriate amount of prestressing to be provided, therefore, cannot be directly determined on the basis of behaviour at ultimate limit state. Supplementary criteria for the design of prestressing in partially prestressed structures are therefore required. These criteria, which will be called *prestressing concepts*, are established to account for aspects of serviceability, economy, and construction in the design of the prestressing as they apply to various types of structures. Regardless of the chosen prestressing concept, however, economic considerations always dictate that the total amount of reinforcement provided (prestressing steel plus mild reinforcing steel) should not exceed what is required for safety.

This aim, however, cannot normally be achieved in structures that are fully prestressed for dead and live load. The long-term prestressing force, allowing for all losses due to friction, relaxation, creep, and shrinkage, is normally small relative to the yield force. Full prestressing consequently requires substantially more prestressing steel than the minimum necessary for safety. In partially prestressed structures, however, the total reinforcement can almost always be determined on the basis of safety alone. Only in exceptional cases must the amount of prestressing be increased above this amount to limit structural deformations.

Careful attention must always be paid to cracking in partially prestressed structures. Cracks due to live load do not normally impair serviceability, since they close immediately after live load is removed. The moments induced by restrained deformations in bridge superstructures usually exceed the cracking moment by only an insignificant amount and result in a small number of narrow cracks. Cracking due to restrained deformations, therefore, can also be tolerated in partially prestressed structures. Moments due to permanent load that are significantly greater than the cracking moment, however, result in a large number of cracks, the widths of which are a function of the stress in the steel. Cracking due to dead load plus prestressing should therefore be restricted or eliminated by a prestressing concept in which at least the deck slab is fully prestressed under dead load plus prestressing.

The likelihood of crack formation in structures that are fully prestressed for dead and live load is small. Even in such structures, however, unexpected tensile stresses and cracks can be produced by restrained deformations and self-equilibrating stresses. It is therefore necessary to provide a sufficiently high reserve of compression in the entire structure or, more reliably, to accept cracking as inevitable and provide mild reinforcement to control and distribute cracks. Minimum mild reinforcement must always be provided in partially prestressed structures. It is also necessary to verify the width and distribution of cracks if the cracking moment is exceeded under permanent load.

Deformations often have a substantial influence on the choice of prestressing concept. Full prestressing can produce unexpected and undesirable upward deflections. Slender, fully prestressed T-girders are particularly susceptible to this type of behaviour. The deformations of partially prestressed structures must always be checked, taking into account the variability inherent in the computed prestressing losses and the reduction of member stiffness due to cracking.

The relative proportion of prestressing and reinforcing steel is of little significance to the economy of the structure. It is normally more economical, however, to reinforce local stress concentrations with mild reinforcing steel as opposed to prestressing steel.

4.7 Long-Term Effects

4.7.1 Fundamentals

The time-varying changes in stresses and deformations induced by creep and shrinkage in concrete and relaxation in prestressing steel are collectively called *long-term effects*. Long term effects must be considered in the calculation of deformations, redistributions of sectional forces due to changes of structural system, and sectional forces due to restrained deformations.

The calculation of long-term effects is based the following analytical models for creep, shrinkage, and relaxation presented in Sections 3.1.1 and 3.1.3:

1. Creep in concrete: $\qquad\qquad\qquad\qquad \varepsilon_{cc}(t, \tau) = \varepsilon_{c,el,28}\, \phi_n k(\tau)\, f(t - \tau)$

2. Shrinkage in concrete: $\qquad\qquad\qquad \varepsilon_{cs}(t, t_0) = \varepsilon_{cs,n}[g(t) - g(t_0)]$

3. Relaxation in prestressing steel: $\qquad\quad \Delta\sigma_P = \Delta\sigma_P(\sigma_{P0})$

a) Analytical Model for Creep

The basic analytical tool for calculating time-varying strains in concrete due to a known stress $\sigma(t)$ is the function $\Phi(t, \tau)$. It is defined as the total (elastic plus creep) strain at time t produced by a unit sustained stress applied instantaneously at time τ. It is related to the *creep function*, $\phi(t, \tau)$, by the following expression:

$$\Phi(t, \tau) = \frac{1}{E_c(\tau)} [1 + \phi(t, \tau)]$$

where $E_c(\tau)$ is the elastic modulus of concrete at time τ. The function $\phi(t, \tau)$ is thus the ratio of creep strain to elastic strain at time of loading. The function Φ is often defined as the ratio of creep strain to elastic strain at 28 days, written $\phi_{28}(t, \tau)$. It is then formulated as

$$\Phi(t, \tau) = \frac{1}{E_c(\tau)} + \frac{\phi_{28}(t, \tau)}{E_{c,28}}$$

where $E_{c,28}$ is the modulus of elasticity at 28 days.

Strains induced by an arbitrary time-varying stress, $\sigma(t)$, can be computed using Boltzmann's principle of superposition. Given a constant stress σ_0, applied at time τ_0, and constant increments of stress $\Delta\sigma_1, \Delta\sigma_2, \ldots, \Delta\sigma_n$, applied at times $\tau_1, \tau_2, \ldots, \tau_n$, the strain $\varepsilon_c(t)$ is obtained from the folowing summation:

$$\varepsilon_c(t) = \sigma_0 \Phi(t, \tau_0) + \sum_{1}^{n} \Delta\sigma_i \Phi(t, \tau_i) \tag{a}$$

For infinitesimally small increments of stress, equation (a) is transformed into

$$\varepsilon_c(t) = \sigma_0 \Phi(t, \tau_0) + \int_{\tau_0}^{t} \frac{\partial\sigma(\tau)}{\partial\tau} \Phi(t, \tau) d\tau \tag{b}$$

where $\sigma(\tau)$ is the total stress at time τ. This equation can be rewritten in terms of $\phi(t, \tau)$

$$\varepsilon_c(t) = \frac{\sigma_0}{E_c(\tau_0)} [1 + \phi(t, \tau_0)] + \int_{\tau_0}^{t} \frac{\partial\sigma(\tau)}{\partial\tau} \frac{1}{E_c(\tau)} [1 + \phi(t, \tau)] d\tau \tag{c}$$

and in terms of $\phi_{28}(t, \tau)$

$$\varepsilon_c(t) = \sigma_0 \left(\frac{1}{E_c(\tau_0)} + \frac{\phi_{28}(t, \tau_0)}{E_{c,28}} \right) + \int_{\tau_0}^{t} \frac{\partial\sigma(\tau)}{\partial\tau} \left(\frac{1}{E_c(\tau)} + \frac{\phi_{28}(t, \tau)}{E_{c,28}} \right) d\tau \tag{d}$$

Equations (b) through (d) are not often used for practical calculations, due to the complexity of the integral. They can, however, be applied to simple problems involving constant increments of stress, for which the integral vanishes. The use of equation (d) in the solution of one such problem is illustrated in the following example:

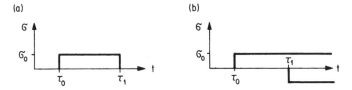

Figure 4.63
Stress function of Example 4.6: a, given function σ; b, equivalent superposition of $+\sigma$ and $-\sigma$

Example 4.6:
Calculation of time-varying strains

The strains induced by a stress σ_0 applied over a finite interval of time $\tau_0 \leq t \leq \tau_1$ are the sum of the strains due to a sustained stress $+\sigma_0$, applied at time τ_0, and a stress $-\sigma_0$, applied at time τ_1 (fig. 4.63). Since both stress functions are horizontal, the derivative $\partial\sigma(\tau)/\partial\tau$ in equation (d) vanishes. The strain $\varepsilon(t)$ is thus given by

$$\varepsilon_c(t) = \sigma_0 \left(\frac{1}{E_c(\tau_0)} + \frac{\phi_{28}(t, \tau_0)}{E_{c, 28}} \right) - \sigma_1 \left(\frac{1}{E_c(\tau_1)} + \frac{\phi_{28}(t, \tau_1)}{E_{c, 28}} \right) \qquad (e)$$

It is assumed that

$$\phi_{28}(t, \tau) = \phi_n k(\tau) f(t - \tau) \qquad \text{(equation 3.1)}$$

where the coefficient ϕ_n and the functions $k(\tau)$ and $f(t-\tau)$ are defined in Section 3.1.1. Substitution into equation (e) yields the following expression:

$$\varepsilon_c(t) = \frac{\sigma_0}{E_c(\tau_0)} + \sigma_0 \frac{\phi_n}{E_{c, 28}} k(\tau_0) f(t - \tau_0)$$

$$- \frac{\sigma_0}{E_c(\tau_1)} - \sigma_0 \frac{\phi_n}{E_{c, 28}} k(\tau_1) f(t - \tau_1)$$

which, for $E_c(\tau_0) \cong E_c(\tau_1)$, simplifies to

$$\varepsilon_c(t) = \sigma_0 \frac{\phi_n}{E_{c, 28}} k(\tau_0) f(t - \tau_0) - \sigma_0 \frac{\phi_n}{E_{c, 28}} k(\tau_1) f(t - \tau_1)$$

Since the function $f(t-\tau)$ tends towards 1 as t increases to infinity, the final long-term strain due to the given stress is

$$\varepsilon_{c, \infty} = \sigma_0 \frac{\phi_n}{E_{c, 28}} (k(\tau_0) - k(\tau_1))$$

Equation (c) can be modified to account for the effects of shrinkage as follows:

$$\varepsilon_c(t) = \frac{\sigma_0}{E_c(\tau_0)} [1 + \phi(t, \tau_0)]$$

$$+ \int_{\tau_0}^{t} \frac{\partial \sigma(\tau)}{\partial \tau} \frac{1}{E_c(\tau)} [1 + \phi(t, \tau)] \, d\tau + \varepsilon_{cs}(t) \tag{f}$$

where the additional term, $\varepsilon_{cs}(t)$, is the shrinkage strain defined in equation (3.2).

b) *Dischinger's Method*

The computation of time-varying strains is made easier by assuming that the function Φ can be written in the following form:

$$\Phi_D(t, \tau_0) = \frac{1}{E_c(\tau_0)} [1 + \phi(t, \tau_0)] \tag{g}$$

for an initial instantaneous application of stress at time τ_0, and

$$\Phi_D(t, \tau_1) = \frac{1}{E_c(\tau_1)} + \frac{1}{E_c(\tau_0)} [\phi(t, \tau_0) - \phi(t, \tau_1)] \tag{h}$$

for a subsequent increment of stress at time τ_1 (Neville, Dilger, and Brooks 1983, 248). The functions $\Phi_D(t, \tau_1)$ are thus parallel to $\Phi_D(t, \tau_0)$, as are the corresponding creep functions ϕ (fig. 4.64). Deformations due to creep are underestimated with these assumptions for Φ_D.

Substituting equations (g) and (h) into equation (f) yields an expression for total strain in the concrete:

$$\varepsilon_c(t) = \frac{\sigma_0}{E_c(\tau_0)} [1 + \phi(t, \tau_0)]$$

$$+ \int_{\tau_0}^{t} \frac{\partial \sigma(\tau)}{\partial \tau} \left(\frac{1}{E_c(\tau)} + \frac{1}{E_c(\tau_0)} [\phi(t, \tau_0) - \phi(t, \tau)] \right) d\tau + \varepsilon_{cs}(t)$$

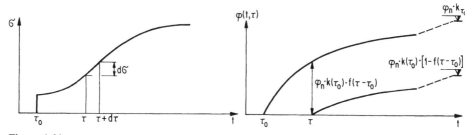

Figure 4.64
Dischinger's method: a, time-varying stresses $\sigma(t)$; b, creep functions ϕ for initial stress $\sigma(\tau_0)$ and subsequent increment of stress $d\sigma$

which can be transformed into the following equation by taking derivatives of both sides with respect to t:

$$\frac{d\varepsilon_c}{dt} = \frac{1}{E_c(t)} \frac{d\sigma}{dt} + \frac{\sigma}{E_c(\tau_0)} \frac{d\phi}{dt} + \frac{d\varepsilon_{cs}}{dt}$$

This expression is known is *Dischinger's equation*.

The inherent inaccuracy of Dischinger's equation, which results from the assumed function, Φ_D, can be improved somewhat by modifications proposed by Rüsch and Jungwirth (1976, 68); its use in practise, however, remains restricted to relatively simple problems. Finite-difference methods can be used to solve Dischinger's equation in more complicated cases. The computational effort required is still, however, considerable.

c) Trost's Method

Equation (f) can be rearranged into the form

$$\varepsilon_c(t) = \frac{\sigma_0}{E_c(\tau_0)} [1 + \phi(t, \tau_0)]$$

$$+ \frac{\sigma(\tau) - \sigma_0}{E_c(\tau_0)} [1 + \mu(t, \tau_0) \phi(t, \tau_0)] + \varepsilon_{cs}(t) \qquad (i)$$

where

$$\mu(t, \tau_0) = \frac{E_c(\tau_0)}{\phi(t, \tau_0)[\sigma(\tau) - \sigma_0]}$$

$$\cdot \int_{\tau_0}^{t} \frac{\partial\sigma(\tau)}{\partial\tau} \frac{1}{E_c(\tau)} [1 + \phi(t, \tau)] d\tau - \frac{1}{\phi(t, \tau_0)}$$

The function $\mu(t, \tau)$, which accounts for the difference in creep response to instantaneously applied stress and to gradually applied stress is called the *aging function*. Trost (1967) found that variation of μ with time after first application of stress, $t - \tau_0$, is small and can be neglected. The function can therefore be taken as a constant; the value of μ depends on ϕ_n (fig. 3.8) and $k(\tau_0)$ (fig. 3.9). Typical values of μ are given in table 4.2.

Table 4.2
Aging Coefficient, μ, for Creep Calculations (adapted from Trost 1967)

ϕ_n	$k(\tau_0)$	
	1.5	0.75
3.0	0.85	0.90
1.5	0.80	0.85

Trost's method is easy to understand and to use. In spite of its simplicity, it gives reliable results in most cases. Equation (i) is often simplified as follows:

$$\varepsilon_c(t) = \frac{\sigma_0}{E_c}[1 + \phi(t)] + \frac{\sigma(t) - \sigma_0}{E_c}[1 + \mu\phi(t)] + \varepsilon_{cs}(t) \qquad (4.10)$$

where $\phi(t)$ and E_c are understood to mean $\phi(t, \tau_0)$ and $E_c(\tau_0)$, respectively.

4.7.2 Calculation of Deformations Due to Permanent Load

Deformations due to permanent load must always be checked, taking into account the effects of creep, shrinkage, and relaxation. The lack of proper attention to these long-term effects can result in excessive deflections, particularly for partially prestressed structures.

Long-term deformations are induced by each of the following increments of stress, $\Delta\sigma$: self-weight and initial prestressing, subsequent stages of prestressing, superimposed dead load, and loss of prestress. When only the final deformations are required, a separate consideration of each $\Delta\sigma$ is usually not necessary. Since the difference between the initial and the final states of stress is normally small, the final concrete strain can be calculated using the final state of stress, σ_∞, and an effective modulus of elasticity, E_c', based on the time of initial loading, τ_0:

$$\varepsilon_{c,\infty} \cong \frac{\sigma_\infty}{E_{c,28}}(1 + \phi_\infty) + \varepsilon_{cs,\infty} = \frac{\sigma_\infty}{E_c'} + \varepsilon_{cs,\infty}$$

where

$$\phi_\infty = \phi_n k(\tau_0) \quad \text{and} \quad E_c' = E_{c,28}/(1 + \phi_\infty)$$

The influence of creep is especially important in second-order stress and stability problems, since the additional deformations due to creep can result in a substantial reduction in ultimate load.

4.7.3 Redistribution of Sectional Forces Due to Change of Structural System

Changes in structural system often occur during the course of construction, particularly when mechanized falsework is used. Typical examples of construction procedures that alter the structural system include the span-by-span construction of girder bridges and cantilever construction with monolithic closure pours at midspan.

Numerical errors can result from calculating the redistributed sectional forces due to dead load and prestressing separately, since both quantities are similar in magnitude and opposite in sign. It is preferable to calculate the associated creep deformations using the combined sectional force due to dead load plus prestressing. It will normally be obvious from inspection of the permanent load moments whether or not redistribution need even be considered. Given the inherent variability of the parameters influencing creep, a refined calculation will be no more reliable than a general estimate. Calculations based on an average creep coefficient for dead load plus prestressing, without considering the actual stress history, are acceptable even for cantilever constructed bridges made continuous at midspan. The following approximation of the final, fully redistributed stress is based on such an average:

$$\sigma_\infty = \sigma^A + 0.8\,(\sigma^E - \sigma^A)$$

where σ^A is the actual stress at time of closure and σ^E is the stress obtained assuming the entire structure was cast simultaneously on conventional falsework.

The redistribution of sectional forces can be quickly and reliably calculated using Trost's method. The compatibility conditions are formulated as a function of time, t. The change of structural system is assumed to occur at time τ_S. A portion of the total creep deformation in at least some of the structural components will normally have occurred before this time. Only the residual creep, $\phi_{res}(t, \tau_0)$, need therefore be considered in the formulation of the compatibility conditions after the change of system:

$$\phi_{res}(t, \tau_0) = \phi(t, \tau_0) - \phi(\tau_S, \tau_0)$$

These concepts are applied in the following example:

Example 4.7:
Redistribution of sectional forces due to a change in structural system

The redistribution of moments is to be calculated for the two-span girder of figure 4.65, consisting of two simply supported spans which are subsequently

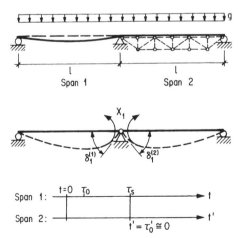

Figure 4.65
System of Example 4.7

made continuous at the intermediate support. The construction sequence of span 1 is as follows:

1. Cast concrete at time $t = 0$
2. Remove falsework and formwork at time $t = \tau_0$
3. Make continuous with span 2 at time $t = \tau_S$

The sequence for span 2 is:

1. Cast concrete at time $t' = 0$
2. Make continuous with span 1 while still supported on falsework
3. Remove falsework and formwork at time $t' = \tau_0' \cong 0$

The compatibility condition at the intermediate support is

$$\delta_{10}^{(1)} + \delta_{10}^{(2)} + X_1(t)\,(\delta_{11}^{(1)} + \delta_{11}^{(2)}) = 0 \tag{a}$$

for all values of t. The unknown redundant moment, $X_1(t)$, can be expressed as the following sum:

$$X_1(t) = X_{10} + \Delta X_1(t)$$

where X_{10} is the redundant moment immediately before the falsework for span 2 is removed ($t' = \tau_0' \cong 0$) and $\Delta X_{11}(t)$ is the change in redundant moment after removal.

The flexibility coefficients in equation (a) are formulated using Trost's method:

1. δ_{10} due to dead load, g:

$$\delta_{10}^{(1)}(t) = \delta_{10,el}^{(1)}\,\phi_{res}^{(1)}(t, \tau_0) = \frac{g\,l^3}{24\,EI}\,\phi_{res}^{(1)}(t, \tau_0)$$

$$\delta_{10}^{(2)}(t) = \delta_{10,el}^{(2)}\,[1 + \phi^{(2)}(t', \tau_0')] = \frac{g\,l^3}{24\,EI}\,[1 + \phi^{(2)}(t', \tau_0')]$$

2. δ_{11} due to $X_{10} = 1$:

$$\delta_{11}^{(1)}(t) = \delta_{11,el}^{(1)}\,[1 + \phi^{(1)}(t, \tau_S)] = \frac{l}{3\,EI}\,[1 + \phi^{(1)}(t, \tau_S)]$$

$$\delta_{11}^{(2)}(t) = \delta_{11,el}^{(2)}\,[1 + \phi^{(2)}(t', \tau_0')] = \frac{l}{3\,EI}\,[1 + \phi^{(2)}(t', \tau_0')]$$

3. δ_{11} due to $\Delta X_1(t) = 1$:

$$\delta_{11}^{(1)}(t) = \delta_{11,el}^{(1)}\,[1 + \mu\phi^{(1)}(t, \tau_S)] = \frac{l}{3\,EI}\,[1 + \mu\phi^{(1)}(t, \tau_S)]$$

$$\delta_{11}^{(2)}(t) = \delta_{11,el}^{(2)}\,[1 + \mu\phi^{(2)}(t', \tau_0')] = \frac{l}{3\,EI}\,[1 + \mu\phi^{(2)}(t', \tau_0')]$$

The creep coefficients used in the above equations are defined as follows:

$$\phi_{res}^{(1)}(t, \tau_0) = \phi_n k(\tau_0)\left[f(t-\tau_0) - f(\tau_S - \tau_0)\right]$$

$$\phi^{(1)}(t, \tau_S) = \phi_n k(\tau_S) f(t - \tau_S)$$

$$\phi^{(2)}(t', \tau_0') = \phi_n k(\tau_0') f(t' - \tau_0')$$

Their final values $(t = \infty)$ are given by

$$\phi_{res, \infty}^{(1)} = \phi_n k(\tau_0)(1 - f(\tau_S - \tau_0))$$

$$\phi_\infty^{(1)} = \phi_n k(\tau_S)$$

$$\phi_\infty^{(2)} = \phi_n k(\tau_0')$$

The compatibility condition for $t = \infty$ is given by

$$\frac{gl^3}{24 EI}(1 + \phi_{res, \infty}^{(1)} + \phi_\infty^{(2)}) + X_{10}\frac{l}{3EI}(2 + \phi_\infty^{(1)} + \phi_\infty^{(2)})$$

$$+ \Delta X_{1, \infty}\frac{l}{3EI}(2 + \mu\phi_\infty^{(1)} + \mu\phi_\infty^{(2)}) = 0 \qquad (b)$$

The value of X_{10} must first be obtained from the compatibility conditions at $t = \tau_S (t' = \tau_0 \cong 0)$, for which $\phi_{res}^{(1)}$, $\phi^{(1)}$, $\phi^{(2)}$, and $\Delta X_1(t)$ are zero. It thus follows that

$$\frac{gl^3}{24 EI} + X_{10}\left(\frac{l}{3EI} + \frac{l}{3EI}\right) = 0$$

and therefore

$$X_{10} = -\frac{gl^2}{16}$$

Substituting this value into equation (b) yields

$$\frac{gl^3}{48 EI}(2\phi_{res, \infty}^{(1)} - \phi_\infty^{(1)} + \phi_\infty^{(2)}) + \Delta X_{1, \infty}\frac{l}{3EI}(2 + \mu\phi_\infty^{(1)} + \mu\phi_\infty^{(2)}) = 0$$

The solution of this equation is

$$\Delta X_{1, \infty} = -\frac{gl^2}{16}\frac{2\phi_{res, \infty}^{(1)} - \phi_\infty^{(1)} + \phi_\infty^{(2)}}{2 + \mu(\phi_\infty^{(1)} + \phi_\infty^{(2)})}$$

Assuming $\phi_{res, \infty}^{(1)} = 1.0$, $\phi_\infty^{(1)} = 1.5$, $\phi_\infty^{(2)} = 2.0$, and $\mu = 0.8$, the following value for $\Delta X_{1, \infty}$ is obtained:

$$\Delta X_{1, \infty} = -0.5\frac{gl^2}{16}$$

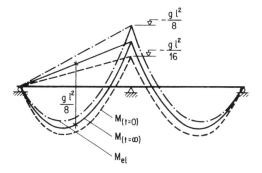

Figure 4.66
Example 4.7: redistribution of moments after removal of falsework for span 2 (M_{el}: elastic moments assuming both spans are cast and loaded simultaneously)

The final bending moments are thus halfway between the moments occurring immediately after falsework for span 2 is removed and the moments of an idealized system for which both spans are cast and loaded simultaneously (fig. 4.66).

4.7.4 Redistribution of Stress Due to Restrained Deformations or Self-Equilibrating States of Stress

Redundant sectional forces induced by restrained deformations are reduced by creep. The reduction is greatest when the deformation occurs instantaneously and somewhat less when the deformation occurs gradually. In both cases, Trost's method can be used to calculate the time-varying stresses and sectional forces. Computations are simplified by assuming that gradually occurring deformations proceed at the same rate as creep.

An expression for the time-varying stresses induced by an instantaneously applied deformation can be derived using Trost's method. The strain $\varepsilon_{c,0}$ is applied at time $t = 0$ and is held constant ($\varepsilon_c(t) = \varepsilon_{c,0}$ for $t \geq 0$). The corresponding elastic stress is denoted $\sigma_0 = E_c \varepsilon_{c,0}$. The stresses are computed using equation (4.10):

$$\varepsilon_c(t) = \varepsilon_{c,0} = \frac{\sigma_0}{E_c}[1 + \phi(t)] + \frac{\sigma(t) - \sigma_0}{E_c}[1 + \mu\phi(t)]$$

(It is assumed that $\varepsilon_{cs}(t) = 0$.) By substituting σ_0/E_c for $\varepsilon_{c,0}$ and rearranging, the following expression for the total stress $\sigma(t)$ is obtained:

$$\sigma(t) = \sigma_0\left(1 - \frac{\phi(t)}{1 + \mu\phi(t)}\right) \tag{a}$$

An analogous equation for the stresses induced by the strain, $\varepsilon_c(t)$, applied gradually to the structure, can also be derived. The initial stress, σ_0, is zero. It is assumed that $\varepsilon_c(t)$ increases at the same rate as creep, $\phi(t)$. The two functions are thus related by the following equation:

$$\varepsilon_c(t) = \varepsilon_{c,\infty} \frac{\phi(t)}{\phi_\infty} \tag{b}$$

The time-varying stresses are calculated using Trost's equation

$$\varepsilon_c(t) = \frac{\sigma_0}{E_c} [1 + \phi(t)] + \frac{\sigma(t) - \sigma_0}{E_c} [1 + \mu\phi(t)]$$

which is transformed into

$$\varepsilon_c(t) = \frac{\sigma(t)}{E_c} [1 + \mu\phi(t)]$$

Substituting equation (b) into this expression yields

$$\varepsilon_{c,\infty} \frac{\phi(t)}{\phi_\infty} = \frac{\sigma(t)}{E_c} [1 + \mu\phi(t)]$$

and thus

$$\sigma(t) = \sigma_{el,\infty} \frac{\phi(t)}{\phi_\infty [1 + \mu\phi(t)]} \tag{c}$$

where $\sigma_{el,\infty} = E_c \varepsilon_{c,\infty}$.

Equations (a) and (c) can also be applied to sectional forces, S. When the deformation is instantaneously applied, the time-varying forces are given by

$$S(t) = S_0 \left(1 - \frac{\phi(t)}{1 + \mu\phi(t)} \right)$$

When the deformation is gradually applied, the following expression can be used:

$$S(t) = S_{e,\infty} \frac{\phi(t)}{\phi_\infty(1 + \mu\phi(t))}$$

where $S_{e,\infty}$ is the elastic sectional force corresponding to the final strain, $\varepsilon_{c,\infty}$. When redundant forces are induced by a combination of an instantaneously deformation, $\varepsilon_{c,0}$, and a gradual deformation, $\varepsilon_c(t)$, the time-varying stresses can be computed using the following equation:

$$\sigma(t) = \sigma_0 \left(1 - \frac{\phi(t)}{1 + \mu\phi(t)} \right) + \sigma_0 \phi_\infty \frac{\phi(t)}{\phi_\infty(1 + \mu\phi(t))} = \sigma_0 \tag{d}$$

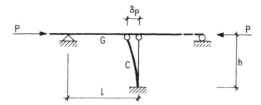

Figure 4.67
Moments in column induced by shortening of girder due to prestressing

provided the final value of the gradually applied strain, $\varepsilon_c(\infty)$, is $\phi_\infty \varepsilon_{c,0}$. The combination of an instantaneous and a gradual deformation thus produces a constant stress equal to the initial elastic stress.

The longitudinal shortening of bridge superstructures induced by prestressing consists of an instantaneous, elastic component and a gradual component due to creep. Equation (d) can thus be used to compute the redundant moments induced in columns by this phenomenon (fig. 4.67). The initial elastic displacement at the tip of the column is denoted δ_{el}; the gradual component, δ_{cc}, is equal to δ_{el} times $\phi(t)$. The total moment at the base of the column, $M(t)$, is thus constant and equal to the elastic moment

$$M_{\text{tot}} = \frac{3\,EI^C}{h^2}\,\delta_{el}$$

where EI^C is the flexural stiffness of the column and h its height. The moment M_{tot} can be decomposed into two components, corresponding to equations (a) and (c), respectively:

$$M(\delta_{el}) = \frac{3\,EI^C}{h^2}\,\delta_{el}\left(1 - \frac{\phi(t)}{1 + \mu\phi(t)}\right)$$

$$M(\delta_{el}\,\phi(t)) = \frac{3\,EI^C}{h^2}\,\delta_{el}\,\phi_\infty\,\frac{\phi(t)}{\phi_\infty(1 + \mu\phi(t))}$$

The contributions of $M(\delta_{el})$ and $M(\delta_{el}\,\phi(t))$ to M_{tot} are plotted in figure 4.68 as functions of t.

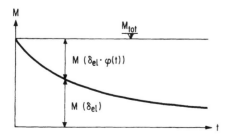

Figure 4.68
Moment at the base of the column, M, due to shortening of girder ($M(\delta_{el})$: instantaneous component; $M(\delta_{el}\,\phi(t))$: gradual component)

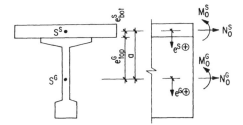

Figure 4.69
Composite section composed of a precast concrete girder and a cast-in-place concrete deck

Load-free states of strain in the cross-section are induced by temperature gradients and, in composite materials, differential creep and shrinkage. Since plane sections must remain plane, the free deformation of the cross-section due to these actions is restrained. This results in self-equilibrating stresses, the evolution of which can be easily calculated using Trost's method.

Cross-sections consisting of a precast girder and a composite, cast-in-place deck slab are frequently used in bridge construction (fig. 4.69). The permanent load stresses are redistributed due to differential creep and shrinkage.

The calculation of the time-varying stresses begins with determining the initial sectional forces in the deck slab and girder due to dead load and prestressing. These forces, which are denoted N_0^S, M_0^S, N_0^G, and M_0^G in figure 4.69, must satisfy equilibrium and, where applicable, compatibility (plane sections). The initial rotations, θ, and strains at the interface, ε^k, are as follows:

$$\varepsilon_0^{S(k)} = \frac{N_0^S}{E^S A^S} + \frac{M_0^S}{E^S I^S} e_{bot}^S \qquad \varepsilon_0^{G(k)} = \frac{N_0^G}{E^G A^G} + \frac{M_0^G}{E^G I^G} e_{top}^G$$

$$\theta_0^S = \frac{M_0^S}{E^S I^S} \qquad\qquad\qquad \theta_0^G = \frac{M_0^G}{E^G I^G}$$

Since the deck slab is normally cast in place without shoring the girders, the parameters N_0^S, M_0^S, $\varepsilon_0^{S(k)}$, and θ_0^S are typically zero.

The free deformations of slab and girder are prevented by time-varying shear forces at the interface, which cause changes in the sectional forces and deformations in each component. Relations between the change in sectional force and change in deformation, valid for arbitrary values of t, are formulated using Trost's method:

$$\Delta\varepsilon^{S(k)}(t) = \varepsilon_0^{S(k)} \phi^S(t)$$
$$+ \left(\frac{\Delta N^S(t)}{E^S A^S} + \frac{\Delta M^S(t)}{E^S I^S} e_{bot}^S\right) (1 + \mu^S \phi^S(t)) + \varepsilon_{cs}^S(t) \qquad (e)$$

$$\Delta\theta^S(t) = \theta_0^S \phi^S(t)$$
$$+ \frac{\Delta M^S(t)}{E^S I^S} (1 + \mu^S \phi^S(t)) \qquad\qquad\qquad (f)$$

$$\Delta \varepsilon^{G(k)}(t) = \varepsilon_0^{G(k)} \phi^G(t)$$

$$+ \left(\frac{\Delta N^G(t)}{E^G A^G} + \frac{\Delta M^G(t)}{E^G I^G} e_{top}^G \right) (1 + \mu^G \phi^G(t)) + \varepsilon_{cs}^G(t) \qquad \text{(g)}$$

$$\Delta \theta^G(t) = \theta_0^G \phi^G(t)$$

$$+ \frac{\Delta M^G(t)}{E^G I^G} (1 + \mu^G \phi^G(t)) \qquad \text{(h)}$$

The sectional forces and deformations thus defined must satisfy the following conditions for all values of t:

1. Equilibrium:

$$\Delta N^S(t) + \Delta N^G(t) = 0 \qquad \text{(i)}$$

$$\Delta M^S(t) + \Delta M^G(t) + a \, \Delta N^G(t) = 0 \qquad \text{(j)}$$

2. Compatibility:

$$\Delta \varepsilon^{S(k)}(t) = \Delta \varepsilon^{G(k)}(t) \qquad \text{(k)}$$

$$\Delta \theta^S(t) = \Delta \theta^G(t) \qquad \text{(l)}$$

The eight equations (e) through (l) can be solved for the four unknown changes in sectional force and four unknown changes in deformation. Because the flexural stiffness of the slab is considerably smaller than that of the girder, $\Delta M^S(t)$ can be taken as zero. Under this assumption, equations (f), (h), and (l) must be dropped; the remaining five equations

$$\Delta \varepsilon^{S(k)}(t) = \varepsilon_0^{S(k)} \phi^S(t) + \frac{\Delta N^S(t)}{E^S A^S} (1 + \mu^S \phi^S(t)) + \varepsilon_{cs}^S(t) \quad \text{(m)}$$

$$\Delta \varepsilon^{G(k)}(t) = \varepsilon_0^{G(k)} \phi^G(t) + \left(\frac{\Delta N^G(t)}{E^G A^G} + \frac{\Delta M^G(t)}{E^G I^G} e_{top}^G \right)$$

$$(1 + \mu^G \phi^G(t)) + \varepsilon_{cs}^G(t) \qquad \text{(n)}$$

$$\Delta N^S(t) + \Delta N^G(t) = 0 \qquad \text{(o)}$$

$$\Delta M^G(t) + a \Delta N^G(t) = 0 \qquad \text{(p)}$$

$$\Delta \varepsilon^{S(k)}(t) = \Delta \varepsilon^{G(k)}(t) \qquad \text{(q)}$$

are solved for $\Delta N^S(t)$, $\Delta N^G(t)$, $\Delta M^G(t)$, $\Delta \varepsilon^{S(k)}(t)$, and $\Delta \varepsilon^{G(k)}(t)$.

The loss of prestress due to shrinkage and creep in concrete can be considered as a redistribution of the self-equilibrating state of stress induced by prestressing and visualized as an increase of the internal lever arm to maintain equilibrium. The prestressing force thus cannot be reduced if the compressive force in the concrete, F_c, is already located close to the upper edge of the cross-section. The initial lever arm for permanent load is greater for partially prestressed sections than for fully

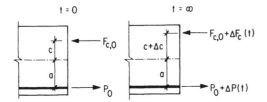

Figure 4.70
Change of internal forces and
lever arm due to loss of prestress

Internal forces : $F_{c,0} = P_0$
Lever arm   : $a + c$

Internal forces : $F_{c,0} + \Delta F_c (t) = P_0 + \Delta P(t)$
Lever arm   : $a + c + \Delta c(t)$

prestressed sections. It thus follows that the potential loss of prestress for the former will be smaller for than the latter.

These concepts are used to derive an equation for the loss of prestress, $\Delta P(t)$, in a flexural member (fig. 4.70). (A decrease in prestressing force corresponds to a positive value for $\Delta P(t)$. It therefore follows that the total prestressing force, $P(t)$, is $P_0 + \Delta P(t)$.) Axial force and moment equilibrium imply that

$$\Delta P(t) = \Delta F_c(t) \tag{r}$$

and

$$[F_{c,0} + \Delta F_c(t)] [a + c + \Delta c(t)] = F_{c,0}(a + c) \tag{s}$$

Expanding equation (s) and neglecting second-order terms, the following expression is obtained:

$$\Delta c(t) = - \frac{\Delta F_c(t)(a+c)}{F_{c,0}} \tag{t}$$

The compatibility condition at the level of the prestressing steel,

$$\Delta \varepsilon_c(t) = \Delta \varepsilon_P(t) \tag{u}$$

is expressed in terms of $\Delta P(t)$ and $\Delta F_c(t)$ with the help of Trost's method:

$$\Delta \varepsilon_c(t) = \frac{\sigma_{c,0}}{E_c} \phi(t) + \frac{\sigma_c(t) - \sigma_{c,0}}{E_c} [1 + \mu \phi(t)] + \varepsilon_{cs}(t) \tag{v}$$

where the concrete stresses are also taken at the level of the prestressing steel. The change in stress $\sigma_c(t) - \sigma_{c,0}$ is decomposed into its axial and flexural components:

$$\sigma_c(t) - \sigma_{c,0} = - \frac{\Delta F_c(t)}{A_c} + \frac{F_{c,0} \Delta c(t) + \Delta F_c(t) c}{I_c} a$$

$$= - \frac{\Delta F_c(t)}{A_c} - \frac{\Delta F_c(t)}{I_c} a^2$$

where second-order terms have been neglected and the expression for $\Delta c(t)$ of equation (t) has been substituted. Equation (v) can now be rewritten

$$\Delta \varepsilon_c(t) = \frac{\sigma_{c,0}}{E_c}\, \phi(t)$$

$$-\left(\frac{\Delta F_c(t)}{E_c A_c} + \frac{\Delta F_c(t)}{E_c I_c}\, a^2\right) [1 + \mu \phi(t)] + \varepsilon_{cs}(t) \tag{w}$$

The time-varying strains in the prestressing steel are given by

$$\Delta \varepsilon_P(t) = \frac{\Delta P(t)}{E_P A_P} \tag{x}$$

Substituting equations (w) and (x) into (u) and letting $\Delta P(t) = \Delta F_c(t)$ yields the following expression for loss of prestress:

$$\Delta P(t) = \frac{n\varrho\,[\sigma_{c,0}\, A_c\, \phi(t) + \varepsilon_{cs}(t)\, E_c\, A_c]}{1 + n\varrho\,[1 + \mu \phi(t)]\left(1 + \dfrac{a^2 A_c}{I_c}\right)}$$

where $n = E_P/E_c$ and $\varrho = A_P/A_c$. If the flexural component of concrete stress is neglected in equation (w), the above expression is transformed into:

$$\Delta P(t) = \frac{n\varrho\,[\sigma_{c,0}\, A_c\, \phi(t) + \varepsilon_{cs}(t)\, E_c\, A_c]}{1 + n\varrho\,(1 + \mu \phi(t))}$$

which is identical to equation (4.9) derived in Section 4.6.8 for axially loaded members.

4.8 Serviceability

4.8.1 Durability

Durable bridges are economical bridges; long service life and low maintenance requirements lead directly to low life-cycle costs. Whereas safety can be "guaranteed" by designs that are in accordance with applicable codes and standards, durability cannot. It is impossible to predict and quantify all of the many causes of deterioration and the performance over time of protective systems. Durability must therefore be addressed not only in the design, but also in the detailing, construction, and inspection of bridges. Design engineers should give careful consideration to construction and inspection, by ensuring that the design and details facilitate proper workmanship and inspection. In addition, designers should prepare a detailed program of periodic inspection, valid for the entire service life of the bridge.

Deterioration of concrete bridges is the result of harmful environmental conditions, the most important of which are as follows:

1. Water in either liquid or gaseous (atmospheric humidity) state
2. Freezing temperatures
3. Deicing chemicals
4. Other aggressive chemicals present in the air, water, or soil
5. Mechanical abrasion

The most serious threats to durability are generally posed by combinations of two or more of these conditions. Water is most hazardous, for example, in conjunction with freeze-thaw cycles or when it serves as a vehicle for dissolved deicing salts.

Among the most effective ways of maximizing durability are high quality concrete and proper detailing. The latter includes measures to limit or eliminate exposure to water and to facilitate inspection. No single countermeasure should be relied on, however, to protect against any of the conditions listed above. Protective systems should always consist of several redundant components to ensure continued protection if any one of them fails.

a) Concrete Quality

Concrete must be of high quality for its own protection and to ensure the protection of the embedded reinforcing steel. Special concrete properties should be specified for all structural components exposed to harsh environmental conditions, such as freezing while saturated or the combined action of freezing and deicing chemicals (see Section 3.1.1). These components include deck slabs, parapets, joints, girder ends, bearing seats, and abutment back walls. Concrete that is resistant to freezing and deicing salt is also required for the soffits of grade separations, which can be easily reached by salt spray from the roadway below for normal clearance heights.

As discussed in Section 3.1.1, a covering layer of concrete of high quality and sufficient thickness is the most effective means of preventing corrosion of the reinforcement. This layer actually protects the steel indirectly; direct protection is provided by an oxide film on the surface of the steel, which forms spontaneously provided the pH of the surrounding concrete is greater than 12. Corrosion is prevented as long as this film remains intact. The oxide can be destroyed, however, by a reduction in the alkalinity of the concrete due to carbonation (a reaction of concrete with atmospheric carbon dioxide) or by contact with chloride ions originating from deicing chemicals. The function of the covering layer of concrete is therefore to safeguard the direct protection mechanism of the steel by restricting the penetration of carbon dioxide and chloride ions.

The rate of penetration of both of these substances is an increasing function of the permeability of the covering layer. Concrete is made more permeable by the presence of capillary pores formed by excess water in the concrete mix (see Section

3.1.1) and microcracks on the surface due to drying shrinkage. Capillary porosity can be effectively reduced by lowering the water-cement ratio. The formation of microcracks is prevented by proper curing conditions. Formwork should not removed less than 3 days after casting. In addition, unformed surfaces should be thermally insulated and protected against water loss by special mats. The importance of low water-cement ratio and proper curing in this regard is illustrated by the following relation: concrete with a water-cement ratio of 0.6, cured for 7 days, has the same capacity for water absorption as concrete with a water-cement ratio of 0.45, cured for 2 days.

The advance of the carbonation front in the covering layer has been found to be proportional to $\sqrt{t}$, where t denotes time. Assuming a relative humidity of 60 to 80 percent and a cement content of 300 kg per cubic metre of concrete, the average depth of carbonation, $\bar{d}$, can be expressed by the following empirical relation:

$$\bar{d} = 0.4\,n\sqrt{t}$$

where $\bar{d}$ is in millimetres, t is in years, and n is capillary porosity in percent. The standard deviation of d is given by the following expression:

$$s_d = 0.2\,n\sqrt{t}$$

Values of d greater than $\bar{d} + s_d$ thus satisfy the inequality

$$d > 0.6\,n\sqrt{t}$$

Assuming d is normally distributed about its mean, $\bar{d}$, it follows that this inequality will be true in less than 16 percent of all observed cases. If this 16 percent fractile value can be accepted after a lifetime of 75 years, the following relation between depth of the covering layer, d, and capillary porosity, n, can be established:

$$\frac{d}{n} > 5 \tag{a}$$

Combinations of d and n satisfying this inequality can be considered effective in preventing corrosion due to carbonation.

Chloride concentrations of less than 0.4 percent of cement weight are generally regarded as insufficient to induce corrosion and thus can be considered acceptable. Concrete covering layers satisfying inequality (a) have been found effective in limiting chloride concentrations at the depth of the reinforcement to 0.4 percent after 75 years, provided the chloride ions are transported to the concrete surface by water vapour. This is normally the case for the soffits of grade separations. A stricter criterion is required, however, for concrete that is in direct contact with liquid salt water. For this type of exposure, d and n must satisfy the following inequality

$$\frac{d}{n} > 15$$

to limit chloride concentrations to 0.4 percent after 75 years. A covering layer with a capillary porosity of 6 percent, corresponding to a water-cement ratio of 0.5, would thus have to be 90 mm thick. Direct contact with salt water should therefore be prevented by the use of waterproofing membranes.

The depths computed from inequality (a) should be increased by at least 10 mm for the top surface of deck slabs, concrete in earth or water, and for all components exposed to particularly aggressive conditions. Additional cover should also be provided as required to accommodate tolerances in bar fabrication. It may be practical to locate large diameter reinforcement 80 mm below the surface and to provide an additional mesh of smaller bars 40 mm below the surface for crack control. Care must be taken to achieve the specified cover during construction. Spacers must always be provided between reinforcement and formwork. Spacers can be pressed into soft materials such as styrofoam, thus reducing the thickness of cover. Only hard materials such as wood or steel should therefore be used for formwork.

Epoxy coating is now commonly used in the U.S.A. and in Canada for the direct protection of mild reinforcement. Although this countermeasure has proven itself to be effective in preventing corrosion, its use must not lead to lowering of concrete quality. Concrete must still be capable of protecting itself against the combined action of freezing and deicing salts and of protecting uncoated prestressing steel against corrosion.

b) Detailing

Many of the details that enhance durability have been known and used for several decades and are no more than simple common sense. The cross-section of the girder should be weatherproof. Structural components located in earth or water should be at least 0.3 m thick. All concrete surfaces not located in water or earth should be accessible for inspection; inaccessible voids and lost forms must therefore never be used. All voids should be provided with openings for ventilation and drainage. Deck slabs must be protected with a high quality waterproofing membrane that has been carefully applied to a clean concrete surface. Wearing surfaces should be at least 80 mm thick to reduce thermal shock in deck slabs exposed to freezing and deicing chemicals.

An efficient deck drainage system is indispensable. The cross-fall of the deck slab should be greater than 2 percent to speed run-off of water to the drainage inlets. Drainage pipes that are not cast into concrete can be easily inspected, repaired, and replaced. In addition, they will not damage the structure by a buildup of ice pressure if blocked. Pipes should therefore be cast into the structure only when: (a) doing otherwise would be aesthetically unacceptable, and (b) a minimum 2 percent slope can be provided. Transverse pipes leading from the edge of the deck to a main pipe between the webs are thus often cast into the structure, normally by locally thickening the deck slab cantilevers. Longitudinal pipes are usually arranged between the webs of box and T girders. The cross-section of slab bridges should be designed so that longitudinal pipes do not detract from the appearance of the bridge.

The deck drainage system must be hydraulically efficient, with adequate cross-section and slope for all pipes. The connection to the main drainage channel and accessibility for inspection and cleaning must be properly detailed. Leaks should be immediately apparent; water from leaking pipes must not be allowed to fill up the voids of box girders, which may lead to serious structural damage. An efficient and accessible drainage system must also be provided at the abutments, since even "waterproof" expansion joints cannot be guaranteed leakproof for the life of the bridge.

Bearings, expansion joints, drainage system components, waterproofing membranes, and wearing surfaces, all of which are exposed to particularly severe environmental conditions, often wear out long before the design service life of bridges. Proper structural behaviour and durability require, however, that each of these components be in good working order at all times. The highest quality materials and workmanship should therefore always be insisted on for these components. Bearings, expansion joints, and drainage pipes must be carefully detailed to ensure ease of inspection, maintenance, and replacement. Bearings should be accessible from at least two sides; expansion joints should be accessible from below. Bearings should be located on raised pedestals to protect them from water and dirt. Sufficient room must be available for the jacks required to lift the bridge for bearing replacement.

Bearings, expansion joints, and other bridge accessories are discussed further in Chapter 6.

4.8.2 Function

As components of a larger transportation system, bridges must provide an acceptable degree of traffic safety and comfort. Roadway width, superelevation, and wearing surfaces must therefore conform to the same standards as highways on grade. Special attention must be given to the design and detailing of guardrails. Rails that can absorb the energy of impact by plastic deformations are the safest. They cannot be used, however, at the edge of the deck slab; additional deck width beyond the limits of the roadway must be provided to allow the guardrails to bend outward on impact. Although this additional width can be put to use as an inspection lane, its cost is normally prohibitive. A safe and economical alternative to metal guardrails are concrete parapets, the interior faces of which are shaped to deflect vehicles back towards the roadway for typical angles of impact.

Proper deck drainage is also indispensable for traffic safety, to prevent flooding and the formation of ice on the roadway. The minimum cross-fall of 2 percent, recommended in Section 4.8.1 for durability, is also the minimum required to prevent flooding. It should thus be provided even where freezing temperatures do not occur. In regions with heavy snowfalls, the roadway should slope down towards both edges to prevent melt water from snowbanks on the high side from flowing over the roadway and freezing. For obvious reasons, drainage inlets must be provided on both edges of the deck when the cross-fall is detailed in this way. Box sections retain heat better than T-sections and are thus less likely to have problems with freezing of the roadway.

Deformations of highway bridges due to live load usually have no influence on rider comfort and thus need not be considered in design. Live load deformations must be limited, however, for bridges carrying high-speed trains. Discontinuities of slope at internal hinges and abutments due to long-term permanent load must be less than 0.4 percent for highways carrying high-speed traffic and 0.8 percent for ordinary roads. Angle breaks caused by differential settlement of backfill behind abutments can be reduced by a sufficiently long approach slab.

Vibrations of highway bridges normally have no influence on rider comfort. Pedestrians, however, are sensitive to vibrations. The dynamic behaviour of slender bridges that carry pedestrians should therefore always be investigated. The human perception of vibration is discussed further in Section 4.8.6. Dynamic investigations must also be performed for railway bridges that carry high-speed trains.

Vehicles and pedestrians should be protected by special shielding devices where extremely strong winds are known to occur.

The primary function of expansion joints and bearings is to permit the movement of the superstructure. In calculating the required movement capacity of these components, the deformations of the structure itself and of the soil must be considered. The accumulation of a series of small annual soil displacements over several decades can be as large as the deformations of the structure that are normally used as a basis for the design of these components. Displacements that exceed the capacity of expansion joints or bearings have serious consequences; their restoration to working order is usually very expensive. It is thus advisable to design bearings and joints conservatively, allowing ample movement capacity. This precautionary measure entails no significant increase in total construction cost.

4.8.3 Appearance

In the context of serviceability, appearance is related to the visible effects of uncontrolled run-off of water, pollution, cracking, and deformations. It is thus

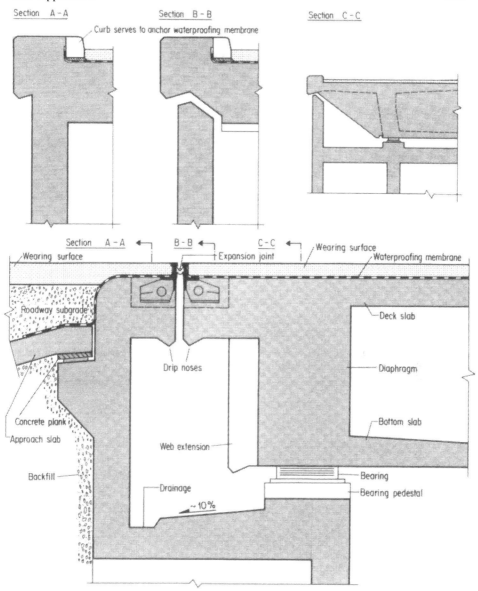

Figure 4.71
Recommended abutment details

not directly related to the aesthetic principles discussed in Chapter 2. Deficiencies due substandard workmanship during construction are also beyond the scope of this section.

Abutments, expansion joints, and the edges of deck slabs must be especially carefully detailed to prevent unsightly stains due to water and dirt (fig. 4.71). It

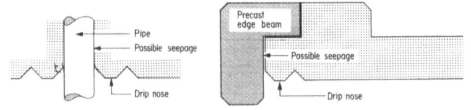

Figure 4.72
Drip nose details

must never be assumed that expansion joints and connections of waterproofing membranes will remain waterproof for the entire life of the bridge. Any water that seeps through at these locations must be collected and carried away. Collection drains and pipes should be designed to be easily accessible for cleaning. A drip nose must be provided whenever a vertical pipe passes through the deck slab, to allow any water that leaks through the waterproofing connection between pipe and concrete to collect and drip away. When the edges of the deck slab are covered by precast components (fig. 4.72), a well detailed drip nose must be provided in the cast-in-place portion to carry away any water that leaks down between the precast and the cast-in-place components.

4.8.4 Cracking

a) Fundamentals

All of the methods used to calculate crack widths and deformations of structures in the cracked state are based on the assumption that the stresses in the reinforcement are known beforehand or can be calculated. In reality, however, it is impossible to know the exact value of the steel stresses under service conditions. Stresses in the steel are a function of many different factors, some of which are subject to considerable variability. The most important of these include prestressing losses, the redistribution of sectional forces, self-equilibrating states of stress, and restrained deformations.

It thus follows that whatever accuracy promised by an "exact" calculation of steel stresses under service conditions is illusory. Simplifications based on rational models of structural behaviour should therefore be used to calculate steel stresses, crack widths, and deformations. It also follows that the criteria used to evaluate cracking behaviour and deformations need not be regarded as "exact" values, but rather as rough, conservative estimates.

Cracking due to live load normally has no influence on durability or appearance and is thus not considered in bridge design. Of much greater importance is cracking caused by the combined effects of permanent load (dead load plus prestressing), restrained deformations, and self-equilibrating stresses. The prestressing concept is usually selected to prevent tensile stresses in the concrete due to permanent load. Concrete stresses in excess of tensile strength are therefore the result of the additional effects of restrained deformations and self-equilibrating stresses.

The sectional forces that must be superimposed onto the permanent load forces to produce the first crack are called the *cracking sectional forces* and are denoted N_r and M_r, or collectively as S_r. In normal cases, an increase of the restrained deformation in the cracked structure results in only a negligible increase in the sectional forces above S_r. The structure responds to such deformations by the formation of more cracks of approximately constant width. Because the steel stresses and crack widths due to an arbitrary restrained deformation differ little from the steel stresses and crack width at the formation of the first crack, the design of minimum reinforcement for crack control can be based on S_r.

The steel stresses associated with the cracking sectional forces can be calculated with reasonable accuracy based on the tensile strength of the concrete in the effective homogeneous cross-section. A calculation based on the external actions and the stiffness of the system in the cracked state is unreliable and should not be attempted.

b) Cracking Due to Flexure and Axial Force

The tensile strength of concrete, f_{ct}, can vary widely from one location to another in a structural element. Tensile strength can be locally reduced in several ways, including inhomogeneities in the concrete, microcracks caused by self-equilibrating stresses, and the presence of transverse reinforcement.

When the sectional forces are constant along the length of an element, the first crack will be produced at the section with the least concrete tensile resistance. Assuming the sectional forces due to permanent load are equal to zero, the cracking moments and axial forces for several typical cases as follows:

1. Pure tension: $\qquad\qquad\qquad\qquad\qquad N_r = f_{ct}\, A_c$

2. Pure bending: $\qquad\qquad\qquad\qquad\qquad M_r = f_{ct}\, W_c$

3. Combined tension and bending: $\qquad N_r + M_r \dfrac{A_c}{W_c} = f_{ct}\, A_c$

where W_c denotes the section modulus of the the homogeneous, uncracked concrete section.

At the first crack, the tensile component of S_r is taken over by the reinforcement. The corresponding stress in the steel at this location is denoted σ_{sr}^{II}:

$$\sigma_{sr}^{II} = \sigma_s^{II}\,(N_r,\, M_r)$$

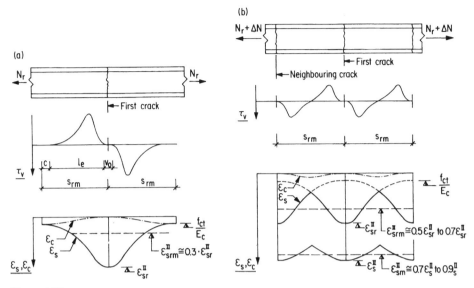

Figure 4.73
Cracking due to axial tension: a, at formation of first crack; b, after several cracks have been
formed

where σ_s^{II} is the general notation for steel stress at a crack, the superscript II
refers to the cracked state (state II in figure 3.18), and the subscript r in σ_{sr}^{II} refers
to the cracking sectional forces. Additional cracks immediately adjacent to the
first crack cannot be formed since the concrete tensile stresses in this region are
less than f_{ct}.

The tensile stress in the steel at the crack is restored gradually to the adjacent
homogeneous concrete sections through bond stresses, τ_v (fig. 4.73). The minimum distance from the first crack beyond which a subsequent crack can form is
denoted s_{rm} and is defined by the following expression:

$$s_{rm} = \frac{v_0}{2} + c + l_e \tag{a}$$

where l_e is the length required to transfer σ_{sr}^{II} to the concrete through bond, v_0 is the
length of destroyed bond at the crack, and c is approximately equal to the
thickness of cover. The parameter s_{rm} thus corresponds to the theoretical
minimum crack spacing.

Provided σ_{sr}^{II} is less than the yield stress, f_{sy}, further cracks will be produced by
increasing the sectional forces only slightly beyond S_r. The cracks eventually
become so closely spaced that concrete stresses in excess of f_{ct} are no longer
possible. This distribution of cracks is called the *stabilized crack pattern*. At this
point, the steel stresses at the cracks are still practically unchanged from the stress

at the cracking load, σ_{sr}^{II}. Increases in sectional forces beyond the formation of this pattern will widen the existing cracks and increase σ_s^{II}, but will produce no new cracks.

The stabilized crack pattern can be modeled as a set of cracks with equal spacing s_{rm}. As shown in figure 4.73b, only half of l_e is available to transfer tension from steel to concrete when the cracks are spaced in this way. This confirms that, even for higher loads, it will be impossible for the concrete stresses between the cracks to exceed f_{ct}. Due to the inherent variability in concrete tensile strength along the length of a member, the actual crack spacing in the stabilized pattern will vary considerably from the theoretical average s_{rm}. If transverse reinforcement is spaced at an interval similar to s_{rm}, the cracks will normally occur at the locations of the transverse reinforcement.

The average stress and strain in the steel over an element of length s_{rm}, centered on a crack, are denoted σ_{sm}^{II} and ε_{sm}^{II}, respectively. Strain in the concrete between cracks can be neglected on the tension side of the member, since elongation due to the sectional forces will be opposite and approximately equal to shortening due to shrinkage. The average theoretical crack width at the extreme tensile fibre can thus be formulated in terms of the steel strain alone:

$$w_m = \varepsilon_{sm}^{II} \, s_{rm}$$

Since increases in σ_s^{II} between formation of the first crack and creation of the stabilized crack pattern are negligible, so are the increases in the corresponding values of w_m. Prior to the formation of the stabilized pattern, therefore, w_m can be taken equal to $w_m(S_r)$, the crack width due to cracking sectional forces:

$$w_m(S_r) = \varepsilon_{srm}^{II} \, s_{rm}$$

(The quantity ε_{srm}^{II} is the average steel strain under cracking sectional force.) After formation of the stabilized crack pattern, average crack width is a function of the average steel strain corresponding to the actual sectional forces, S:[1]

$$w_m(S) = \varepsilon_{sm}^{II} \, s_{rm}$$

The parameters ε_{sm}^{II} and s_{rm} can be calculated as follows:

1. *Average Steel Strain* ε_{sm}^{II}. The average steel strain ε_{sm}^{II} is a function of steel strain at the crack location ε_s^{II}, the ratio $\sigma_s^{II}/\sigma_{sr}^{II}$, bond strength, and type of load (initial, long-term, or repeated). Figure 4.74 shows the increase of ε_{sm}^{II} and the average concrete strain, ε_{cm}^{II}, with increasing sectional force.

---

[1] The cracking behaviour described above is based on the assumption of constant sectional forces along the length of the member. When the distribution of sectional forces varies along the member, the first crack is produced at the most highly stressed section. Increases in sectional forces will create additional cracks and will widen the initial crack.

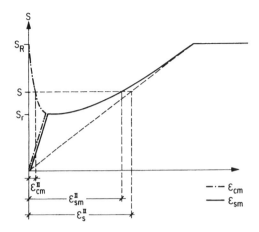

Figure 4.74
Average steel and concrete strains (S: sectional force; S_r: cracking sectional force; S_R: ultimate sectional force)

The following expressions can be used to calculate ε_{sm}^{II} (Comité Euro-International du Béton 1985):

$$\varepsilon_{sm}^{II} = \lambda \varepsilon_s^{II}$$

where

$$\lambda = \left[1 - \beta_1 \beta_2 \left(\frac{\sigma_{sr}^{II}}{\sigma_s^{II}} \right)^2 \right]$$

The parameter β_1 is taken as 1.0 for deformed reinforcing steel and 0.5 for smooth reinforcing steel; β_2 is taken as 1.0 for initial loading and 0.5 for long-term load or many cycles of a repeated load.

For the typical case of deformed reinforcing steel and long-term load, therefore, the average steel strain is given by

$$\varepsilon_{sm}^{II} = \left[1 - 0.5 \left(\frac{\sigma_{sr}^{II}}{\sigma_s^{II}} \right)^2 \right] \varepsilon_s^{II} = \lambda \varepsilon_s^{II}$$

from which it follows that $0.5 \leq \lambda \leq 1.0$.

Prior to the formation of the stabilized crack pattern, for sectional forces between 1.0 and 1.5 times the cracking forces, ε_{sm}^{II} is a function of strain at the crack due to S_r, and lies between $0.5\varepsilon_{sr}^{II}$ and $0.7\varepsilon_{sr}^{II}$. After formation of the stabilized crack pattern, for sectional forces greater than 1.5 times the cracking forces, ε_{sm}^{II} is a function of the strain at the crack due to the actual sectional forces, and lies between $0.7\varepsilon_s^{II}$ and $0.9\varepsilon_s^{II}$.

2. Average Crack Spacing s_{rm}. The average crack spacing, s_{rm}, corresponds to the length required to develop f_{ct} in the concrete at a section adjacent to an existing crack. It is calculated using equation (a):

$$s_{rm} = \frac{v_0}{2} + c + l_e$$

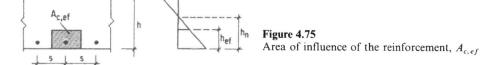

Figure 4.75
Area of influence of the reinforcement, $A_{c,ef}$

where v_0 is the length of destroyed bond at the crack, c the length required for stress to spread out from the reinforcement to the outer fibres of the section, and l_e is the length required to transfer the tensile stress from the reinforcement to the homogeneous section through bond.

An approximate expression for l_e can be formulated based on the following equation of equilibrium:

$$l_e \, d\pi \, \tau_{vm} = f_{ct} \, k \, A_{c,ef} \tag{b}$$

where the left-hand side represents the force transferred from steel to concrete through bond and the right-hand side represents the force necessary to crack the concrete in the zone of influence of the reinforcement. The quantities d and τ_{vm} denote bar diameter and average bond stress, respectively.

The effectiveness of reinforcement in limiting crack spacing and width is limited to a zone in the vicinity of the bars. The area of this zone of influence, denoted $A_{c,ef}$, is defined by the following equation:

$$A_{c,ef} = s \, h_{ef}$$

where s is bar spacing. The effective depth, h_{ef}, is a function of the shape of the concrete stress diagram before cracking and the depth of the member (fig. 4.75). Approximate values for h_{ef} can be taken from table 4.3. The tensile force in the zone of influence immediately before cracking is thus $f_{ct} \, k \, A_{c,ef}$, where the factor k is 1 for rectangular stress distributions and 0.5 for triangular distributions.

Table 4.3
Effective Depth, h_{ef}, of the Zone of Influence of the Reinforcement, $A_{c,ef}$ (fig. 4.75)

$h_n \leq 0.75h$ (Tension at one edge of the section)		$1.00h \leq h_n$ (Tension at both edges of the section)	
h_n	h_{ef}	$h/2$	h_{ef}
≤ 250 mm	h_n	≤ 250 mm	$h/2$
> 250 mm	250 mm	> 250 mm	250 mm

(Linear interpolation can be used for values of h_n between $0.75h$ and $1.00h$.)

By rearranging equation (b), the following expression for l_e is obtained:

$$l_e = \frac{f_{ct} k A_{c,ef}}{\tau_{vm} \pi d} \tag{c}$$

The quotient f_{ct}/τ_{vm} can be assumed constant and taken as 0.4 for deformed reinforcing steel. Equation (c) can therefore be simplified as follows:

$$l_e \cong 0.4 \frac{k A_{c,ef}}{\pi d} \cong \frac{k A_{c,ef}}{8 d} \tag{d}$$

The length of disturbed bond, v_0, can be taken between $4d$ and $7d$. The length required to spread the stress from the end of the developed bar to the outer edge of the member, c, is approximately equal to the thickness of the covering layer.

Example 4.8:
Reinforcement to limit crack width

The width of cracks produced by the cracking tensile force, N_r, is to be calculated for the bottom slab of a box section. The following dimensions and parameters are given:

Slab thickness	$h = 180$ mm
Reinforcement ratio	$\varrho = 0.6$ percent
Bar spacing	$s = 150$ mm
Thickness of covering layer	$c = 40$ mm
Concrete tensile strength	$f_{ct} = 2$ N/mm^2

Reinforcement consists of 10 mm diameter bars top and bottom.

The cracking tensile force per metre of slab width is given by

$$N_r = f_{ct} A_c = (2) (180) (1\,000) = 360 \text{ kN}$$

The given area of steel per metre of width is

$$A_{s,tot} = \varrho A_c = (0.006) (180) (1\,000) = 1\,080 \text{ mm}^2$$

The stress and strain in the steel under cracking load are thus

$$\sigma_{sr}^{II} = \frac{N_r}{A_{s,tot}} = 330 \text{ N/mm}^2 \qquad \varepsilon_{sr}^{II} = \frac{\sigma_{sr}^{II}}{E_s} = 1.6 \times 10^{-3}$$

The average steel strain can be taken as

$$\varepsilon_{srm}^{II} \cong 0.6 \varepsilon_{sr}^{II} = 0.96 \times 10^{-3}$$

which is in the range of possible values corresponding to the cracking force, N_r. It is assumed that $v_0 = 4d = 40$ mm. The transition length l_e is calculated using equation (d):

$$l_e = \frac{kA_{c,ef}}{8d} = \frac{(1.0)\ (150)\ (90)}{(8)\ (10)} = 170 \text{ mm}$$

The average spacing between cracks is thus

$$s_{rm} = \frac{v_0}{2} + c + l_e = 20 + 40 + 170 = 230 \text{ mm}$$

The average crack width can now be computed:

$$w_m = s_{rm}\ \varepsilon_{srm}^{II} = (230)\ (0.96 \times 10^{-3}) = 0.22 \text{ mm}$$

The model of cracking behaviour derived above is based on the assumption that bond is the only mechanism by which tensile stresses are developed in concrete sections adjacent to existing cracks. This assumption is valid for members subjected to axial tension. For flexural members, however, tensile stresses are developed through bond, as described above, and through the eccentricity of the compressive force in the concrete at the crack, F_c. As shown in figure 4.76, F_c is equilibrated in the uncracked section by a linear stress distribution, which is considered to be fully developed at a distance h from the crack, where h is the depth of the member. The eccentricity of F_c is normally large enough to produce tensile stresses at the edge of the uncracked section. Additional tensile stresses are created by whatever portion of the tension in the reinforcement at the crack, F_t, has been transferred to the concrete over this distance.

For bridge girders and other flexural members of relatively great depth, values of s_{rm} computed considering only the transfer of tension to concrete through bond

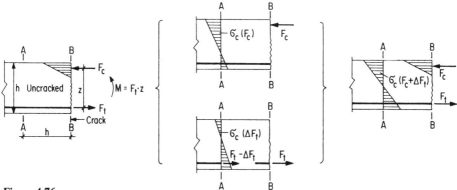

Figure 4.76
Development of tensile stress due to eccentricity of flexural compressive force, F_c

(equation (a)) are smaller than h. The effect of the eccentricity of F_c can thus be conservatively neglected. For shallow beams and slabs, however, crack spacing computed using equation (a) will be of similar size to or greater than h. In such cases, crack spacing will be not be determined by bond, but rather by the eccentricity of the flexural compressive force and can thus be chosen between $1.0h$ and $1.5h$ for design.

c) Minimum Reinforcement and Reinforcement to Limit Crack Width

The progressive formation of cracks of constant width, from the first crack to the stabilized pattern, is only possible when the stress in the steel at the first crack is less than the yield stress:

$$\sigma_{sr}^{II} < f_{sy} \tag{e}$$

Otherwise, only one crack can be formed, the width of which increases with the crack-producing restrained deformations. *Minimum reinforcement*, sufficient to ensure that σ_{sr}^{II} is less than f_{sy}, must therefore always be provided to guarantee the proper distribution of cracks.

Under certain circumstances, however, it may be necessary to limit the width of cracks to a specified value. The reinforcement required for this purpose is typically greater than the minimum reinforcement designed to satisfy inequality (e). In simple cases, reinforcement to limit crack width can be designed on the basis of section and material properties only. In general, however, this reinforcement must be designed considering the actual distribution of the cracking sectional forces and the sectional forces due to restrained deformations.

Example 4.9:
Reinforcement to limit crack width in a slab strip subjected to restrained shrinkage

The relevant dimensions and material properties of the slab strip are as follows:

Length	$l = 10.0$ m
Width	$w = 1.0$ m
Thickness	$h = 0.2$ m
Concrete tensile strength	$f_{ct} = 2$ N/mm²
Modulus of elasticity of concrete (including the influence of creep)	$E_c = 20$ kN/mm²
Shrinkage coefficient	$\varepsilon_{cs} = -0.0002$
Modulus of elasticity of steel	$E_s = 200$ kN/mm²

Cracks must not be wider than 0.2 mm.

The tensile stress in the uncracked concrete section required for complete restraint of ε_{cs} is

$$\sigma_c = -\varepsilon_{cs} E_c = 4 \text{ N/mm}^2$$

Since σ_c is greater than f_{ct}, the slab strip will crack. The reinforcement is designed to ensure that the maximum allowable crack width, 0.2 mm, is not exceeded under the cracking tensile force, N_r. It is assumed that $\varepsilon_{srm}^{II} = 0.6\varepsilon_{sr}^{II}$ and $s_{rm} = 200$ mm.

The cracking tensile force is

$$N_r = A_c f_{ct} = 400 \text{ kN}$$

The allowable steel stress is determined from the allowable crack width:

$$w_m = \varepsilon_{srm}^{II} s_{rm} = 0.6 \frac{\sigma_s^{II}}{E_s} s_{rm}$$

$$\sigma_s^{II} = \frac{w_m E_s}{0.6 s_{rm}} = 330 \text{ N/mm}^2$$

The required area of steel, therefore, is

$$A_{s,\text{req}} = \frac{N_r}{\sigma_s^{II}} = 1200 \text{ mm}^2$$

which corresponds to the following reinforcement ratio:

$$\varrho_{\text{req}} = \frac{A_{s,\text{req}}}{A_c} = 0.6\%$$

By taking $\varepsilon_{srm}^{II} = 0.6\varepsilon_{sr}^{II}$, it was implicitly assumed that the stabilized crack pattern had not yet been formed. This assumption must now be verified. The shortening of the unrestrained plate strip due to shrinkage is

$$\Delta l = l\varepsilon_{cs} = -2 \text{ mm}$$

The number of cracks, n, is obtained from the following compatibility condition:

$$nw_m + \frac{f_{ct}}{E_c}(l - ns_{rm}) = -\Delta l$$

where the second term on the left-hand side accounts for the deformation of the concrete between the cracks. The solution $n = 5.6$ is obtained; 6 cracks are therefore required for compatibility. This number is considerably less than the maximum possible in the stabilized crack pattern:

$$n_0 = \frac{l}{s_{rm}} = 50$$

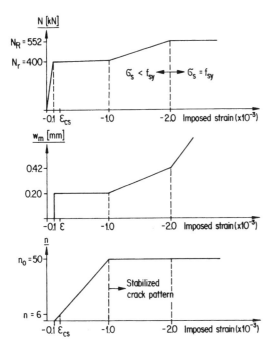

Figure 4.77
Example 4.9: summary of cracking
behaviour

The given deformation is thus not sufficient to produce the stabilized cracked pattern. The assumption for the average steel strain is therefore valid.

Since n is considerably less than n_0, the stresses and crack widths would not be significantly changed if the shrinkage strain ε_{cs} were doubled; only the number of cracks would increase. The behaviour of the slab strip under progressively larger imposed strains is shown in figure 4.77.

The calculation of this example is approximate, since the restraint of shrinkage provided by the reinforcement and the effects of creep in the concrete were not directly considered.

Example 4.10:
Reinforcement to limit crack width in a column subjected to imposed deformation

A lateral displacement $u^T = 400$ mm is imposed at the tip of the column shown in figure 4.78. The sectional forces thus produced are a function of the stiffness of the column and thus cannot easily be calculated. The design of the reinforcement should therefore not be based on the bending moment diagram, but rather on the deformations of the system. The reinforcement is designed to limit crack width to a given allowable value. It must then be verified that the deformations corresponding to the chosen reinforcement are compatible with the given tip displacement.

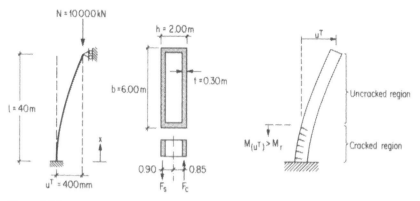

Figure 4.78
Example 4.10: imposed displacement at tip of column

The superimposed axial load at the top of the column, N, is 10000 kN. The following section and material properties are assumed:

Area	$A_c = 4.44 \text{ m}^2$
Moment of inertia	$I_c = 2.76 \text{ m}^4$
Section modulus	$W_c = 2.76 \text{ m}^3$
Concrete tensile strength	$f_{ct} = 2 \text{ N/mm}^2$
Modulus of elasticity of concrete (including the influence of creep)	$E_c = 20 \text{ kN/mm}^2$
Shrinkage coefficient	$\varepsilon_{cs} = -0.0002$
Modulus of elasticity of steel	$E_s = 200 \text{ kN/mm}^2$

The allowable average crack width is 0.2 mm.

It is assumed that $\varepsilon_{srm}^{II} = 0.8\,\varepsilon_{sr}^{II}$ and $s_{rm} = 200$ mm. The allowable steel stress, σ_s^{II}, is calculated to limit crack width:

$$w_m = \varepsilon_{srm}^{II}\, s_{rm} = 0.8\, \frac{\sigma_s^{II}}{E_s}\, s_{rm} = 0.2 \text{ mm}$$

It therefore follows that

$$\sigma_{s,\text{allow}}^{II} = 250 \text{ N/mm}^2$$

The curvature in the uncracked portion of the column is calculated using the following equation:

$$\frac{1}{r} = \frac{\Delta\varepsilon_c}{h} \tag{f}$$

where $\Delta\varepsilon_c$ is the algebraic difference in the strains at the extreme tension and compression fibres. In the cracked portion of the column, curvature is determined from the concrete strain at the centroid of the compression flange and the steel strain at the centroid of the reinforcement:

$$\frac{1}{r} = \frac{\varepsilon_{srm}^{II} - \varepsilon_c^{II}}{h_0} \tag{g}$$

where

$$\varepsilon_c^{II} = \frac{\sigma_c^{II}}{E_c} + \varepsilon_{cs}$$

and

$$\varepsilon_{srm}^{II} = 0.8\,\varepsilon_{sr}^{II} = 0.8\,\frac{\sigma_s^{II}}{E_s} = 1.0 \times 10^{-3}$$

The parameter h_0 denotes the internal lever arm, which is approximately equal to 1.75 m.

A bending moment diagram must now be assumed. Second-order moments will be neglected for this example; the moment diagram due to the lateral displacement at the tip will thus be linear. It is assumed that the moment at the lower third of the column is equal to the cracking moment, M_r. At this point, the axial force N due to self-weight plus superimposed load is 12960 kN. The value of M_r is obtained by setting the tensile stress in the concrete equal to f_{ct}:

$$-\frac{N}{A_c} + \frac{M_r}{W_c} = 2 \text{ N/mm}^2$$

from which it follows that $M_r = 13\,580$ kN $\cdot$ m.

The curvature in the uncracked section at the lower third point is calculated using equation (f):

$$\frac{1}{r} = \frac{\Delta\varepsilon_c}{h} = 0.25 \times 10^{-3} \text{ m}^{-1}$$

The curvature in the cracked section at this point is computed using equation (g). The strain at the centroid of the compression flange is

$$\varepsilon_c^{II} = \frac{\sigma_c^{II}}{E_c} + \varepsilon_{cs} = -0.6 \times 10^{-3}$$

and the strain at the centroid of the reinforcement in the tension flange is

$$\varepsilon_{srm}^{II} = 1.0 \times 10^{-3}$$

The corresponding curvature is therefore

$$\frac{1}{r} = \frac{\varepsilon_{srm}^{II} - \varepsilon_c^{II}}{h_0} = 0.91 \times 10^{-3}$$

The sectional forces at the base of the column are $N = 14400$ kN and $M = 20370$ kN · m. Equation (g) is used to calculate the curvature at this location. The compressive strain in the concrete is slightly higher than at the lower third point. (It should not be allowed to exceed -0.002, however, to ensure a suitable margin of safety against crushing.) The average strain in the steel is set equal to the value of ε_{srm}^{II} at the lower third point to limit crack width. The curvature at the base is therefore

$$\frac{1}{r} = 0.99 \times 10^{-3}$$

The tip displacement corresponding to the computed curvatures (fig. 4.79) is calculated using the method of virtual work:

$$u_{corr}^T = \int_0^l \frac{1}{r} \bar{M} dx = 480 \text{ mm}$$

In other words, the column tip can displace up to 480 mm when crack widths are limited to 0.2 mm. The imposed displacement of 400 mm will therefore produce crack widths of less than 0.2 mm.

The sectional forces are resisted by a resultant tensile force, F_s, and a resultant compressive force, F_c, which can be computed from the following equilibrium equations:

$$F_c - F_s = N$$
$$0.85 F_c + 0.90 F_s = M$$

The required reinforcement can be calculated from F_s and the allowable steel stress σ_s^{II}:

$$A_s = \frac{F_s}{\sigma_s^{II}}$$

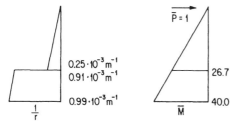

0.25·10⁻³ m⁻¹
0.91·10⁻³ m⁻¹

0.99·10⁻³ m⁻¹

$\frac{1}{r}$

$\bar{P} = 1$

26.7

40.0

$\bar{M}$

Figure 4.79
Example 4.10: curvatures $1/r$ due to imposed tip displacement and moments $\bar{M}$ due to virtual load at the tip

from which it follows that:

$A_s = 5860 \text{ mm}^2$ ($\varrho = 0.32\%$) at the lower third point
$A_s = 18500 \text{ mm}^2$ ($\varrho = 1.03\%$) at the base

(The reinforcement ratio, ϱ, is defined relative to gross flange area.)

Example 4.11:
Cracking in a continuous prestressed girder due to temperature gradient

This example focuses on the effect of deck slab reinforcement on cracking, for members of constant flexural resistance. The girder is infinite, with all spans of length 40 m. The distribution of dead load g and live load q, prestressing layout, and cross-section dimensions are given in figure 4.80. The section and material properties are as follows:

Gross area of cross-section	$A_c = 5.40 \text{ m}^2$
Area of bottom slab at midspan	$A_{c,\text{bot}} = 0.90 \text{ m}^2$
Moment of inertia	$I_c = 3.64 \text{ m}^4$
Section moduli	$W_{c,\text{top}} = 4.80 \text{ m}^3$
	$W_{c,\text{bot}} = 4.50 \text{ m}^3$
Concrete tensile strength	$f_{ct} = 2 \text{ N/mm}^2$
Modulus of elasticity of concrete (including the effects of creep)	$E_c = 20 \text{ kN/mm}^2$
Yield stress of reinforcing steel	$f_{sy} = 460 \text{ N/mm}^2$
Modulus of elasticity of reinforcing steel	$E_s = 200 \text{ kN/mm}^2$
Yield stress of prestressing steel	$f_{Py} = 1520 \text{ N/mm}^2$
Modulus of elasticity of prestressing steel	$E_P = 200 \text{ kN/mm}^2$

The area of mild reinforcing steel in the deck slab at the supports, $A_{s,\text{top}}$, is 18000 mm². This corresponds to a reinforcement ratio, ϱ_{top}, of 0.6 percent relative

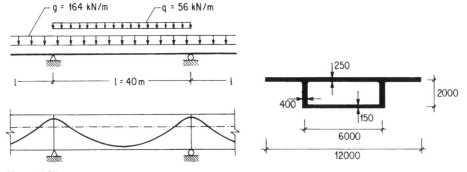

Figure 4.80
Example 4.11: structural system and loads (cross-section dimensions in mm)

to the gross area of the deck slab. The reinforcement ratio at midspan, defined as the area of mild reinforcement divided by the gross area of the bottom slab,

$$\varrho_{bot} = \frac{A_{s,\,bot}}{A_{c,\,bot}}$$

is varied from 0 to 0.01. For a given value of ϱ_{bot}, the area of prestressing is calculated to ensure safety at ultimate limit state:

$$|M_d^M| + |M_d^S| = 1.4 \frac{(g+q)\,l^2}{8} \leq \frac{1}{1.3}\,(M_R^M + M_R^S)$$

where the superscripts M and S denote midspan and supports, respectively, and

$$M_R^M = (A_P\,f_{Py} + \varrho_{bot}\,A_{c,\,bot}\,f_{sy})\,z \qquad M_R^S = (A_P\,f_{Py} + A_{s,\,top}\,f_{sy})\,z$$

(Identical lever arms $z = 2.00$ m are assumed for prestressing and reinforcing steel at midspan and at the supports.) The ultimate resistance of the girder is thus constant for all combinations of bottom slab reinforcement and prestressing.

It is assumed that identical areas of prestressing steel and prestressing forces are provided at midspan and at the supports, and that the effective long-term prestressing force, $\sigma_{P,\,\infty}$, is $0.67 f_{Py}$.

The number, n, and width, w_m, of cracks in the bottom slab were calculated for permanent load $(g + P_\infty)$ plus a uniform warming of the deck slab, ΔT. The results are presented in figure 4.81 as functions of ΔT and of the reinforcement ratio of the bottom slab, ϱ_{bot}. The procedure used to calculate n and w_m is illustrated for the following case:

Prestressing: $P_\infty = 9650$ kN
Bottom slab reinforcement: $A_{s,\,bot} = 5400$ mm^2
                                                                 $\varrho_{bot} = 0.6\,\%$
Temperature change in the deck slab: $\Delta T = 30\,°$C

Assuming that the cracking sectional forces are resisted by mild reinforcing steel alone, w_m can be computed using the following equation:

$$w_m = \varepsilon_{srm}^{II}\,s_{rm} \qquad\qquad\qquad\qquad\qquad\qquad\text{(h)}$$

Crack spacing, s_{rm}, is taken as 200 mm. The average steel strain, ε_{srm}^{II}, is assumed equal to $0.8\varepsilon_{sr}^{II}$, where $\varepsilon_{sr}^{II} = \sigma_{sr}^{II}/E_s$. The stress σ_{sr}^{II} is a function of the cracking tensile force in the bottom slab, N_r:

$$N_r = A_{c,\,bot}\,f_{ct} = (0.90)\,(2)\,(10^3) = 1\,800\ \text{kN}$$

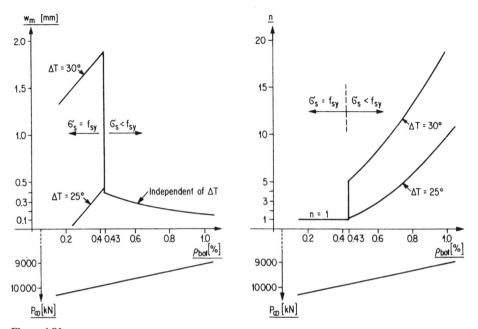

Figure 4.81
Example 4.11: crack width, w_n, and number of cracks, n, as a function of reinforcement ratio in the bottom slab, ϱ_{bot}

It therefore follows that

$$\sigma_{sr}^{II} = \frac{N_r}{A_{s,\,bot}} = \frac{1800}{5400}\,(10^3) = 330\;\text{N/mm}^2 < f_{sy}$$

(Since the reinforcement provided is greater than the required minimum, the use of equation (h) is valid.) The strain ε_{srm}^{II} is given by

$$\varepsilon_{srm}^{II} = 0.8\,\frac{330}{200}\,(10^{-3}) = 0.00133$$

The width of the cracks is therefore

$$w_m = (0.00133)\,(200) = 0.27\;\text{mm}$$

The value of w_m is independent of the imposed deformation due to ΔT, provided $\sigma_{sr}^{II} < f_{sy}$.

The number of cracks, n, is calculated from the compatibility conditions of the cracked system. A statically determinate primary system of simply supported spans is assumed. The thermal gradient ΔT produces a rotation at the supports, $\phi\,(\Delta T)$. If $\phi\,(\Delta T)$ is less than the rotation due to the minimum cracking moment of

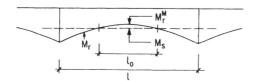

Figure 4.82
Example 4.11: cracking moment diagram with redundant moment M_S (M_r^M: cracking moment at midspan; l_0: extent of cracking)

the system, $\phi(M_r)$, compatibility can be restored elastically. (This case is primarily of theoretical interest, since the actual values of imposed deformations and cracking moments can vary considerably from the computed values. It should therefore not be assumed that $\phi(M_r) > \phi(\Delta T)$ guarantees that no cracks will be produced.) When $\phi(M_r) < \phi(\Delta T)$, compatibility must be restored by rotations due to cracking and due to flexural deformation of the uncracked portions of the girder. The latter component is produced by the actual redundant moment, M_s, that produces the crack pattern corresponding to the former component. The moment M_s is normally only slightly greater than M_r (fig. 4.82).

The positive cracking moment, M_r, is obtained from the following equation, defined for the bottom fibres of the section:

$$\sigma(g) + \sigma(P_\infty) + \sigma_0(\Delta T) + \sigma(M_r) = f_{ct}$$

where $\sigma(g)$ and $\sigma(P_\infty)$ are stresses due to dead load and prestressing, respectively, in the continuous system, $\sigma_0(\Delta T)$ is the self-equilibrating stress due to thermal gradient in the statically determinate primary system, and $\sigma(M_r)$ is the stress due to the cracking moment. The M_r diagram obtained for the values of P_∞ and ΔT given above is shown in figure 4.83. The minimum value of M_r occurs at midspan, and is denoted M_r^M.

The angles of rotation due to $\phi(\Delta T)$ and $\phi(M_r^M)$ are shown in figure 4.84. Since $\phi(\Delta T) > \phi(M_r^M)$, cracks are required to restore compatibility. The compatibility condition is formulated as follows:

$$\phi(\Delta T) + \phi(M_s) + \frac{n}{2}\phi(w_m) = 0 \tag{i}$$

where $\phi(w_m)$ is the support rotation due to two cracks of width w_m, located symmetrically on either side of midspan. The angle $\phi(M_s)$ is defined by the following equation:

$$\phi(M_s) = -\frac{M_s}{E_c I_c}\frac{l - n s_{rm}}{2}$$

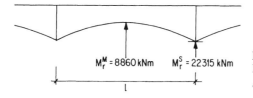

$M_r^M = 8860\ \text{kNm}$ $\quad M_r^S = 22315\ \text{kNm}$

Figure 4.83
Example 4.11: positive cracking moment for $\Delta T = 30\,°C$, $\varrho_{bot} = 0.6\%$

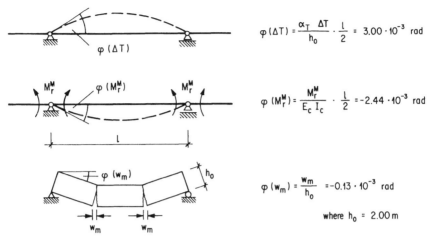

$$\varphi(\Delta T) = \frac{\alpha_T \, \Delta T}{h_o} \cdot \frac{l}{2} = 3.00 \cdot 10^{-3} \text{ rad}$$

$$\varphi(M_r^M) = \frac{M_r^M}{E_c I_c} \cdot \frac{l}{2} = -2.44 \cdot 10^{-3} \text{ rad}$$

$$\varphi(w_m) = \frac{w_m}{h_o} = -0.13 \cdot 10^{-3} \text{ rad}$$

where $h_o = 2.00$ m

Figure 4.84
Example 4.11: angles of rotation ϕ due to ΔT, M_r, and two cracks of width w_m

As an initial approximation, M_s can be taken equal to the cracking moment M_r^M in the above equation:

$$\phi(M_s) \cong -\frac{M_r^M}{E_c I_c} \frac{l - n s_{rm}}{2} = -\frac{8860}{(20 \times 10^6)\,(3.64)} \frac{40 - 0.2n}{2}$$

$$= -2.44 \times 10^{-3} + (1.22 \times 10^{-5})\,n$$

This expression for $\phi(M_s)$ is then substituted into equation (i):

$$3.00 \times 10^{-3} - 2.44 \times 10^{-3} + (1.22 \times 10^{-5})\,n - (0.13 \times 10^{-3})\frac{n}{2} = 0$$

from which it follows that $n = 11$.

4.8.5 Deformations

a) Fundamentals

The long-term deformations of the main structural elements due to permanent load must always be checked, taking into account the effects of creep and shrinkage. The deformations that occur immediately after the removal of falsework should also be calculated and monitored during construction. This permits the early detection of design errors, such as incorrect assumptions for prestressing losses due to friction.

Long-term deflections due to permanent load should be limited to roughly 1/750 of the span length. This can normally be accomplished by a proper choice of prestressing concept. Special care must be taken in the design of the prestressing for cantilever-constructed bridges, long-span partially-prestressed girder bridges, rigid frame bridges, and slab bridges, which are particularly prone to excessive deformations. Long deck slab overhangs should always be transversely pre-stressed. Any computed residual deformation due permanent load in excess of 1/750 of the span length should be compensated by camber of the formwork.

Reliable values of material properties are essential for all deformation calcu-lations. Average values that have been confirmed by experience are sufficient in most cases. Only in exceptional situations where high accuracy is required, for example in cantilever construction, are special material tests necessary. Even material properties obtained from special tests, however, are subject to consider-able variability. Rigorous analyses of the effects of creep and shrinkage on structural deformations are thus usually of dubious value.

Deformations are most commonly calculated using the method of virtual work:

$$1 \cdot \delta = \int \bar{S} \, d\lambda \tag{a}$$

where δ is the deformation to be calculated, $\bar{S}$ is the sectional force due to a virtual load of magnitude 1 applied in the direction of δ and $d\lambda$ is the real deformation of the differential element dx due to permanent load and any strains due to creep, shrinkage, and change of temperature.

Equation (a) is valid for both uncracked and cracked structures.

b) *Deformations in the Uncracked State*

In the uncracked state, deformations of differential elements are calculated using the appropriate material and section properties. The short-term, static defor-mations are listed below for the principal sectional forces:

1. Axial Strain:
$$\varepsilon_x = \frac{N}{E_c A} + \alpha_T \, \Delta T^\circ$$

2. Curvature:
$$\frac{d\phi}{dx} = \frac{M}{E_c I} + \alpha_T \frac{T_{bot}^\circ - T_{top}^\circ}{h}$$

3. Shear strain:
$$\gamma_m = \frac{V}{G_c \, \alpha A_c}$$

4. Twist:
$$\frac{d\theta}{dx} = \frac{T}{G_c K_c}$$

where T° denotes temperature and T denotes torsional moment.

Axial strains and curvatures are normally calculated using the section properties of a *transformed section*, in which the area of reinforcement is increased by the

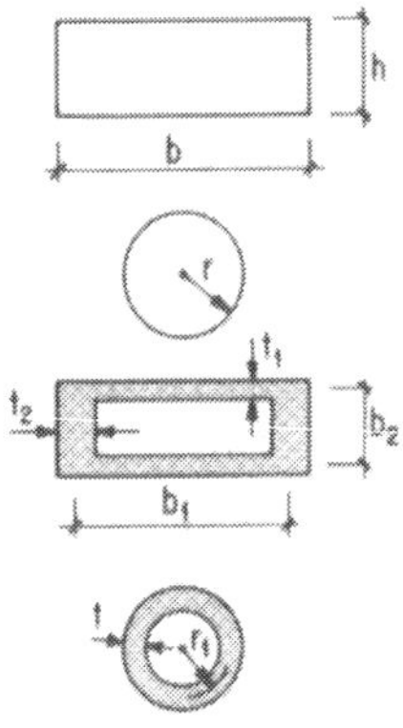

$$K \cong \frac{b \cdot h^3}{3} \cdot (1 - 0.63 \cdot \frac{h}{b})$$

$$K = \frac{\pi \cdot r^4}{2}$$

$$K = 2 \cdot \frac{b_1^2 \cdot b_2^2}{b_1/t_1 + b_2/t_2}$$

$$K = 2 \cdot \pi \cdot r_1^3 \cdot t$$

Figure 4.85
Torsional constants K for common
cross-sections

factor $n = E_s/E_c \cong 5$ to convert it into an equivalent area of concrete. The reinforcement is not normally considered, however, in the calculation of shear strains and twists. For box and T-sections, the shear area αA_c can be taken as the effective web area. Values of the torsional constant, K, are given in figure 4.85 for commonly used cross-section types.

Long-term deformations can be calculated using the methods presented in Section 4.7. In simple cases, for example constant stress, deformations can be calculated approximately using an average creep coefficient ϕ:

$$\delta_c = \delta_{c,el}(1 + \phi) + \delta_{cs}$$

Concrete strains due to creep and shrinkage are restrained by the reinforcement. This effect can be considered by reducing the creep and shrinkage coefficients by the factor

$$k = 1 - 10\varrho$$

where ϱ denotes reinforcement ratio.

c) Deformations in the Cracked State

In the cracked state, deformations of differential elements are calculated from the associated axial strains in the concrete and steel. The average steel strain, ε_{sm}^{II} is approximately 0.8 times the steel strain at the crack location, ε_s^{II}. The instantaneous static deformations are as follows:

1. Axial strain:
$$\varepsilon_x = \frac{\varepsilon_{bot} + \varepsilon_{top}}{2}$$

2. Curvature:
$$\frac{d\phi}{dx} = \frac{\varepsilon_{sm,bot}^{II} - \varepsilon_{c,top}}{d}$$

where $\varepsilon^{II}_{sm, \, bot}$ is the average strain in the reinforcement, $\varepsilon_{c, \, top}$ is concrete strain at the extreme compression fibre, and d is the distance from the extreme compression fibre to the centroid of reinforcement.

3. Shear strain is calculated approximately using a truss model, assuming that the stirrups are vertical and that the compression diagonals are inclined at roughly $45°$:

$$\gamma_m \cong V \left[\frac{4}{E_c A_0} + \frac{0.8s}{E_s A_s^w z} + \frac{1}{4(EA)_{top}} + \frac{1}{4(EA)_{bot}} \right]$$

where the following parameters are defined:

V: shear force
t: wall thickness
z: distance between centroids of upper and lower longitudinal reinforcement
$A_0 = tz$
A_s^w: stirrup area
s: stirrup spacing
EA: axial stiffness of longitudinal reinforcement

4. Twist: $\dfrac{d\theta}{dx} = \sum \bar{v} z_i \gamma_{m, i}$

where

$\gamma_{m, i}$: shear strain in the ith element of the section
z_i: distance between centroids of longitudinal steel in the ith element
$\bar{v}$: torsional shear flow due the virtual load $\bar{T} = 1$

Shear strains in the cracked state can be many times greater than in the uncracked state. The shear component of deformation can nevertheless normally be neglected, provided cracking is restricted to a relatively small region in the member. Shear deformations must be considered whenever shear cracks extend over a significant length. This is often the case with torsion; significant deformations can occur after the torsional cracking load has been exceeded.

Long-term deformations in the cracked state can be calculated by increasing the concrete strains in the above expressions to account for the effects of creep and shrinkage.

4.8.6 Vibrations

a) Fundamentals

Vibrations are normally of secondary importance in prestressed concrete bridges. Dynamic analyses are therefore only required in exceptional cases, for example

Table 4.4
Human Perception of Structural Vibration (Rausch 1973, 178)

K	Class	Human Perception
$K < 0.10$	A	Imperceptible
$0.10 \leq K < 0.25$	B	Barely perceptible
$0.25 \leq K < 0.63$	C	Perceptible
$0.63 \leq K < 1.60$	D	Easily perceptible
$1.60 \leq K < 4.00$	E	Strongly perceptible
$4.00 \leq K < 10.00$	F	Very strongly perceptible
$10.00 \leq K < 25.00$	G	Very strongly perceptible
$25.00 \leq K < 63.00$	H	Very strongly perceptible
$63.00 \leq K$	I	Very strongly perceptible

very slender bridges, heavily loaded bridges with pedestrian traffic, or actual pedestrian bridges.

Human sensitivity to structural vibration is primarily a function of acceleration. It is commonly quantified in terms of amplitude and frequency using a *sensitivity factor*, K, defined as follows (Rausch 1973, 178):

$$K = d \, \frac{f^2}{2\sqrt{1 + (f/f_0)^2}}$$

where d is amplitude of vibration in mm, f is frequency in Hz, and $f_0 = 10$ Hz. Table 4.4 relates values of K and the perceived intensity of vibration. The range of sensitivity has been divided into nine classes, A through I. These classes can be correlated with the psychological and physiological effects of vibrations on humans, and thus can be used as design criteria for structures. Motion in classes A, B, C, and D can generally be regarded as acceptable; vibrations in classes E and F may be unpleasant but can still be considered endurable. Vibrations in classes G, H, and I are unendurable and should be prevented. The sensitivity classes are presented graphically in figure 4.86.

b) Single Degree of Freedom Oscillator

The simplest model of a vibrating system is the *single degree of freedom oscillator*, consisting of a single concentrated mass, a viscous damper, and a linear spring (fig. 4.87). The corresponding parameters of the system are mass m, damping coefficient c, and spring constant k. It is assumed that the system is forced into motion by a load that varies periodically as a function of time, t. Displacements away from the position of static equilibrium are denoted y; derivatives with respect to t are indicated with dots $(dy/dt = \dot{y})$. The equation of motion of the system, the solution of which gives y as a function of time, is derived from dynamic equilibrium of the forces acting on the mass:

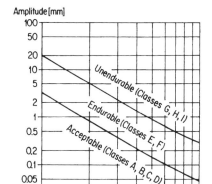

Figure 4.86
Human sensitivity to vibration (adapted from Rausch 1973)

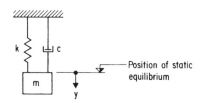

Position of static equilibrium

Figure 4.87
Single degree of freedom oscillator (m: mass; c: damping; k: stiffness)

1. Inertial force: $\qquad F_I = m\ddot{y}$

2. Damping force: $\qquad F_D = c\dot{y}$

3. Elastic force: $\qquad F_E = ky$

4. Load: $\qquad F_L = a_0 \sin \bar{\omega} t$

Equilibrium of forces requires that

$$m\ddot{y} + c\dot{y} + ky = a_0 \sin \bar{\omega} t \qquad\qquad (a)$$

c) Undamped Free Vibration

When the loading function and damping force are equal to zero for all values of t, the motion of the system is referred to as *undamped free vibration*. Equation (a) thus simplifies to

$$m\ddot{y} + ky = 0$$

or, equivalently,

$$\ddot{y} + \omega^2 y = 0 \qquad\qquad (b)$$

where

$$\omega^2 = \frac{k}{m} \tag{c}$$

The general solution of equation (b) is

$$y = C_1 \sin \omega t + C_2 \cos \omega t \tag{d}$$

The parameter ω is called the *circular frequency* of the system. It is related to the period, T_0, and the *cyclic frequency*, f_0, of the motion by the following equations:

$$T_0 = \frac{2\pi}{\omega} \qquad f_0 = \frac{1}{T_0} = \frac{\omega}{2\pi}$$

The constants of integration are determined from the initial conditions. Substituting

$$y(t=0) = y_0 \quad \text{and} \quad \dot{y}(t=0) = 0$$

into equation (d) yields the constants

$$C_1 = 0 \quad \text{and} \quad C_2 = y_0$$

The solution is therefore

$$y = y_0 \cos \omega t$$

d) Damped Free Vibration

When a damping force proportional to velocity is assumed in addition to the elastic and inertial forces, equation (a) is transformed into

$$m\ddot{y} + c\dot{y} + ky = 0 \tag{e}$$

or, equivalently,

$$\ddot{y} + 2\xi\omega\dot{y} + \omega^2 y = 0 \tag{f}$$

where ω is defined by equation (c) and

$$\xi = \frac{c}{2m\omega} \tag{g}$$

When a solution of the form

$$y = Ce^{\lambda t}$$

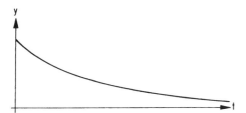

Figure 4.88
Overdamped oscillation

is substituted into equation (f), the following pair of values for λ is obtained:

$$\lambda_{1,2} = -\xi\omega \pm i\omega\sqrt{1-\xi^2}$$

The general solution can thus be expressed in the form

$$y = C_1 e^{\lambda_1 t} + C_2 e^{\lambda_2 t}$$

If $\xi > 1$, then λ_1 and λ_2 are both real; the corresponding motion is called "overdamped" (fig. 4.88). If $\xi < 1$, then λ_1 and λ_2 are complex numbers with nonzero imaginary components:

$$\lambda_{1,2} = -\xi\omega \pm i\omega_D$$

where

$$\omega_D = \omega\sqrt{1-\xi^2} \tag{h}$$

In this case, the general solution is given by the following equation:

$$y = C_1 e^{-\xi\omega t} \cos\omega_D t + C_2 e^{-\xi\omega t} \sin\omega_D t \tag{i}$$

The motion is referred to as "underdamped".

As shown in figure 4.89, underdamped oscillations are not periodic. The interval of time between two successive passes of the mass through the position of static

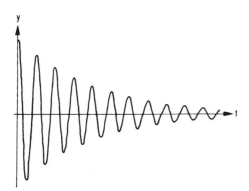

Figure 4.89
Underdamped oscillation

equilibrium is nevertheless constant. This interval, equal to $2\pi/\omega_D$, is the period of the underdamped vibration, T_0; the parameter ω_D is its circular frequency. The values of ξ normally encountered in structures are small, typically less than 0.10. It follows from equation (h) that $\omega_D = 0.995\omega$ for $\xi = 0.10$. The natural frequency of damped structures can therefore be calculated for the corresponding undamped system with negligible error.

The constants of integration are determined from the initial conditions. Assuming

$$y(t = 0) = y_0 \quad \text{and} \quad \dot{y}(t = 0) = 0$$

it follows that

$$C_1 = y_0 \quad \text{and} \quad C_2 = 0$$

These constants correspond to the following solution:

$$y = y_0\, e^{-\xi\omega t} \cos\omega_D t \tag{j}$$

The peak displacements of the system are denoted y_n. For the oscillations described by equation (j), peaks occur at times $\tau = 2n\pi/\omega_D$, for all positive integers n:

$$y_n = y_0\, e^{-2\pi n\xi\omega/\omega_D}$$

Peak y_{n+1}, which occurs at time $\tau + T_0$, is

$$y_{n+1} = y_0\, e^{-2\pi n\xi\omega/\omega_D}\, e^{-2\pi\xi\omega/\omega_D}$$

The ratio of the two successive peaks is therefore

$$\frac{y_n}{y_{n+1}} = e^{2\pi\xi\omega/\omega_D} \tag{k}$$

which is a constant. Taking the logarithm of both sides of this equation yields the following expression:

$$\ln\frac{y_n}{y_{n+1}} = \frac{2\pi\xi\omega}{\omega_D} = \delta$$

The parameter δ is called the *logarithmic decrement of damping*. An approximate formulation of equation (k), valid for small values of ξ, is obtained by linearizing e^δ:

$$\frac{y_n}{y_{n+1}} = e^\delta \cong 1 + \delta$$

e) Forced Vibration

Two separate cases of forced vibration are considered:

1. Forced Vibration of an Undamped Oscillator. The equation of motion is obtained by setting c equal to zero in equation (a):

$$m\ddot{y} + ky = a_0 \sin\bar{\omega}t \tag{l}$$

Making the substitution $\omega^2 = k/m$ yields the following equivalent expression:

$$\ddot{y} + \omega^2 y = \frac{a_0}{m} \sin\bar{\omega}t \tag{m}$$

The loading function is harmonic, with circular frequency $\bar{\omega}$.

The solution to equation (m) is of the form $y(t) = y_0(t) + y_1(t)$, where y_0 is the general solution of the homogeneous differential equation and y_1 is a particular solution to the complete equation. The solution y_0 is obtained directly from equation (d):

$$y_0 = C_1 \sin\omega t + C_2 \cos\omega t$$

The function y_1 is chosen as

$$y_1 = A \sin\bar{\omega}t$$

with

$$A = \frac{a_0}{k} \frac{1}{1 - \dfrac{\bar{\omega}^2}{\omega^2}}$$

The complete solution is thus

$$y = C_1 \sin\omega t + C_2 \cos\omega t + \frac{a_0}{k} \frac{1}{1 - \dfrac{\bar{\omega}^2}{\omega^2}} \sin\bar{\omega}t$$

For the initial conditions $y(t = 0) = 0$ and $\dot{y}(t = 0) = 0$, the solution is:

$$y = \frac{a_0}{k} \frac{1}{1 - \dfrac{\bar{\omega}^2}{\omega^2}} \left(\sin\bar{\omega}t - \frac{\bar{\omega}}{\omega} \sin\omega t \right)$$

The corresponding motion is shown in figure 4.90 for $\bar{\omega}/\omega = 2$, 4, and 10. It is apparent from these curves that the primary frequency of the oscillations is very close to the natural frequency of the system in free vibration. This phenomenon can be observed whenever $\bar{\omega}/\omega$ is greater than 1, for instance when bridges that are loaded dynamically by trucks.

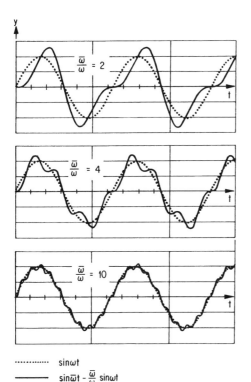

Figure 4.90
Forced vibrations

.... $\sin \omega t$

———— $\sin \bar\omega t - \dfrac{\bar\omega}{\because} \sin \omega t$

2. Forced Vibration of an Underdamped Oscillator. The equation of motion

$$m\ddot{y} + c\dot{y} + ky = a_0 \sin \bar\omega t \qquad\qquad \text{(equation (a))}$$

is again transformed into the form

$$\ddot{y} + 2\xi\omega\dot{y} + \omega^2 y = \frac{a_0}{m} \sin \bar\omega t$$

using the substitutions $\omega^2 = k/m$ and $\xi = c/2\omega m$. Its solution, $y(t)$, can also be expressed as the sum $y_0(t) + y_1(t)$, where y_0 is the general solution to the homogeneous equation and y_1 is a particular solution to the complete equation. The function y_0 is obtained directly from equation (i):

$$y_0 = C_1 e^{-\xi\omega t} \cos \omega_D t + C_2 e^{-\xi\omega t} \sin \omega_D t$$

The solution y_1 is chosen as

$$y_1 = B_1 \sin \bar\omega t + B_2 \cos \bar\omega t \qquad\qquad \text{(n)}$$

where

$$B_1 = \frac{a_0}{k} \frac{1-\beta^2}{(1-\beta^2)^2 + (2\xi\beta)^2} \qquad \text{and} \qquad B_2 = \frac{a_0}{k} \frac{-2\xi\beta}{(1-\beta^2)^2 + (2\xi\beta)^2}$$

The parameter β is defined as $\bar{\omega}/\omega$. The particular solution can also be expressed as

$$y_1 = A \sin(\bar{\omega}t - \theta)$$

where

$$A = \sqrt{B_1^2 + B_2^2} \quad \text{and} \quad \tan\theta = -\frac{B_2}{B_1}$$

For the assumed initial conditions

$$y(t=0) = 0 \quad \text{and} \quad \dot{y}(t=0) = 0$$

the constants of integration are

$$C_1 = A\sin\theta \quad \text{and} \quad C_2 = A\frac{\xi\omega\sin\theta - \bar{\omega}\cos\theta}{\omega_D}$$

The solution is thus

$$y = Ae^{-\xi\omega t}\left(\sin\theta\cos\omega_D t + \frac{\xi\omega\sin\theta - \bar{\omega}\cos\theta}{\omega_D}\sin\omega_D t\right) + A\sin(\bar{\omega}t - \theta)$$

f) Natural Frequency of Beams (Simplified Calculation)

The single degree of freedom oscillator can be used as a simplified model for calculating the natural frequency of beams. The procedure will be formulated for a simply supported beam of length l, flexural stiffness EI, and mass per unit length m. (As stated in Part (d), above, the natural frequency can be calculated from the undamped system with negligible error.) Only one half of the total mass is considered to be dynamically effective; it is assumed that a mass of $ml/2$ is lumped at midspan and a mass of $ml/4$ is lumped at each support (fig. 4.91). Letting y denote the deflection of the beam at midspan, the inertial force is as follows:

$$F_I = \frac{ml}{2}\ddot{y}$$

The flexural stiffness of the beam assumes the role of the spring. The spring constant is defined as the concentrated force at midspan required to produce a unit deflection at the same location:

$$k = \frac{48\,EI}{l^3}$$

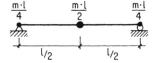

Figure 4.91
Lumped masses for the simplified calculation of the natural frequency of a simply supported beam

The elastic force is thus

$$F_E = ky = \frac{48\,EI}{l^3}\,y$$

The equation of motion is obtained from dynamic equilibrium of the lumped mass at midspan:

$$\ddot{y} + \frac{96\,EI}{ml^4}\,y = 0 \qquad\qquad\text{(equation (b))}$$

Its general solution is

$$y = C_1 \sin\omega t + C_2 \cos\omega t \qquad\qquad\text{(equation (d))}$$

where

$$\omega = \frac{9.8}{l^2}\sqrt{\frac{EI}{m}} \qquad\qquad\text{(equation (c))}$$

The natural period and natural frequency of the vibrating beam are thus

$$T_0 = \frac{2\pi}{\omega} = 0.64l^2\sqrt{\frac{m}{EI}} \quad\text{and}\quad f_0 = \frac{1}{T_0} = \frac{1.56}{l^2}\sqrt{\frac{EI}{m}}$$

g) Natural Frequencies of Beams (Exact Calculation)

The exact calculation of natural frequency is based on the dynamic equilibrium of each differential element of the beam, dx. The deflections of the beam are described by the function $y(x, t)$, where x is the longitudinal coordinate along the axis of the beam. As in Part (f), damping will be neglected. The length of the beam, l, flexural stiffness $EI(x)$, and mass per unit length m are given. The inertial force acting on a typical differential element can thus be formulated as

$$dF_I = m\,\frac{\partial^2 y}{\partial t^2}\,dx = m\ddot{y}\,dx$$

and the elastic force as

$$dF_E = \frac{\partial^2}{\partial x^2}\left(\frac{\partial^2 y}{\partial x^2}\,EI\right)dx = (EIy'')''\,dx$$

(Dots and primes denote partial derivatives with respect to t and x, respectively.) The equation of motion is obtained directly from the dynamic equilibrium of these two forces:

$$m\ddot{y} + (EIy'')'' = 0 \qquad\qquad\text{(o)}$$

The solution of this equation is based on the assumption that y can be expressed as the product of a function of x only and a function of t only:

$$y(x, t) = u(x)\, g(t)$$

It is assumed that $g(t)$ is periodic:

$$g(t) = C_1 \sin \omega t + C_2 \cos \omega t$$

It thus follows that

$$y(x,\, t) = u(x)\, (C_1 \sin \omega t + C_2 \cos \omega t)$$

and

$$\ddot{y}(x, t) = -u(x)\, \omega^2\, (C_1 \sin \omega t + C_2 \cos \omega t)$$

The functions of time drop out when these expressions are substituted into equation (o). The following differential equation for the deflected shape is obtained:

$$(EIu'')'' - m\omega^2\, u = 0 \tag{p}$$

Assuming EI is constant, the general solution of this equation is

$$u(x) = B_1 \sin \alpha x + B_2 \cos \alpha x + B_3 \sinh \alpha x + B_4 \cosh \alpha x$$

where

$$\alpha = \frac{m\omega^2}{EI} \tag{q}$$

For a simply supported beam, the boundary conditions are as follows:

$$u(0) = 0 \qquad u(l) = 0$$
$$u''(0) = 0 \qquad u''(l) = 0$$

It follows that

$$u(x) = B_1 \sin \frac{n\pi x}{l}$$

for any positive integer n. The natural frequencies of the beam are obtained by substituting $\alpha = n\pi/l$ into equation (q):

$$\omega_n = \frac{n^2 \pi^2}{l^2} \sqrt{\frac{EI}{m}}$$

The positive integer n defines the *mode* of vibration; $n = 1$ corresponds to the fundamental mode of vibration, $n > 1$ to higher modes. The corresponding functions $u_n(x) = B_1 \sin(n\pi x/l)$ are called the *mode shapes* of the structure. The fundamental frequency is

$$\omega_1 = \frac{\pi^2}{l^2} \sqrt{\frac{EI}{m}}$$

with corresponding cyclic frequency

$$f_1 = \frac{\omega_1}{2\pi} = \frac{\pi}{2l^2} \sqrt{\frac{EI}{m}} \cong \frac{1.57}{l^2} \sqrt{\frac{EI}{m}}$$

h) Fundamental Frequency of Continuous Beams (Iterative Method)

An iterative solution of equation (p)

$$(EIu'')'' = m\omega^2 u \qquad\qquad\qquad (r)$$

can be readily obtained by substituting the equivalent static load $q = m\omega^2 u$ for the right-hand side:

$$(EIu'')'' = q$$

The load q is approximated by $q_0 = m\omega^2 u_0$, where the assumed displacement u_0 must be compatible with the boundary conditions. The displacements u_1 due to q_0 can then be calculated:

$$(EIu_1'')'' = q_0 \qquad\qquad\qquad (s)$$

The solution of equation (s) can be simplified by using the conjugate beam method, in which u_1 is equal to the moment diagram in the conjugate beam due to the load $q_0^* = M(q_0)/EI$.

The fundamental frequency ω is then solved from the condition

$$u_1(x) = u_0(x)$$

Example 4.12:
Fundamental frequency of a simply supported beam (Iterative method)

It is assumed that the mode shape u_0 is a parabola that satisfies the given boundary conditions (fig. 4.92). The moment at midspan due to $q_0 = m\omega^2 u_0$ is thus given by

$$M_m(q_0) = \frac{m\omega^2 u_{0,m} l^2}{9.6}$$

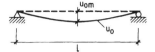

Figure 4.92
Example 4.12: fundamental mode shape of a simply supported beam

where $u_{0,m}$ is the value of the function u_0 at midspan. The conjugate beam is now loaded with $q_0^* = M(q_0)/EI$. The calculation is simplified by assuming that q_0^* is also parabolic. The resulting moment at midspan is equal to u_1:

$$u_{1,m} = M_m(q_0^*) \cong \left(\frac{m\omega^2 u_{0,m} l^2}{9.6}\right) \frac{l^2}{9.6} = \frac{m\omega^2 u_{0,m} l^4}{92.2\,EI}$$

The value of ω is obtained from the condition $u_1 = u_0$ at midspan:

$$\omega^2 = \frac{92.2\,EI}{ml^4}$$

The fundamental cyclic frequency is thus

$$f = \frac{\omega}{2\pi} = \frac{1.53}{l^2}\sqrt{\frac{EI}{m}}$$

Example 4.13:
Fundamental frequency of a non-prismatic continuous beam (Iterative method)

The span lengths of the girder and the maximum and minimum values of EI and m are given in figure 4.93. The first approximation to the fundamental mode shape, u_0, is taken as the deflected shape due to the load q shown in figure 4.94a. The deflected shape u_1 due to $q_0 = mw^2 u_0$ is shown in figure 4.94b. The accuracy of the approximation can be checked by comparing the ratio of u_0 to u_1 at midspan of the side spans and main span:

$$\frac{u_0}{u_1} = 1.74 \times 10^{-4}\,\frac{EI}{m\omega^2} \qquad \text{(side spans)}$$

$$\frac{u_0}{u_1} = 1.51 \times 10^{-4}\,\frac{EI}{m\omega^2} \qquad \text{(main span)}$$

| Flexural stiffness : EI | * | 1 | 20 | 1 | 20 | 1 |
| Mass | : m | * | 1 | 2 | 1 | 2 | 1 |

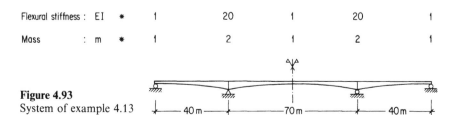

Figure 4.93
System of example 4.13

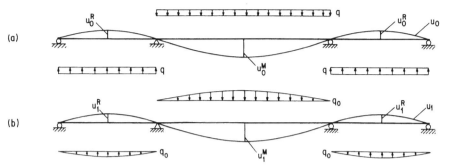

Figure 4.94
Example 4.13: a, load q and deflection u_0; b, load q_0 and deflection u_1

The condition $u_0 = u_1$ thus cannot be satisfied at all points in the structure. Further iterations are performed until agreement between u_n and u_{n+1} is acceptable at all locations along the girder. The fundamental frequency ω is then obtained by setting $u_n = u_{n+1}$. The final result is:

$$\omega = 1.26 \times 10^{-2} \sqrt{\frac{EI}{m}}$$

with corresponding cyclic frequency:

$$f = \frac{\omega}{2\pi} = 2.0 \times 10^{-3} \sqrt{\frac{EI}{m}}$$

i) Estimation of Parameters

The vibrational behaviour of simple systems can often be modeled using single degree of freedom oscillators. The reliability of the model depends on the accuracy with which the natural frequency and damping of the oscillator and the characteristics of the loading function can be estimated.

The natural frequency can be calculated with relative ease using the methods described in this section. The following expression can be used as an initial approximation for fundamental cyclic frequency, in Hertz, of girder bridges with longest span length L, in metres:

$$f = \frac{100}{L} + 0.5$$

The damping coefficient of bridges will always be less than the critical value $c_{cr} = 2m\omega$. Measurements reported by Cantieni (1983) show that the logarithmic decrement, δ, normally lies between 0.02 and 0.35, with an average value of approximately 0.08 (fig. 4.95). This corresponds roughly to values of ξ ranging between 0.003 and 0.058, with an average of 0.013. Damping in long bridges is

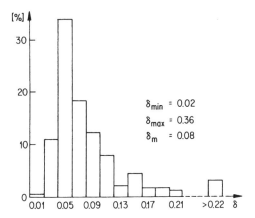

Figure 4.95
Histogram of logarithmic decrements of damping measured at 198 concrete bridges in Switzerland (adapted from Cantieni 1983)

δ_{min} = 0.02
δ_{max} = 0.36
δ_{m} = 0.08

usually less than in short bridges, and less in straight bridges than in curved or skew bridges. The damping of actual structures can be estimated from simple measurements of the attenuation of free vibrations. A logarithmic decrement of 0.05 and a natural frequency of 3 Hz, typical values for highway bridges, correspond to an attenuation of the fundamental mode of roughly 50 percent in 5 seconds. Higher modes are attenuated more quickly.

The loading function has a strong effect on vibrational behaviour. Resonant vibrations occur when the frequency of the loading, $\bar{\omega}$, is equal to the natural frequency of the structure, ω. Vibrations can be induced simply by the horizontal movement of a load across a beam. These oscillations are small, however, in comparison with the vibrations produced by heavy trucks, which themselves vibrate due to unevenness in the pavement. The frequency spectrum of these dynamic wheel loads exhibits two pronounced peaks: the first between 2 and 5 Hz, corresponding to the vibration of the truck chassis and body, and the second between 10 and 15 Hz, corresponding to the vibration of the truck axles. The dynamic effects of truck loads can be treated quasi-statically by increasing the design live load by a factor of $1 + \phi$, where the dynamic increment ϕ is defined as follows:

$$\phi = \frac{A_{dyn} - A_{stat}}{A_{stat}}$$

where A denotes deflection or strain.

The danger of resonance is normally small, since a simultaneous excitation by several trucks travelling over the bridge is highly unlikely. The dynamic increment for a single truck load can, however, be substantial. For natural frequencies lying in the 2 Hz to 5 Hz region, the dynamic increment can be as large as 70%, depending on the roughness of the pavement (fig. 4.96). An average dynamic increment of 30 to 40 percent should be considered in this frequency range assuming normal pavement conditions. In other frequency ranges, the dynamic increment is roughly 10 to 20 percent.

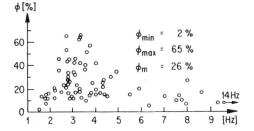

Figure 4.96
Dynamic increments for a single truck and normal pavement conditions, measured at 92 bridges in Switzerland (adapted from Cantieni 1983)

The circular frequency of loads due to pedestrian traffic varies between 10.5 rad/s and 12.5 rad/s, assuming 100 to 120 steps per minute. Resonant vibrations can occur when the fundamental frequency is also in this range, especially when the bridge is loaded by soldiers marching in step.

References

Cantieni, R. 1983. *Dynamische Belastungsversuche an Strassenbrücken in der Schweiz* (Dynamic load tests of highway bridges in Switzerland). Forschungs- und Arbeitsberichte Abteilung 116, Massivbau, Bericht Nr. 116/1. Dübendorf: Eidgenössische Materialprüfungs- und Versuchsanstalt.

Comité Euro-International du Béton. 1985. *CEB Design Manual on Cracking and Deformations* (Bulletin d'information no. 158-E). Lausanne: Comité Euro-International du Béton.

Neville, A.M., W.H. Dilger, and J.J. Brooks. 1983. *Creep of Plain and Structural Concrete.* London and New York: Construction Press.

Rausch, E. 1973. Maschinenfundamente und andere dynamisch beanspruchte Baukonstruktionen (Machine foundations and other dynamically loaded structures). In vol. 2 of *Beton-Kalender 1973.* Berlin, Munich, Düsseldorf: Verlag von Wilhelm Ernst & Sohn.

Rüsch, H. 1960. Researches Toward a General Flexural Theory for Structural Concrete. *Proceedings of the ACI* 57: 1 – 28.

Rüsch, H. and D. Jungwirth. 1976. *Stahlbeton – Spannbeton. Band 2: Berücksichtigung der Einflüsse von Kriechen und Schwinden auf das Verhalten der Tragwerke* (Reinforced concrete – prestressed concrete. Vol. 2: Consideration of the effects of creep and shrinkage on structural behaviour). Düsseldorf: Verner-Verlag.

SIA 160. 1989. *Norm 160: Einwirkungen auf Tragwerke* (Standard 160: Actions on structures). Zürich: Schweizerischer Ingenieur- und Architekten-Verein.

SIA 162. 1989. *Norm 162: Betonbauten* (Standard 162: Concrete structures). Zürich: Schweizerischer Ingenieur- und Architekten-Verein.

SIA 162/1. 1989. *Norm 162/1: Betonbauten, Materialprüfung* (Standard 162/1: Concrete structures, testing of materials). Zürich: Schweizerischer Ingenieur- und Architekten-Verein.

Thürlimann, B., P. Marti, P. Ritz, and J. Pralong. 1983. *Anwendung der Plastizitätstheorie auf Stahlbeton* (Use of the theory of plasticity in reinforced concrete). Zürich: Eidgenössische Technische Hochschule Zürich, Institut für Baustatik und Konstruktion.

Trost, H. 1967. Auswirkungen des Superpositionsprinzips auf Kriech- und Relaxationsprobleme bei Beton und Spannbeton (Effects of the principle of superposition on creep and relaxation problems in concrete and prestressed concrete). Parts 1, 2. *Beton- und Stahlbetonbau.* 62: 230 – 238, 261 – 269.

Wolfensberger, R. 1964. *Traglast und optimale Bemessung von Platten* (Ultimate load and optimal design of slabs). Wildegg: Technische Forschungs- und Beratungsstelle der Schweizerischen Zementindustrie.

5 Analysis and Design of Bridge Superstructures

5.1 Structural Models and Load Distribution

5.1.1 General Ideas

a) The Structural Model

Concrete bridge superstructures typically consist of relatively thin slabs mono-lithically connected to each other along longitudinal joints. Although these systems satisfy the geometrical conditions for classification as thin shell structures, it is not appropriate to analyse them according to elastic shell theory. The use of elastic shell theory requires linear elastic, isotropic and homogeneous material behaviour, which is never the case in prestressed concrete bridge superstructures. At ultimate limit state, the superstructure is transformed into an inhomogeneous, fully cracked composite member. Moreover, self-equilibrating stresses of un-known magnitude and distribution are always present under service conditions, a result of differential creep and shrinkage as well as temperature gradients.

The restrictions on material behaviour imposed by elastic shell theory can be overcome through the use of simplified models that satisfy equilibrium and suitably chosen compatibility conditions. These models normally consist of an assemblage of beam elements. The entire superstructure can be modeled as a single beam, provided the superstructure can deflect under load without distortion of the cross-section. Otherwise, it must be modeled as an assemblage of several beams. A plane grid, for example, can be used to model a superstructure whose cross-section undergoes significant deformation due to transverse bending.

The deformability of the cross-section, and hence the choice of model, is a function of the following factors: form of the cross-section, additional stiffening elements such as diaphragms, ratio of span length to cross-section width, support conditions, and load arrangement. The role played by the form of the cross-section is illustrated in figure 5.1. It can be observed that transverse bending is generally present in T-girders with three or more webs and, by analogy, in multiple-cell box girders. These cross-sections must therefore be modeled as plane grids. Single-cell box girders and double-T girders, however, can always be modeled as single beams provided the span length is large in relation to the cross-

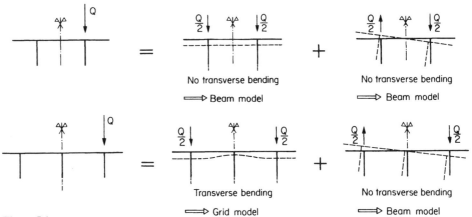

Figure 5.1
Relation between form of the cross-section and transverse deformation

section dimensions. This condition can be expressed quantitatively by the inequality

$$\frac{l_0}{2(b_0 + h_0)} > 1$$

where the idealized cross-section dimensions b_0 and h_0 are defined in figure 5.2. The effective span length l_0 is defined as follows:

1. For simply supported girders, l_0 is equal to the span length
2. For cantilever girders, l_0 is equal to twice the cantilever length
3. For continuous girders, l_0 is equal to the length between inflection points

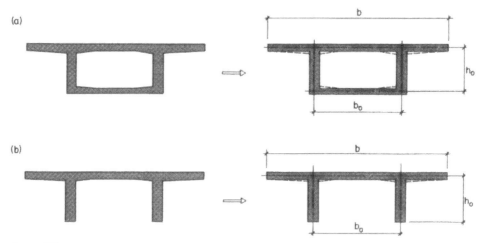

Figure 5.2
Idealized cross-section dimensions: a, box girder; b, T-girder

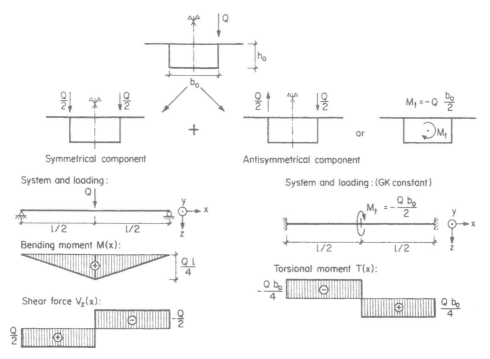

Figure 5.3
Sectional forces due to an eccentrically applied concentrated load

When the superstructure is modeled as a single beam, the external loads are equilibrated by bending moments M, shear forces V, and torsional moments T, denoted collectively as *sectional forces*. An eccentrically applied load can be resolved into symmetrical and antisymmetrical components, as shown in figures 5.3 and 5.4. The symmetrical component induces bending moments and shear forces. The antisymmetrical component, a force couple, is equivalent to an external torque M_t (concentrated loads) or m_t (distributed loads). Torque induces torsional moments in the structure.

Cross-sections can be classified into two groups, according to the mechanism by which torsional moments are resisted. *Closed sections* (e.g. hollow boxes) resist torsional moments by means of a closed shear flow; *open sections* (e.g. T-sections) resist torsion primarily by bending moments in the webs.

In plane grid models, a longitudinal beam is used to model each cell of the box girder or web of the T-girder. Transverse beams, whose properties are derived from those of the top and bottom slabs, link the longitudinal beams together. The number of transverse beams can, in general, by freely chosen; three to five per span are normally sufficient.

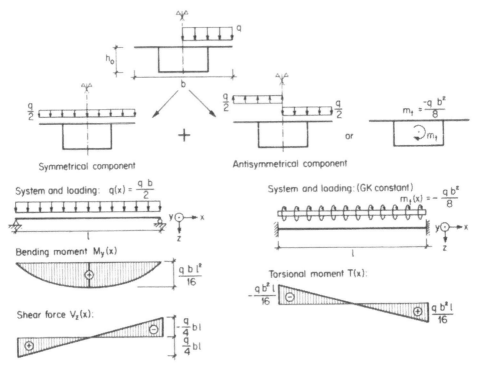

Figure 5.4
Sectional forces due to an eccentrically applied distributed load

b) Effective Flange Width

Ultimate Limit State. As shown in figure 4.30, the force in the compression flange at ultimate limit state is built up from the horizontal component of the compression diagonals in the webs. Transverse tension is required to deviate these forces from the web-flange junction to the centroid of the effective compression flange. The angle of deviation, and hence the effective flange width, will thus depend on the amount of transverse reinforcement provided in the compression flange. The interaction of effective flange width, transverse reinforcement, and flange thickness at ultimate limit state is discussed further in Section 5.3.3.

Concrete tensile strength is neglected in calculating the axial resistance of the tension flange. The longitudinal reinforcement distributed across its width may be considered, provided the shear resistance of the flange is sufficient to transfer the yield force of the reinforcement to the webs.

Service Conditions. The flanges of box and T-sections respond to longitudinal bending of the superstructure primarily in a state of plane stress. A flange can be isolated as a free body loaded by the horizontal shear stresses at the flange-web junctions. The load distributes laterally into the flange through in-plane shear.

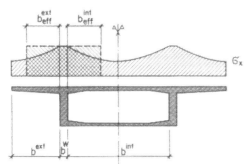

Figure 5.5
Effective flange width (service conditions)

The associated shear deformations cause a gradual decrease in normal stress away from the web. The variation of normal stress along the width of the flange is, in general, nonlinear (fig. 5.5).

Calculations are simplified by defining an *effective flange width* b_{eff}, over which the normal stresses are assumed to be uniformly distributed and equal to the maximum stress in the nonlinear distribution. The value of b_{eff} follows from equality of the resultant flange forces obtained from the nonlinear stress distribution and from the equivalent uniform distribution. Investigations into stress distributions in T-girders have shown that b_{eff} is primarily a function of the following factors (Rüsch 1972, 151): (1) type of loading (distributed or concentrated); (2) structural system (simply supported, cantilever, or continuous beam); and (3) dimensions (ratio of flange thickness to depth of section, ratio of flange span to girder span).

The flange stresses under service conditions are normally required only for the calculation of deformations. Due to variability in material properties, the accuracy of the computed deformations will not be significantly increased by refinement in the calculation of flange stresses. The following approximate expressions for b_{eff} can therefore be used for both box girders and T-girders. (The symbols are defined in figure 5.5):

1. Top slab:

$$b_{eff}^{ts} = b^w + b_{eff}^{int} + b_{eff}^{ext}$$

where

$$b_{eff}^{int} = \text{the lesser of } l_0/6 \text{ and } b^{int}/2$$

$$b_{eff}^{ext} = \text{the lesser of } l_0/6 \text{ and } b^{ext}$$

The effective span length l_0 is defined as the distance between inflection points of the girder, and can be assumed equal to $0.8l$ for end spans and $0.6l$ for interior spans.

2. Bottom slab:

$$b_{eff}^{bs} = b^w + b_{eff}^{int}$$

where

$$b_{\text{eff}}^{\text{int}} = \text{the lesser of } l_0/10 \text{ and } b^{\text{int}}/2$$

and l_0 is as defined for the top slab.

5.1.2 Torsion and Introduction of Loads in Single-Cell Box Girders

a) Torsion

Torsional moments T in a single-cell box girder are normally considered to be equilibrated by a state of pure shear. This phenomenon is called *St. Venant torsion*. The torsional shear stresses in the deck slab cantilevers are small relative to those in the walls of the box and are therefore neglected. Because the webs and slabs are thin relative to the overall box dimensions, the shear stresses can be assumed constant across the wall thickness. The torsional stresses can thus be expressed as a constant closed shear flow around the box, v [N/m] (fig. 5.6). The value of v is obtained from the equation of moment equilibrium about one of the corners of the box:

$$T = (vb_0)\,h_0 + (vh_0)\,b_0$$

where b_0 and h_0 are as defined in figure 5.2. Defining

$$A_0 = b_0\,h_0$$

and solving for v yields the following expression:

$$v = \frac{T}{2A_0}$$

The equations

$$\tau^{ts} = v/t^{ts} \qquad \tau^{bs} = v/t^{bs} \qquad \tau^{w} = v/b^{w}$$

can then be used to calculate the shear stresses in each of the girder elements.

Strictly speaking, torsion in box girder bridges is rarely equilibrated by shear stresses alone. The closed shear flow v is normally accompanied by both transverse and longitudinal bending. This can be demonstrated by deriving the necessary

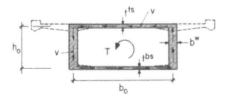

Figure 5.6
St. Venant torsion in a closed cross-section

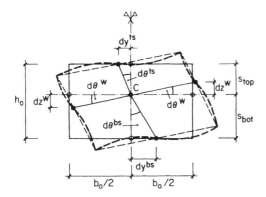

Figure 5.7
General torsional deformation of a box
cross-section

conditions for pure torsion in a box section symmetrical about its vertical axis
(fig. 5.7). In the absence of transverse and longitudinal bending, torsional
moments will produce only twist about the shear centre C. The angle of twist will
be constant at all points in the section:

$$d\theta^{ts} = d\theta^{bs} = d\theta^{w} = d\theta$$

The in-plane displacement of the web can be formulated in terms of its individual
shear deformation (fig. 5.8):

$$dz^{w} = \gamma^{w}\, dx = \frac{v}{b^{w} G}\, dx \tag{a}$$

where G is the shear modulus of concrete. It can also be expressed in terms of the
twist $d\theta$:

$$dz^{w} = \frac{b_{0}}{2}\, d\theta \tag{b}$$

Substituting (b) into (a) yields the following equation:

$$\frac{b_{0}}{2}\, b^{w} = \frac{v}{G}\frac{dx}{d\theta} \tag{c}$$

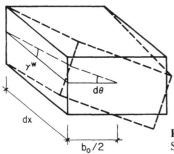

Figure 5.8
Shear deformation in a box girder web due to pure torsion

Similar expressions can be obtained for the top and bottom slabs:

$$s_{\text{top}} \, t^{ts} = \frac{v}{G} \frac{dx}{d\theta} \qquad\qquad\qquad (d)$$

$$s_{\text{bot}} \, t^{bs} = \frac{v}{G} \frac{dx}{d\theta} \qquad\qquad\qquad (e)$$

Substituting (d) into (c) and (e) into (c) yields the equations

$$s_{\text{top}} \, t^{ts} = \frac{b_0}{2} \, b^w \qquad s_{\text{bot}} \, t^{bs} = \frac{b_0}{2} \, b^w$$

The necessary condition for pure torsion is obtained by adding these two equations together:

$$b_0 \left(\frac{1}{t^{ts}} + \frac{1}{t^{bs}} \right) = h_0 \, \frac{2}{b^w}$$

This condition is rarely satisfied by cross-sections typically used in box girder bridges. The box width is normally greater than its depth and the webs are normally thicker than both the top slab and the bottom slab. The closed shear flow normally associated with box girders will thus, strictly speaking, always be accompanied by transverse and longitudinal bending.

The girder deformations due to torsionally induced transverse and longitudinal bending are nevertheless small relative to the deformations due to torsional shear flow. The total deformation due to torsion can thus be assumed equal to the deformation in pure torsion, θ, defined by the following equation:

$$\theta(x) = \int \frac{T(x)}{GK} \, dx + C \qquad\qquad\qquad (5.1)$$

where C is obtained from the boundary conditions and K is defined as the *torsional constant* of the section.

An expression for K can be derived for an uncracked box section using the method of virtual work. A girder segment of length dx is considered. The length and thickness of section element i are denoted b_i and t_i, respectively. The real torsional moment T is assumed to produce the twist $d\theta$. The corresponding shear deformation of section element i is

$$\gamma_i \, dx = \frac{T}{2A_0 G t_i} \, dx \qquad\qquad\qquad \text{(cf. equation (a))}$$

The virtual torsional moment $\bar{T} = 1$ produces the shear flow $\bar{v} = 1/2A_0$. The corresponding shear force in section element i is

$$\bar{v} b_i = \frac{b_i}{2A_0}$$

The external and internal work of the virtual forces and real displacements can be equated:

$$1 \cdot d\theta = \sum_i (\bar{v} b_i) (\gamma_i dx) = \sum_i \left(\frac{b_i}{2A_0}\right) \left(\frac{T}{2A_0 G t_i} dx\right)$$

It follows that

$$\frac{d\theta}{dx} = \sum_i \frac{b_i}{2A_0} \frac{T}{2A_0 G t_i} = \sum_i \frac{T}{4G A_0^2} \frac{b_i}{t_i} = \frac{T}{G} \frac{\sum_i (b_i/t_i)}{4A_0^2}$$

This equation can be rewritten as

$$\frac{d\theta}{dx} = \frac{T}{GK} \tag{5.2}$$

where the torsional constant K is defined as

$$K = \frac{4A_0^2}{\sum_i b_i/t_i}$$

b) Eccentric Loads

The sectional forces shown in figures 5.3 and 5.4 were calculated from the equilibrium and compatibility conditions of a prismatic bar with constant flexural and torsional stiffness. The actual cross-section geometry was not considered in the calculation, and the given load was replaced by statically equivalent loads. Although the sectional forces shown are correct, they are insufficient for a complete calculation of the stresses in the member. Self-equilibrating stresses, which are a consequence of the actual cross-section geometry and loading arrangement, will also be present. Even though they make no net contribution to the sectional forces, they will affect the local behaviour of the deck slab, bottom slab and webs. The actual stress at a given point in the cross-section will be equal to the sum of the self-equilibrating stresses and the stresses obtained from the sectional forces.

The self-equilibrating stresses due to the antisymmetrical component of the load will be considered in further detail. Figure 5.9 shows the *diagonal forces R* produced by an antisymmetrical couple of concentrated loads Q applied in the plane of the webs. Likewise, distributed diagonal forces r will be induced by a distributed couple q. (As shown in figure 5.10, the diagonal forces are reduced when the couple Q is applied to the deck slab cantilevers.) These self-equilibrating forces will produce transverse and longitudinal bending of the cross-section elements.

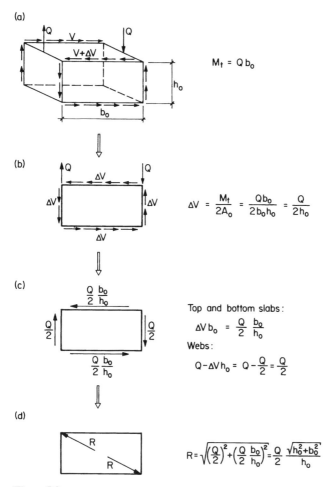

Figure 5.9
Introduction of couple Q applied in the plane of the webs: a, girder element; b, equilibrium of couple Q and shear flow difference Δv; c, shear forces; d, self-equilibrating diagonal forces R

The response of the girder to the diagonal forces can be investigated using the model shown in figure 5.11. The model consists of two components: (1) a *hinged box girder*, in which the monolithic connections along the longitudinal joints have been replaced by hinges; and (2) a series of *transverse frames*. The deck slab, webs and bottom slab of the hinged box girder deflect as longitudinal beams due to the diagonal forces, resulting in the deformation of the cross-section into a parallelogram. This deformation is partially restrained by bending in the transverse frames. Assuming linear elastic material behaviour, the restraining force will be proportional to the deflection of the cross-section elements. The behaviour of the deck slab, webs and bottom slab under the action of the diagonal forces is thus equivalent to the behaviour of a beam on an elastic foundation.

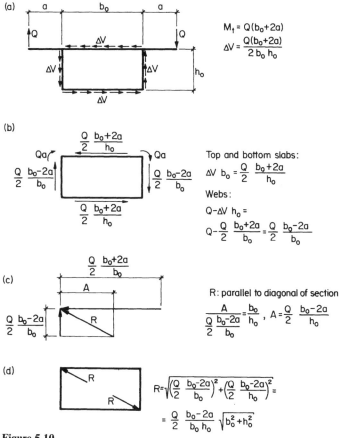

(a)

$$M_t = Q(b_0 + 2a)$$

$$\Delta V = \frac{Q(b_0 + 2a)}{2 b_0 h_0}$$

(b)

Top and bottom slabs:

$$\Delta V \; b_0 = \frac{Q}{2} \frac{b_0 + 2a}{h_0}$$

Webs:

$$Q - \Delta V \; h_0 =$$

$$Q - \frac{Q}{2} \frac{b_0 + 2a}{b_0} = \frac{Q}{2} \frac{b_0 - 2a}{b_0}$$

(c)

R: parallel to diagonal of section

$$\frac{A}{\frac{Q}{2}\frac{b_0-2a}{b_0}} = \frac{b_0}{h_0} \;, \quad A = \frac{Q}{2} \frac{b_0 - 2a}{h_0}$$

(d)

$$R = \sqrt{\left(\frac{Q}{2}\frac{b_0-2a}{b_0}\right)^2 + \left(\frac{Q}{2}\frac{b_0-2a}{h_0}\right)^2} =$$

$$= \frac{Q}{2} \frac{b_0 - 2a}{b_0 h_0} \sqrt{b_0^2 + h_0^2}$$

Figure 5.10
Introduction of couple Q applied to the cantilevers: a, equilibrium of Q and Δv; b, shear forces and moments; c, forces at upper left-hand corner; d, self-equilibrating diagonal forces R (component of shear force parallel to diagonal of cross-section)

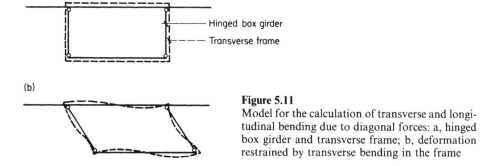

(a)

Hinged box girder

Transverse frame

(b)

Figure 5.11
Model for the calculation of transverse and longitudinal bending due to diagonal forces: a, hinged box girder and transverse frame; b, deformation restrained by transverse bending in the frame

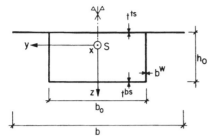

Figure 5.12
Box section with vertical axis of symmetry; origin of coordinates at centroid S

The fourth-order differential equation of the elastic curve will be formulated for one of the webs of the cross-section shown in figure 5.12. (The origin of the coordinate system is located at the centroid S of the section.) The equation is based on the moment-curvature relationship of the web in the hinged box girder, modified to account for the elastic restraint of the transverse frames.

Elastic Curve of a Web in the Hinged Box. The force method is used to compute the normal stresses in the hinged girder due to the distributed diagonal force $r(x)$. The section is assumed symmetrical about its vertical axis. The statically determinate primary system consists of four simply supported beams. Its cross-section is shown in figure 5.13. Force r is resolved into horizontal and vertical components r_h and r_v:

$$r_h = r \frac{b_0}{\sqrt{(b_0^2 + h_0^2)}} \qquad r_v = r \frac{h_0}{\sqrt{(b_0^2 + h_0^2)}} \tag{f}$$

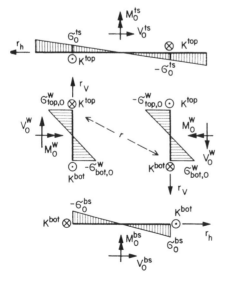

Figure 5.13
Longitudinal bending in the hinged box girder due the diagonal forces r: primary system with stresses and edge shear forces

which are applied as loads to the girder elements. Since the two force components are proportional $(r_h(x) = (b_0/h_0)\, r_v(x)$ for all $x)$, the corresponding sectional forces will also be proportional:

$$M_0^{ts}(x) = -M_0^{bs}(x) = M_0^w(x)\,\frac{b_0}{h_0} \tag{g}$$

and

$$V_0^{ts}(x) = -V_0^{bs}(x) = V_0^w(x)\,\frac{b_0}{h_0}, \tag{h}$$

for all x.

Edge shear forces $K^{top}(x)$ and $K^{bot}(x)$ are required for compatibility of longitudinal strain. They are defined as the integral of the corresponding edge shear stresses:

$$K^{top}(x) = \int_0^x \tau^{top}(s)\,b^w\,ds \qquad K^{bot}(x) = \int_0^x \tau^{bot}(s)\,b^w\,ds$$

The compatibility conditions can be expressed as the following system of equations:

$$A_1\,M_0^{ts}(x) + A_2\,M_0^w(x) = A_3\,K^{top}(x) + A_4\,K^{bot}(x)$$
$$A_5\,M_0^{bs}(x) + A_6\,M_0^w(x) = A_7\,K^{top}(x) + A_8\,K^{bot}(x)$$

where the strains at the slab-web interfaces have been converted into linear functions of the sectional forces using the constants A_i. The forces $K^{top}(x)$ and $K^{bot}(x)$ can be rewritten in terms of $M_0^w(x)$ alone, using equation (g):

$$K^{top}(x) = B_1\,M_0^w(x) \qquad K^{bot}(x) = B_2\,M_0^w(x)$$

for all x. It is therefore necessary to solve for $K^{top}(x)$ and $K^{bot}(x)$ at only one arbitrary location along the length of the girder.

The constants B_1 and B_2 are functions of the cross-section dimensions. The distribution of stresses in the statically indeterminate system that correspond to these constants (fig. 5.14) is conveniently expressed in terms of the parameters a and k^w. The parameter a is defined as the distance from the neutral axis to the middle surface of the top slab:

$$a = h_0\,\frac{1 + 3\,\dfrac{h_0}{b_0}\,\dfrac{b^w}{t^{bs}}}{1 + \dfrac{t^{ts}}{t^{bs}}\left(\dfrac{b}{b_0}\right)^3 + 6\,\dfrac{h_0}{b_0}\,\dfrac{b^w}{t^{bs}}} \tag{5.3}$$

The quantity k^w relates M_0^w to the web moment in the statically indeterminate system, M^w:

$$M^w = k^w\,M_0^w \tag{5.4}$$

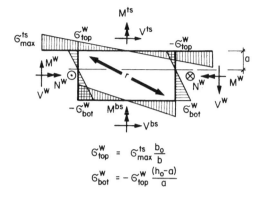

$$\sigma^{w}_{top} = \sigma^{ts}_{max} \frac{b_0}{b}$$

$$\sigma^{w}_{bot} = -\sigma^{w}_{top} \frac{(h_0-a)}{a}$$

Figure 5.14
Longitudinal bending in the hinged box girder due the diagonal forces r: total stresses in the statically indeterminate system

where

$$k^{w} = \frac{\dfrac{1}{3}\left[1 + \dfrac{t^{bs}}{t^{ts}}\left(\dfrac{b_0}{b}\right)^3\right] + 2\dfrac{h_0}{b_0}\dfrac{b^w}{t^{ts}}\left(\dfrac{b_0}{b}\right)^3}{\dfrac{1}{3}\dfrac{b_0}{h_0}\dfrac{t^{bs}}{b^w} + \dfrac{2}{3}\left[1 + \dfrac{t^{bs}}{t^{ts}}\left(\dfrac{b_0}{b}\right)^3\right] + \dfrac{h_0}{b_0}\dfrac{b^w}{t^{ts}}\left(\dfrac{b_0}{b}\right)^3} \tag{5.5}$$

The associated bending stresses in the web of the statically indeterminate system are obtained from the expressions

$$\sigma^{w}_{top} = \frac{M^{w}\,a}{I^{w}} \tag{i}$$

$$\sigma^{w}_{bot} = \frac{M^{w}(a-h_0)}{I^{w}} \tag{j}$$

where

$$I^{w} = \frac{b^w\,h_0^3}{12}$$

The moment-curvature relationships of the web in the hinged box girder can now be formulated in terms of M_0^w and w_v, the vertical deflection of the web. Substituting equation (5.4) into the familiar equation

$$EI^{w}\frac{d^2 w_v}{dx^2} = -M^{w} \tag{5.6}$$

results in

$$EI^{w}\frac{d^2 w_v}{dx^2} = -k^{w}\,M_0^{w} \tag{k}$$

The differential equation of the elastic curve of the web is obtained by taking two successive derivatives of equation (k):

$$EI^w \frac{d^3 w_v}{dx^3} = -k^w V_0^w \tag{5.7}$$

$$EI^w \frac{d^4 w_v}{dx^4} = k^w r_v \tag{1}$$

where r_v is the vertical component of the diagonal force (equation (f)).

It can be shown that the vertical shear forces in the web of the hinged box, V^w, are equal to the web shear forces in the primary system V_0^w. The edge shear forces $K^{top}(x)$ and $K^{bot}(x)$ thus produce only a redistribution of the web shear stresses, leaving the total shear force unchanged. The shear forces and moments are related by the following expression:

$$V^w = V_0^w = \frac{dM_0^w}{dx} = \frac{1}{k^w} \frac{dM^w}{dx} \tag{m}$$

where $dM^w/dx \neq V^w$. Recalling equation (h), the shear forces in the top and bottom slab can be computed from V^w as follows:

$$V^{ts} = -V^{bs} = \frac{b_0}{h_0} V^w \tag{n}$$

Elastic Restraint of the Transverse Frame. The deformations of the hinged girder are partially restrained by the transverse frames. When the corners of the cross-section are displaced as shown in figure 5.15, transverse bending moments are induced. The distribution of these moments is identical to the moments produced by a diagonal force r^Q. The elastic restraint due to transverse frame action can be accounted for in equation (1) by adding a term corresponding to the vertical component of the diagonal force, r_v^Q, expressed as a function of the vertical displacement w_v.

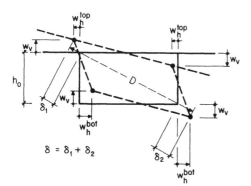

$$\delta = \delta_1 + \delta_2$$

Figure 5.15
Total displacements of the section due to longitudinal and transverse bending

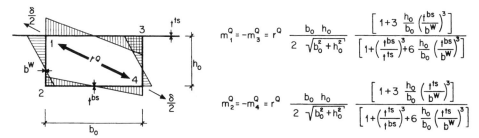

Figure 5.16
Transverse bending moments per unit girder length due to r^Q

The increase in length of the diagonal of the section, δ, is computed from the moments due to r^Q (fig. 5.16) using the method of virtual work:

$$\delta = r^Q \frac{h_0^3 b_0^2}{E(b^w)^3 (b_0^2 + h_0^2)} C_1 \qquad (o)$$

where

$$C_1 = \frac{2\left[(t^{ts})^3 + (t^{bs})^3 + \frac{b_0(b^w)^3}{2h_0}\right] + 3\frac{h_0}{b_0}\left(\frac{t^{ts}\,t^{bs}}{b^w}\right)^3}{(t^{ts})^3 + (t^{bs})^3 + 6\frac{h_0}{b_0}\left(\frac{t^{ts}\,t^{bs}}{b^w}\right)^3} \qquad (5.8)$$

The relation between δ and the joint displacements follows from the geometry of figure 5.15:

$$\delta = \frac{h_0}{\sqrt{(b_0^2 + h_0^2)}}\left[2w_v + \frac{b_0}{h_0}(w_h^{top} + w_h^{bot})\right]$$

This expression can be reduced to a function of w_v alone using the relation

$$\frac{2w_v}{b_0} = \frac{w_h^T + w_h^B}{h_0}$$

which is derived from the stress distribution of figure 5.14 and the moment-curvature relations of the webs, top slab, and bottom slab. It follows that

$$w_v = \delta\frac{\sqrt{(b_0^2 + h_0^2)}}{4h_0} \qquad (p)$$

Combining equations (o) and (p) yields an expression relating diagonal force and vertical displacement

$$r^Q = w_v\frac{4E(b^w)^3\sqrt{(b_0^2 + h_0^2)}}{C_1 h_0^2 b_0^2} \qquad (q)$$

This equation is equivalent to the following relation between the vertical component of r^Q and w_v:

$$r_v^Q = c_f\, w_v$$

where $r_v^Q = r^Q h_0 / \sqrt{(b_0^2 + h_0^2)}$, and

$$c_f = \frac{4E(b^w)^3}{C_1\, h_0\, b_0^2} \tag{5.9}$$

Equilibrium Equation of the Web. Equation (l)

$$EI^w \frac{d^4 w_v}{dx^4} = k^w\, r_v$$

which was defined for webs of the hinged box, can now be modified to account for the elastic restraint due to bending of the transverse frame. The total load acting on the web will no longer be r_v, but rather the algebraic sum of r_v and the vertical force due to elastic restraint r_v^Q (fig. 5.17). It therefore follows that

$$EI^w \frac{d^4 w_v}{dx^4} = k^w (r_v - r_v^Q) = k^w (r_v - c_f\, w_v)$$

This expression can be rearranged into the familiar equation of a beam on an elastic foundation

$$\frac{EI^w}{k^w} \frac{d^4 w_v}{dx^4} + c_f\, w_v = r_v \tag{5.10}$$

whose solution w_v can be obtained from published tables and charts (see, for example, Hetenyi 1964).

The solution w_v can then be used to obtain the longitudinal and transverse bending stresses and the longitudinal shear forces in the section components:

1. The longitudinal bending moments in the web M^w are obtained from equation (5.6). The corresponding flexural stresses can then be calculated from equations (i) and (j).

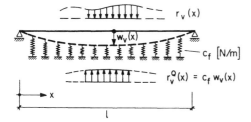

Figure 5.17
Web modeled as a beam on an elastic foundation

2. The transverse bending moments are calculated from the equations given in figure 5.16, where r^Q is obtained from equation (q).
3. The shear in the web $V^w(r)$ is obtained from equation (m). The corresponding shear forces in the top and bottom slabs, $V^{ts}(r)$ and $V^{bs}(r)$, are obtained from equation (n). The total shear force in one web is equal to the sum of $V^w(r)$ and the shear due to the torsional sectional force, $V^w(T)$:

$$V^w_{tot} = V^w(T) + V^w(r) \tag{5.11}$$

The total shear forces in the deck slab and bottom slab are obtained in an identical manner.

The effect of the self-equilibrating diagonal force r on the stresses in a box girder is investigated in the following example.

Example 5.1:
Symmetric and eccentric loads in a simply-supported box girder

A simply supported box girder of length l and cross-section dimensions shown in figure 5.18 is given. The girder is torsionally fixed at both ends. Two load cases are considered: (1) two trucks at midspan, equidistant from the longitudinal axis of the bridge, and (2) one truck at midspan eccentric to the bridge axis. Both cases are modeled using simplified loads (fig. 5.19). The web shear forces V^w and the web bending moments M^w are to be calculated.

For the symmetric load, V^w is obtained by dividing the total shear V by 2. The web bending moment M^w is obtained from the total bending moment M in an analogous manner to equation (5.4):

$$M^w = k^w_{sym} M^w_0 = k^w_{sym} \frac{M}{2}$$

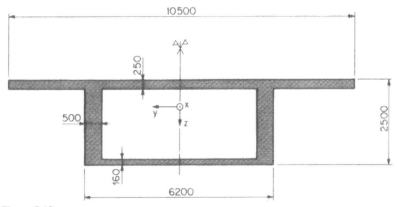

Figure 5.18
Cross-section of example 5.1 (dimensions in mm)

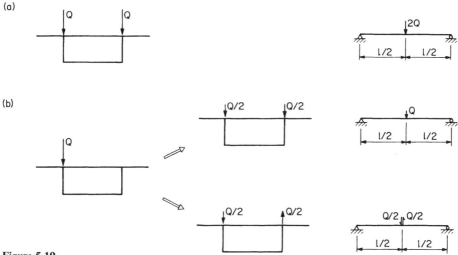

Figure 5.19
Load cases of example 5.1: a, symmetric; b, eccentric

where $M_0^w = M/2$ is the moment which would be resisted by one web, assuming no shear transfer between the webs and the top and bottom slabs. For the given cross-section, $k_{\text{sym}}^w = 0.2$.

The antisymmetric loading results in shear due to torsion T as well as shear and bending due to the self-equilibrating diagonal force R. The relevant quantities are

External torque:
$$M_t = \frac{Q b_0}{2}$$

Torsional moment:
$$T = \pm \frac{Q b_0}{4}$$

Torsional shear flow:
$$v = \pm \frac{Q b_0}{4} \frac{1}{2 b_0 h_0} = \frac{Q}{8 h_0}$$

Diagonal force:
$$R = \frac{Q}{4 h_0} \sqrt{(b_0^2 + h_0^2)}$$

Vertical component of R:
$$R_v = \frac{Q}{4}$$

where $b_0 = 5.70$ m and $h_0 = 2.295$ m. The web shear due to torsion is obtained directly from the shear flow v:

$$V^w(T) = \pm \frac{Q}{8 h_0} h_0 = \frac{Q}{8}$$

The web shears and moments due to R are obtained from the deflections of the elastically supported webs, w_v (equation (5.10)), where

$$I^w = \frac{(0.5)\,(2.30)^3}{12} = 0.504 \text{ m}^4$$

$$k^w = 0.554 \qquad\qquad\qquad\qquad\qquad\qquad\text{(equation (5.5))}$$

$$C_1 = 16.7 \qquad\qquad\qquad\qquad\qquad\qquad\text{(equation (5.8))}$$

$$c_f = 0.000401\,E \qquad\qquad\qquad\qquad\qquad\text{(equation (5.9))}$$

(E is in N/mm².) Equations (5.6) and (5.7) are used to solve for $M^w(r)$ and $V^w(r)$, respectively. $V^w(T)$ and $V^w(r)$ are then combined according to equation (5.11).

The results are shown in figures 5.20 and 5.21. The bending moments due to eccentrically applied load (case 4 of figure 5.21) are less than the moments due to full symmetrical load. The maximum web shear due to eccentric load is equal to the maximum shear due to full symmetric load.

It can be concluded that full symmetrical load will govern the design of the superstructure for longitudinal bending. The shear forces resulting from eccentric live load are nevertheless considerable. With regard to the longitudinal response of the system, therefore, eccentric loads need only be considered for shear design.

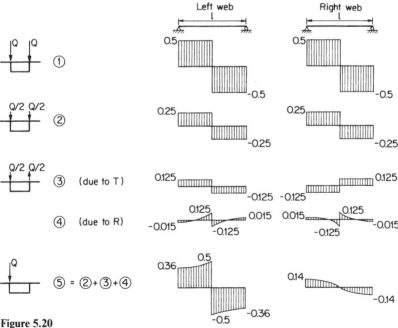

Figure 5.20
Web shear forces V^w of example 5.1

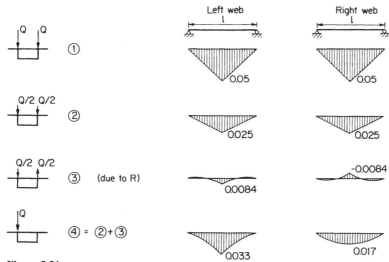

Figure 5.21
Web bending moments M^w of example 5.1

5.1.3 Torsion and Eccentric Loads in Double-T Girders

Torsional moments in open sections are resisted by a combination of St. Venant torsion and differential web bending. The latter phenomenon, illustrated schematically in figure 5.22, is known as *warping torsion*. The torsional moment at a given section can be expressed as

$$T(x) = T^{SV}(x) + T^W(x)$$

where $T^{SV}(x)$ and $T^W(x)$ denote the St. Venant and warping components, respectively. The ratio $T^{SV}(x)/T^W(x)$ varies along the length of the superstructure, and is also a function of cross-section dimensions and span length. (The warping component predominates in most commonly used open sections.) The components $T^{SV}(x)$ and $T^W(x)$ can be obtained at each point along the length of the superstructure from the compatibility condition of equal twist due to St. Venant and warping torsion:

$$\theta^W(x) = \theta^{SV}(x) \tag{5.12}$$

The angle $\theta^W(x)$ is defined by the equation

$$\theta^W(x) = 2w_v(x)/b_0 \tag{5.13}$$

where w_v is the vertical deflection of the web due to flexure and b_0 is as shown in figure 5.23. The angle $\theta^{SV}(x)$ is computed from equation (5.1):

$$\theta^{SV}(x) = \frac{1}{GK} \int T^{SV}(x)\, dx + C \tag{5.14}$$

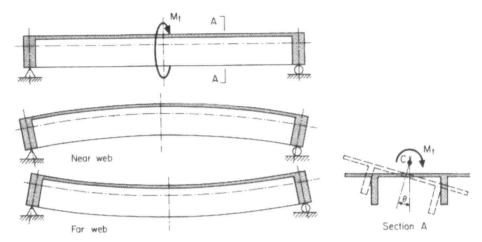

Figure 5.22
Deformation of a double-T girder due to warping torsion

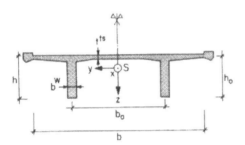

Figure 5.23
Double-T cross-section: symbols

where

$$K \cong \tfrac{1}{3}[(t^s)^3 b + 2(b^w)^3 h_0] \tag{5.15}$$

for a double-T girder and C is determined from the boundary conditions.

It is convenient to express the external loads as the sum of a component resisted by warping torsion and a component resisted by St. Venant torsion. External torque M_t is thus written as

$$M_t = M_t^W + M_t^{SV} = Q^W b_0 + Q^{SV} b_0$$

where $Q^W b_0$ and $Q^{SV} b_0$ are the corresponding couples applied in the plane of the webs.

An exact calculation of the torsional response of an open section can be complicated and time-consuming. The computational effort required can be substantially reduced by assuming that the ratio of St. Venant torsion to warping torsion is constant over the entire girder length, that is,

$$\frac{T^{SV}(x)}{T^{W}(x)} = k$$

where k is a constant. The compatibility condition of equation (5.12) now needs only be solved at one arbitrary point along the length of the superstructure. The loss of accuracy resulting from this simplification is insignificant. Design at ultimate limit state can normally be based on a constant ratio $T^{SV}/T^{W} = 1/2$ for spans greater than 50 m and $1/3$ for spans less than 50 m. A similar simplification can be made for the external load components:

$$\frac{Q^{SV}}{Q^{W}} = k \tag{a}$$

It is convenient to compute the warping torsional response in terms of the unknown parameter Q^{W} and the St. Venant torsional response in terms of kQ^{W}. The value of k can be obtained from the compatibility condition of equation (5.12). The components Q^{W} and Q^{SV} are then calculated from

$$Q^{W} = Q/(1+k)$$

and equation (a).

For girders with a vertical axis of symmetry, the calculation of the warping component of torsional response can be further simplified using the model shown in figure 5.24. The superstructure is separated along its centreline into two equal halves. Each of the half-girders is loaded by $Q^{W} = M_{t}^{W}/b_{0}$ and is analysed separately as a continuous beam. (Only torsionally fixed supports of the original girder can be considered as vertical supports of the half-girders.) The bending moments and shear forces thus produced are denoted M_{y} and V_{z}, respectively. Stresses can be calculated without significant error by assuming that the principal axes of the half-girder sections are parallel to the principal axes of the original section.

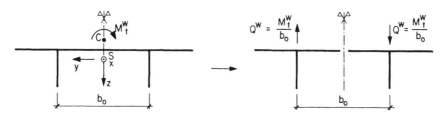

Figure 5.24
Simplified model for the calculation of warping torsional response

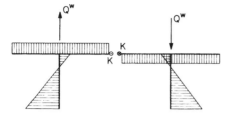

Figure 5.25
Refined model for the calculation of warping torsional response: primary system with normal stresses and redundant shear force K

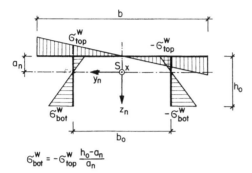

$$\sigma^w_{bot} = -\sigma^w_{top} \frac{h_0 - a_n}{a_n}$$

Figure 5.26
Refined model for the calculation of warping torsional response: total normal stresses in the statically indeterminate system

The main advantage of this model is its simplicity. It does not satisfy compatibility of longitudinal strain at the centreline nor does it satisfy equilibrium ($M_z \neq 0$). These shortcomings occur because the shear stresses in the deck slab at the cut are neglected.

The model can be refined to account for the shear stresses at the centreline. The primary system of figure 5.25 is identical to the model of figure 5.24, except for the unknown redundant shear forces K which have been added. The stress distribution in the full section (fig. 5.26) is computed from the condition of equal longitudinal strain at the centreline. The solution can be characterized by two parameters: the dimension a_n, defined as the distance from the middle surface of the top slab to the neutral axis, and $\bar{I}_n$, the moment of inertia satisfying the equation

$$\sigma^w = \frac{M_y z}{\bar{I}_n}$$

where σ^w is the stress in the web of the original (uncut) section, M_y is the bending moment due to Q^W in one of the half-girders and z is vertical distance from the neutral axis defined by the parameter a_n. Expressions for a_n and $\bar{I}_n$ can be formulated in terms of the cross-section dimensions:

$$a_n = \frac{3 h_0^2 b_0^2 b^w}{b^3 t^s + 6 h_0 b_0^2 b^w} \tag{5.16}$$

$$\bar{I}_n = \frac{b^w a_n^3}{3} \left[1 + \frac{3 t^s b^2}{2 b^w b_0 a_n} + \left(\frac{h_0}{a_n} - 1 \right)^3 \right] \tag{5.17}$$

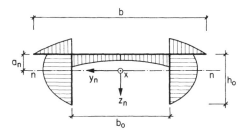

Figure 5.27
Shear flow due to warping torsion (schematic)

The stress distribution of figure 5.26 is completely defined by these two parameters and the moment M_y. The normal stresses are computed using the equations

$$\sigma^w = \frac{M_y z}{\overline{I}_n} \qquad \text{(web)}$$

$$\sigma^s = \sigma^w_{top} \frac{2y}{b_0} \qquad \text{(deck slab)}$$

The shear flow shown in figure 5.27 is obtained using the following equations:

1. Webs:

$$\tau^w(z) b^w = \frac{V_z b^w}{\overline{I}_n} \int\limits_z^{h_0 - a_n} s\,ds$$

2. Deck slab:

$$\tau^s(y) t^s = \frac{V_z t^s}{\overline{I}_n} \frac{2 a_n}{b_0} \int\limits_y^{b/2} s\,ds \qquad (b)$$

Between the webs, the constant shear flow

$$\tau^w_{top} b^w = \frac{V_z b^w (h_0^2 - 2 a_n h_0)}{2 \overline{I}_n}$$

must be added to the shear calculated from equation (b).

Example 5.2:
Torsion and eccentric loads in a simply-supported double-T girder

A simply-supported double-T girder of length 40 m is loaded by a force couple Q applied in the plane of the webs (fig. 5.28). Both ends of the girder are torsionally fixed. The torsional moments will be resisted by a combination of St. Venant and warping torsion.

The stresses due to warping torsion are calculated in terms of the unknown load component Q^W. The sectional forces in the left half-girder are shown in

(a)

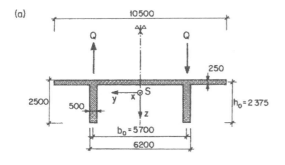

(b)

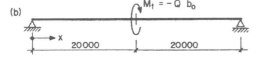

Figure 5.28

Example 5.2: a, cross-section and loads Q; b, girder and equivalent torque M_t; c, torsional moments T (dimensions in mm)

(c)

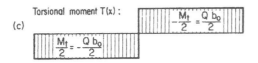

figure 5.29. The bending stresses and shear flow at midspan, assuming no shear transfer between the two half-girders at the centreline, are shown in figure 5.30. These are obtained from the centroid a_s and moment of inertia I of the half-section, neglecting skew bending:

$$a_s = 0.575 \text{ m}$$
$$I = 1.43 \text{ m}^4$$

The equilibrium error, M_z, is equal to $1.28 Q^W$ [m].

Figure 5.31 shows the exact stress distribution at midspan, based on the parameters

$$a_n = 0.528 \text{ m} \qquad\qquad\qquad\qquad\qquad \text{(equation 5.16)}$$
$$\bar{I}_n = 1.75 \text{ m}^4 \qquad\qquad\qquad\qquad\qquad \text{(equation 5.17)}$$

The compatibility condition (equation 5.12) is formulated at midspan. The rotations θ^{SV} and θ^W are obtained as follows:

1. $$\theta^{SV}(x = 20 \text{ m}) = \frac{1}{GK} \int_0^{20} \frac{kQ^W b_0}{2} \, dx = \frac{10kQ^W b_0}{GK} \qquad \text{(equation (5.14))}$$

where $G = 0.4E$, $b_0 = 5.70$ m and

$$K = \tfrac{1}{3}[(0.25)^3 \ 10.5 + 2(0.50)^3 \ 2.375] = 0.25 \text{ m}^4 \qquad \text{(equation (5.15))}$$

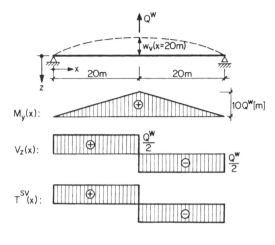

$M_y(x)$:

$V_z(x)$:

$T^{SV}(x)$:

Figure 5.29
Sectional forces in the left half-girder

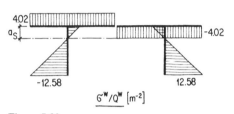

$\bar{\sigma}^W/Q^W \; [m^{-2}]$

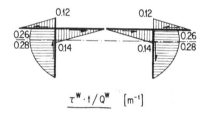

$\tau^W \cdot t / Q^W \; [m^{-1}]$

Figure 5.30
Bending stresses and shear flow at midspan due to Q^w (no shear transfer between the two half-girders)

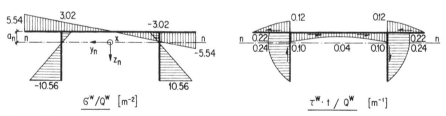

$\bar{\sigma}^W/Q^W \; [m^{-2}]$

$\tau^W \cdot t / Q^W \; [m^{-1}]$

Figure 5.31
Bending stresses and shear flow at midspan due to Q^W (including shear transfer between the two half-girders)

It therefore follows that $\theta^{SV}(x = 20 \text{ m}) = 570 k (Q^W/E) \; [m^{-2}]$.

2.
$$w_v(x = 20 \text{ m}) = \frac{Q^W l^3}{48 E \bar{I}_n} = \frac{Q^W (40)^3}{48 E (1.75)} = 762 (Q^W/E) \; [m^{-1}]$$

$$\theta^W (x = 20 \text{ m}) = 2 (762) (Q^W/E)/5.70 = 267 (Q^W/E) \; [m^{-2}]$$
$$\text{(equation (5.13))}$$

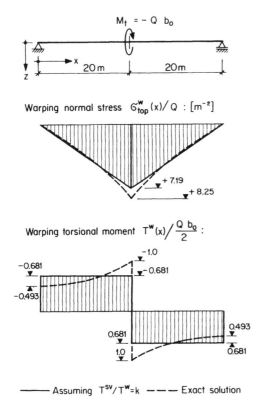

$$M_t = -Q\, b_o$$

Warping normal stress $G_{top}^W(x)/Q$: $[m^{-2}]$

+ 7.19
+ 8.25

Warping torsional moment $T^W(x)/\dfrac{Q\, b_o}{2}$:

-1.0
-0.681
-0.681
-0.493
0.681
0.493
1.0
0.681

——— Assuming $T^{SV}/T^W = k$ — — — Exact solution

Figure 5.32
Warping normal stresses and warping
torsional moment

Setting $\theta^W = \theta^{SV}$, the solution $k = 0.47$ is obtained. It follows that $Q^W = 0.681\,Q$ and $Q^{SV} = 0.319\,Q$.

The flexural stresses due to warping and the associated warping torsional moments are presented in figure 5.32. Also included are the results of a more exact calculation, in which the ratio $T^{SV}(x)/T^W(x)$ was not assumed constant. It is apparent that the assumption of a constant ratio $T^{SV}(x)/T^W(x)$ leads to satisfactory results.

5.1.4 Structural Models for Bridge Superstructures

Multiple-cell box girders, T-girders with three or more webs, and slabs can be modeled as plane grids of longitudinal and transverse beams.

The longitudinal beams in a grid model of a box girder correspond to the cells of the box (fig. 5.33). The transverse beams model the top and bottom slabs spanning between the webs, which are assumed to behave as a Vierendeel girder. An equivalent shear stiffness can thus be assigned to the transverse beams based on the flexural stiffness of the top and bottom slabs.

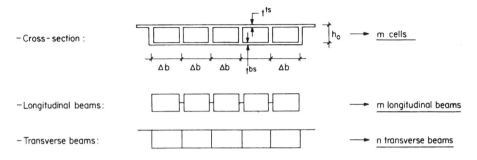

Stiffness	Longitudinal Beam (LB)	Transverse Beam (TB) $\quad (\Delta l = \dfrac{l}{n+1}$: Beam width$)$
Flexural	$EI^{LB} \cong \dfrac{EI_{tot}}{m}$	$M=1 \overset{\Delta\varphi}{\curvearrowright} M=1 \qquad \Delta\varphi = \dfrac{1}{EI^{TB}} \Delta b$ $$EI^{TB} = \dfrac{\Delta b}{\Delta\varphi} \cong E \Delta l \, h_o^2 \dfrac{t^{ts} \, t^{bs}}{t^{ts}+t^{bs}}$$
Shear	$\alpha GA^{LB} \cong \infty$	$V=1 \qquad V=1 \qquad \gamma \, \Delta b = \dfrac{1}{\alpha GA^{TB}} \Delta b$ $$\alpha GA^{TB} = \dfrac{1}{\gamma} \cong E \dfrac{\Delta l}{\Delta b^2} \left[(t^{ts})^3 + (t^{bs})^3 \right]$$
Torsional	$GK^{LB} \cong \dfrac{GK_{tot}}{m}$	$GK^{TB} \cong 0$

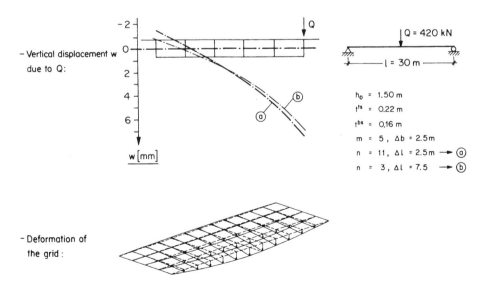

$Q = 420$ kN

$l = 30$ m

$h_o = 1.50$ m
$t^{ts} = 0.22$ m
$t^{bs} = 0.16$ m
$m = 5$, $\Delta b = 2.5$ m
$n = 11$, $\Delta l = 2.5$ m → ⓐ
$n = 3$, $\Delta l = 7.5$ → ⓑ

Figure 5.33
Grid models: multiple-cell box girder

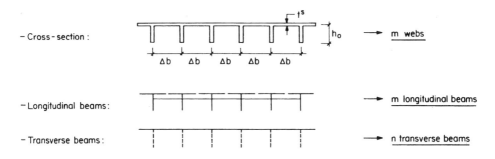

Stiffness	Longitudinal beam (LB)	Transverse beam (TB) $(\Delta = \frac{l}{n+1}$: Beam width)
Flexural	$EI^{LB} \cong \dfrac{EI_{tot}}{m}$	$EI^{TB} \cong \dfrac{1}{12} E \Delta l (t^s)^3$
Shear	$\alpha G A^{LB} \cong \infty$	$\alpha G A^{TB} \cong \infty$
Torsional	$GK^{LB} \cong \dfrac{GK_{tot}}{m}$	$GK^{TB} \cong 0$

- Vertical displacement w
  due to Q :

- Deformation of
  the grid :

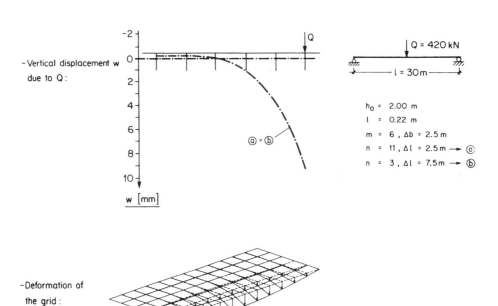

Figure 5.34
Grid models: multiple-web T-girder

- Cross-section :

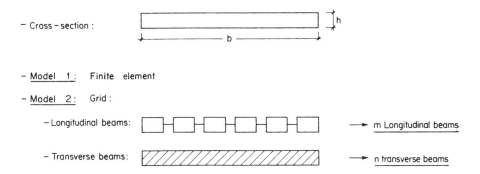

- <u>Model 1</u> :   Finite   element

- <u>Model 2</u> :   Grid :

    - Longitudinal beams: ⟶ m Longitudinal beams

    - Transverse beams: ⟶ n transverse beams

Stiffness	Longitudinal beam (LB)	Transverse beam (TB) $(\Delta l = \frac{l}{n+1}$: Beam width)
Flexural	$EI^{LB} \cong \dfrac{EI_{tot}}{m}$	$EI^{TB} \cong \dfrac{1}{12} E \Delta l\, h^3$
Shear	$\alpha G A^{LB} \cong \infty$	$\alpha G A^{TB} \cong \infty$
Torsional	$GK^{LB} \cong \dfrac{GK_{tot}}{m}$	$GK^{TB} \cong \dfrac{1}{3} G \Delta l\, h^3$ for $h << \Delta l$

- Vertical displacement w
  due to Q :

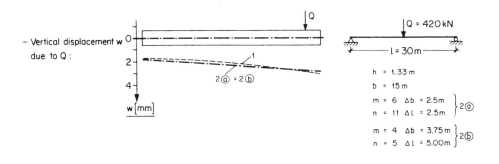

$Q = 420\,kN$

$l = 30\,m$

$h\ =\ 1.33\,m$

$b\ =\ 15\,m$

$m = 6 \quad \Delta b = 2.5m$
$n = 11 \quad \Delta l = 2.5m$ } 2ⓐ

$m = 4 \quad \Delta b = 3.75m$
$n = 5 \quad \Delta l = 5.00m$ } 2ⓑ

- Deformation of
  the grid :

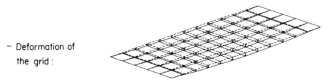

Figure 5.35
Grid models: solid slab

Grid models for T-girders include one longitudinal beam for each web (fig. 5.34). Three to five transverse beams per span are normally sufficient; further refinement will not significantly increase the accuracy of the model results. The stiffness of the transverse beams is obtained directly from the section properties of slab strips of width $l/(n+1)$, where l is the longitudinal span length and n is the number of transverse beams per span. Diaphragms can be considered as individual transverse T-beams. Their section properties are calculated assuming an effective deck slab width $b_m = \Delta b/2$, where Δb denotes web spacing.

Solid slabs can be modeled with a grid consisting of a relatively small number of elements (fig. 5.35). Four longitudinal beams are sufficient when the ratio of span length to cross-section width (l/b) lies between 2 and 3. The number of beams can be reduced to three when $l/b \geq 3$.

Grid models will yield reasonably accurate sectional forces and deformations provided the member stiffnesses in flexure, shear, and torsion have been properly chosen. An exact calculation of member stiffness is complex and time-consuming due to cracking and plastification of concrete. Results of acceptable accuracy are always obtained when the member stiffnesses are calculated from the properties of the homogeneous uncracked section. The calculation of a member stiffness, K, can be further simplified by considering its associated sectional force S_K:

1. When S_K is not required for equilibrium, K can be assigned a very small value $(\cong 0)$. The member can then be considered perfectly flexible with regard to the deformations associated with K. The computed values of S_K will be very small.
2. When S_K is required for equilibrium and K has no significant influence on the system deformations, K can be assigned a very large value $(\cong \infty)$. The member can be considered perfectly rigid.
3. When S_K is required for equilibrium and K has a substantial influence on the system deformations, K should be calculated from the homogeneous uncracked section properties.

5.2 Structural Function of Cross-Section Components

Each of the cross-section components of a bridge superstructure performs several structural functions simultaneously. These functions can be classified as *longitudinal* or *transverse*. The longitudinal structural function is associated with the beam sectional forces M, V, and T; the transverse function is associated with transverse bending of the cross-section.

A definite advantage of beam statics as opposed to shell theory is that it makes possible the separate calculation of stresses due to separate structural functions. This increases the clarity of the analysis and simplifies the detection of numerical errors. Furthermore, any details required to improve structural behaviour can be more efficiently designed when beam statics is used.

The calculation of peak structural response of cross-section components must consider structural function. The loading arrangement that produces the peak response in one function will not normally produce a peak response in the other functions. For example, maximum compression (longitudinal function) and maximum transverse bending (transverse function) in the bottom slab of a box girder will each be caused by two different load arrangements. Stresses resulting from the transverse function of cross-section components are normally confined to a relatively small region. The transverse function thus has a correspondingly small influence on the resistance to in-plane stresses resulting from longitudinal function. It is therefore normally sufficient to design a cross-section component for the least favourable of the following two combinations:

1. Maximum response of the longitudinal function plus permanent load response of the transverse function.
2. Maximum response of the transverse function plus permanent load response of the longitudinal function.

Stresses and sectional forces must be calculated for each of the principal structural functions of a given cross-section component. These functions are as follows:

a) Deck Slab

1. *Transverse* function as a slab, to carry its self-weight and live load to the webs. Structural response: transverse bending, longitudinal bending, and vertical shear.
2. *Transverse* function as a frame component, to resist distortion of the section. Structural response: transverse bending and vertical shear.
3. *Longitudinal* function as a tension or compression chord, to resist bending of the entire cross-section about its horizontal axis. Structural response: longitudinal axial force and in-plane shear.
4. *Longitudinal* function as a shear wall, to resist torsion of the entire cross-section. Structural response: longitudinal axial force and in-plane shear.
5. *Longitudinal* function as a web, to resist bending of the entire cross-section about its vertical axis. Structural response: longitudinal axial force and in-plane shear.

The design of the deck slab is normally governed by function (1), for which the peak response is produced by dead load plus specified wheel loads. The transverse reinforcement over the webs will therefore be fully utilized only over a short distance. There is usually sufficient reinforcement available away from this area of peak steel stress to resist the in-plane shear forces due to functions (3) and (4). The cross-sectional area of the slab is normally greater than required to resist the longitudinal compression due to function (3).

b) Webs

1. *Longitudinal* function as a web, to resist bending of the entire cross-section about its horizontal axis.

Structural response: vertical shear and longitudinal axial force. (Vertical axial forces must also be considered for the webs of bridges in which loads are superimposed onto the bottom slab.)
2. *Longitudinal* function as a shear wall, to resist torsion of the entire cross-section.
   Structural response: longitudinal axial force and vertical shear.
3. *Longitudinal* function as a tension or compression chord, to resist bending of the entire section about its vertical axis.
   Structural response: longitudinal axial force and vertical shear.
4. *Transverse* function as a frame component, to provide rotational restraint to the deck slab and (when applicable) bottom slab, as well as to resist distortion of the section.
   Structural response: in closed sections, transverse bending and shear perpendicular to the plane of the web. In open sections, torsion.

The design of the webs is governed by the least favourable of the following two combinations: maximum shear (functions (1) and (2)) plus transverse bending due to permanent load (function (4)), or maximum transverse bending (function (4)) plus shear due to permanent load (functions (1) and (2)).

c) Bottom Slab

1. *Transverse* function as a slab, to carry its self-weight and any superimposed loads to the webs.
   Structural response: primarily transverse bending, as well as vertical shear.
2. *Transverse* function as a frame component, to resist distortion of the cross-section.
   Structural response: transverse bending and vertical shear.
3. *Longitudinal* function as a compression or tension chord, to resist bending of the entire cross-section about its horizontal axis.
   Structural response: longitudinal axial force and in-plane shear.
4. *Longitudinal* function as a shear wall, to resist torsion of the entire cross-section.
   Structural response: longitudinal axial force and in-plane shear.
5. *Longitudinal* function as a web, to resist bending of the entire cross-section about its vertical axis.
   Structural response: longitudinal axial force and in-plane shear.

The design of the bottom slab is fundamentally different from the design of the deck slab. Function (1) is of little significance for the bottom slab. The design is governed by a combination of functions: shear forces due to functions (3), (4), and (5), superimposed with transverse bending from functions (1) and (2). The thickness of the bottom slab at the slab-web connection is determined from the maximum in-plane shear stress.

5.3 Analysis and Design of Cross-Section Components

5.3.1 Deck Slab

a) Fundamental Considerations

Deck slabs are exposed to a very complex environment of external influences. These include static loads, dynamic loads due to vehicle impact, and severe temperature gradients. In addition, decks may be exposed to deicing chemicals and freeze-thaw action due to possible deficiencies in the waterproofing system. The models used for the analysis and design of deck slabs are based on rather crude simplifications of loads, environmental conditions, and structural behaviour. For this reason, the design should not be based on the theoretical ultimate limit state, in which the capacity of the slab has been completely exhausted through full moment redistribution and mobilization of membrane forces. A portion of its capacity should be reserved for actions other than static loads which are difficult to include in the analysis. The sectional forces should be calculated using elastic methods, to ensure satisfactory cracking behaviour under service conditions. The additional costs of prestressing the deck slab can always be justified in terms of improved structural behaviour and durability, even for short spans.

Deck slab thickness should never be chosen as small as possible, to facilitate placement of concrete and to improve dynamic behaviour. Thin, transversely prestressed slabs require an extremely dense arrangement of reinforcement. For typical box girders with 6 m spacing between webs and 3 m cantilevers, deck slab thickness should not be less than 200 mm between the webs and 220 mm at the edge of the cantilevers. An additional thickness of about 100 mm is required at the webs, where the combined web and deck slab reinforcement is particularly dense (fig. 5.36).

Increasing the slab thickness by only 10 mm results in a considerable improvement in durability with negligible additional costs. The additional thickness increases superstructure dead load by about 2 percent. The added cost of concrete and longitudinal prestressing, combined with the small saving in deck slab reinforcement, results in a net increase in total construction cost of only 0.01 percent for bridges spanning 50 m.

Deck slabs require a relatively large amount of reinforcement. The typical consumption of steel in the deck slab is about 140 kg per m^3 of concrete or about

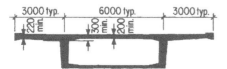

Figure 5.36
Minimum deck slab thicknesses in mm for a two-lane highway bridge

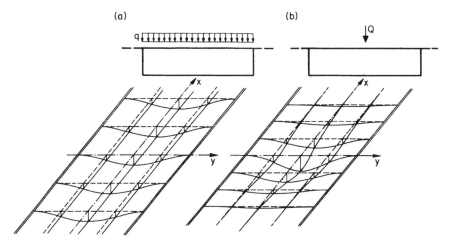

Figure 5.37
Deflected shape of a slab of infinite length fixed along two parallel edges: a, due to a uniform load over the entire slab surface; b, due to a concentrated load

65 percent of the entire mild reinforcement in the superstructure. (This figure was obtained by converting the area of deck slab prestressing to an area of mild steel of equal ultimate tensile force.) The deck slab reinforcement should therefore be designed and detailed with special care.

The deck slab can be treated as a one-way slab for most of its length. It need only be considered as a slab supported on three sides in the immediate vicinity of diaphragms, which are normally located only at abutments, piers, and internal hinges.

Loads that are uniformly distributed over the surface of a one-way slab produce a cylindrical deflected shape with no curvatures or moments in the longitudinal direction (fig. 5.37a). The slab can therefore be analysed as a beam for uniform loads. The deflected shape due to a concentrated load, however, has curvatures in the x-direction which are almost as large as those in the y-direction (fig. 5.37b). The bending moments in the x- and the y-directions thus induced will therefore also be of similar magnitude.

b) Fundamentals of Plate Theory

The theory of elastic plates is based on the following assumptions:

1. The material is homogeneous, isotropic, and linear elastic
2. The principle of plane sections applies, that is, lines normal to the middle surface of the plate remain normal to the deflected middle surface after deformation
3. Plate thickness is small relative to the other plate dimensions
4. Plate deflections are small relative to plate thickness

The equilibrium conditions will be formulated assuming Poisson's ratio is equal to zero. This simplification has no effect on the accuracy of sectional forces computed at ultimate limit state. Under service conditions, the additional steel stress due to Poisson's ratio is normally less than 5 percent of yield stress.

Normal stresses are designated σ_x and σ_y, where the subscript corresponds to the coordinate axis parallel to the stress. Tensile normal stresses are considered positive; compressive normal stresses are considered negative. Shear stresses are designated by the letter τ and two subscripts, for example τ_{xy}. The first subscript denotes the coordinate axis normal to the plane under consideration. The second subscript denotes the direction of stress. Positive shear stress is defined by the positive coordinate direction when positive normal stress acts in the positive coordinate direction. Otherwise, positive shear stress is defined by the negative coordinate direction.

When all loads are perpendicular to the middle surface, no net axial or shear forces parallel to the middle surface are produced. This phenomenon is called pure plate bending. The normal stresses σ_x and σ_y, and the shear stresses parallel to the middle surface, τ_{xy} and τ_{yx}, thus have an antisymmetrical triangular distribution (fig. 5.38).

The sectional forces are designated as follows:

Bending moments m_x, m_y
Twisting moments m_{xy}, m_{yx}
Shear forces v_x, v_y

When Poisson's ratio is neglected, the sectional forces, stresses, and deflections w of a plate of thickness h are related as follows:

$$m_x = \int_{-h/2}^{h/2} \sigma_x \, z \, dz = -D \frac{\partial^2 w}{\partial x^2} \tag{5.18}$$

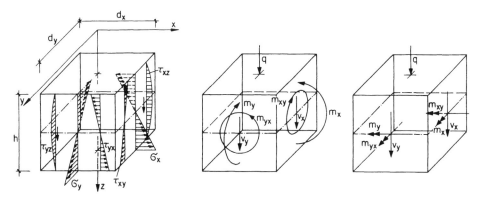

Figure 5.38
Stresses and sectional forces for pure plate bending

$$m_y = \int_{-h/2}^{h/2} \sigma_y\, z\, dz = -D\frac{\partial^2 w}{\partial y^2} \tag{5.19}$$

$$m_{xy} = \int_{-h/2}^{h/2} \tau_{xy}\, z\, dz = -D\frac{\partial^2 w}{\partial x\, \partial y} \tag{5.20}$$

$$m_{yx} = \int_{-h/2}^{h/2} \tau_{yx}\, z\, dz = -D\frac{\partial^2 w}{\partial x\, \partial y}$$

The parameter $D = Eh^3/12$ is defined as the flexural stiffness of the plate.

The above equations are analogous to the moment-curvature relationship for flexure in beams. The moments m_x and m_y are proportional to their respective curvatures, equal to the second partial derivatives of w. The twisting moments $m_{xy} = m_{yx}$ are proportional to the twist, defined by the second mixed partial derivative of w. Using these relations, a qualitative picture of the moments in the plate can be obtained from its deflected shape.

The equilibrium conditions are derived as follows for the differential plate element of figure 5.39:

$$\left(\frac{\partial m_x}{\partial x}\, dx\right) dy + \left(\frac{\partial m_{yx}}{\partial y}\, dy\right) dx - (v_x\, dx)\, dy = 0 \tag{a}$$

$$\left(\frac{\partial m_y}{\partial y}\, dy\right) dx + \left(\frac{\partial m_{xy}}{\partial x}\, dx\right) dy - (v_y\, dy)\, dx = 0 \tag{b}$$

$$\left(\frac{\partial v_x}{\partial x}\, dx\right) dy + \left(\frac{\partial v_y}{\partial y}\, dy\right) dx - q\,(dx\, dy) = 0 \tag{c}$$

Solving equations (a) and (b) for v_x and v_y, respectively, and substituting these expressions into equation (c), the following equation is obtained:

$$\frac{\partial^2 m_x}{\partial x^2} + 2\frac{\partial^2 m_{xy}}{\partial x\, \partial y} + \frac{\partial^2 m_y}{\partial y^2} = -q \tag{d}$$

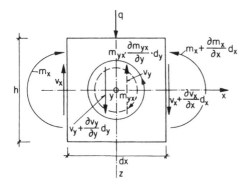

Figure 5.39
Equilibrium of a differential plate element

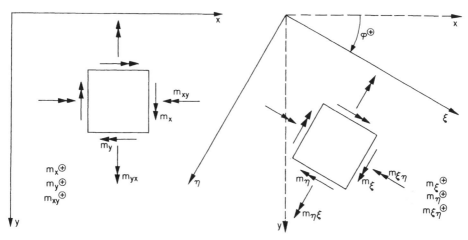

Figure 5.40
Sign convention for moments in the transformed (ξ, η) coordinate system

The differential equation for the elastic curve of the plate results from substitution
of equations (5.18), (5.19), and (5.20) into equation (d):

$$\frac{\partial^4 w}{\partial x^4} + 2 \frac{\partial^4 w}{\partial x^2 \partial y^2} + \frac{\partial^4 w}{\partial y^4} = \frac{q}{D} \tag{5.21}$$

The bending and twisting moments can be determined for a surface of arbitrary
orientation using the transformation formulas of Section 4.4.2 (fig. 5.40):

$$m_\xi = m_x \cos^2 \phi + m_y \sin^2 \phi + m_{xy} \sin 2\phi$$

$$m_\eta = m_x \sin^2 \phi + m_y \cos^2 \phi - m_{xy} \sin 2\phi$$

$$m_{\xi\eta} = \tfrac{1}{2}(m_y - m_x) \sin 2\phi + m_{xy} \cos 2\phi$$

c) Calculation of Sectional Forces

The sectional forces can be considered as the sum of two components: (1) forces
due to the loads, assuming the webs support the slab rigidly in the vertical
direction, and (2) forces due to the vertical deflections of the webs. The latter
component is only significant for multiple-cell box girders and multiple-web T
girders. The sectional forces in single-cell box girders and double-T girders can
thus be calculated assuming the webs do not displace in the vertical direction. This
makes possible the use of simplified finite element models and influence surfaces in
deck slab analysis.

d) Statical Models for Slabs on Rigid Supports

Deck slabs can be normally be modeled as plates of constant thickness, neglecting
any haunches or thickening of the cantilever tips (fig. 5.41). Equilibrium will

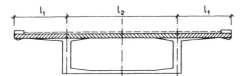

Figure 5.41
Statical model for slabs on rigid supports

always be satisfied by this simplified model. Any differences in sectional forces resulting from variation in slab thickness will be small, and can be accommodated by a slight moment redistribution at ultimate limit state. The effects of edge stiffening need only be considered when a continuous concrete parapet is used.

The geometrical boundary conditions to be used in the model depend on the type of cross-section:

1. Closed sections:
   The slab rotations at the webs are normally sufficiently small that the slab can be assumed fully fixed at this location, regardless of the type of loading (fig. 5.42).
2. Open sections:
   The geometrical boundary conditions should correspond to the type of loading and the stiffening effect of the webs. Except in the immediate vicinity of diaphragms, the torsional stiffness of the webs is normally insufficient to restrain slab rotations under uniform load. Hinge supports are therefore assumed for this case (fig. 5.43). The degree of fixity is relatively high for concentrated loads, however, since torsional moments are transferred to the webs only over a short length (fig. 5.44). Away from the immediate vicinity of the load, twist in the web is restrained by the unloaded slab. The end moments for concentrated loads can be assumed equal to about one-half of the full fixed-end moments. In all cases, the torsional moments received from the slab must be considered in the design of the webs of double-T girders.

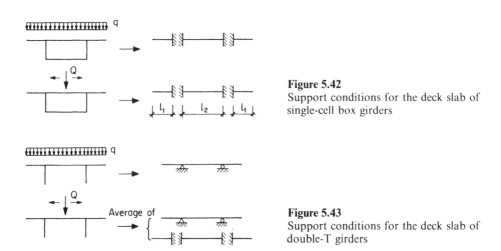

Figure 5.42
Support conditions for the deck slab of single-cell box girders

Figure 5.43
Support conditions for the deck slab of double-T girders

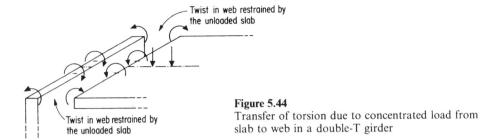

Figure 5.44
Transfer of torsion due to concentrated load from slab to web in a double-T girder

These simplified models satisfy equilibrium and are adequate in many typical cases. It may nevertheless be worthwhile to determine the degree of fixity of the deck slab more exactly, by taking into account the actual stiffness of the individual cross-section elements. Open and closed cross-sections must again be distinguished:

1. Closed sections:
   Since the lower end of the web is restrained by the bottom slab, moments transferred to the web from the deck slab cantilever can be resisted in transverse bending. In commonly used cross-sections, the flexural stiffness of the web is greater than that of the deck slab. Most of the cantilever moment will therefore be transferred to the web. The cross-section elements can be modeled as beams. The moments in the web, deck slab, and bottom slab can then be calculated according to their respective flexural stiffnesses using moment distribution (fig. 5.45).
2. Open sections:
   It can be assumed that one half of the cantilever moment is transferred to the deck slab, while the other half is transferred to the webs (fig. 5.46). The torsion in the web thus produced is transferred back to the neighbouring unloaded regions of the slab.

A more refined calculation can be made using the model shown in figure 5.47. The slab is represented by four identical beam elements. The width of the beams, denoted a, is equal to half the influence length of the given truck load at the web, assuming the load spreads out at an angle of 45°. The total cantilever moment is assumed to be divided into two equal halves, M^C, which are applied at beams 2 and 3.

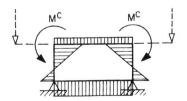

Figure 5.45
Distribution of the cantilever moment M^C to the box section components according to beam theory

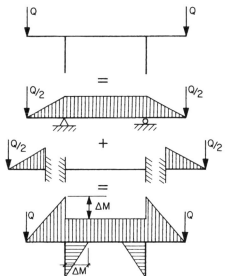

Figure 5.46
Estimation of the web moment ΔM due to live load on the cantilevers in a double-T section

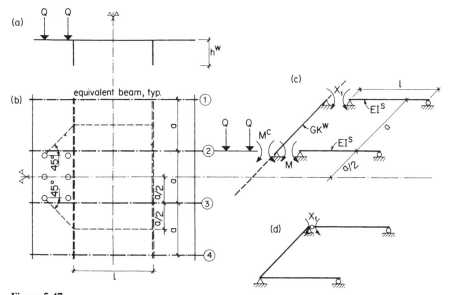

Figure 5.47
Model for the calculation of the elastic restraint of the deck slab cantilevers in an open section: a, cross-section; b, plan; c, simplified model; d, primary system with redundant moment X_1

Assuming the load is symmetrically applied, the calculation of the moments in beams 1 through 4 can be carried out for one half of the model (fig. 5.47c). The primary system is shown in figure 5.47d. The compatibility condition is formulated as

$$\delta_1 = \delta_{10} + X_1 \, \delta_{11} = 0$$

The rotations δ_{10} and δ_{11} are obtained as functions of the flexural stiffness of the beams EI^s and the torsional stiffness of the web GK^w:

$$\delta_{10} = -\frac{M^C l}{3 EI^s} \qquad \delta_{11} = \frac{2l}{3 EI^s} + \frac{a}{GK^w}$$

The redundant moment is

$$X_1 = M^C \, \frac{1}{2 + 3 \dfrac{EI^s a}{GK^w l}} \tag{5.22}$$

The moment in beam 2 is therefore equal to:

$$M = M^C \left(1 - \frac{1}{2 + 3 \dfrac{a EI^s}{l GK^w}} \right) \tag{5.23}$$

Example 5.3:
Elastic restraint of the deck slab cantilever in an open section

The following geometrical parameters are given:

Web depth	$h^w = 2.0$ m
Web width	$b^w = 0.5$ m
Slab thickness	$t^s = 0.25$ m
Interior slab span	$l = 5.0$ m

It is assumed that $a = 3.5$ m, and that $G = 0.4 E$. It therefore follows that

$$K^w = 0.28 \, (0.5)^3 \, (2.0) = 0.070 \text{ m}^4$$
$$EI^s = 3.5 \, (0.25)^3/12 = 0.0046 \text{ m}^4$$

The solution is obtained from equations (5.22) and (5.23):

$$X_1 = M^C \, \frac{1}{2 + 3 \dfrac{3.5 \, (0.0046)}{5 \, (0.4) \, (0.070)}} = 0.43 \, M^C$$

$$M = 0.57 \, M^C$$

The resulting bending and torsional moment diagram is shown in figure 5.48.

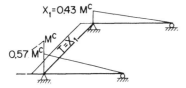

Figure 5.48
Bending moments in the equivalent beams due to the cantilever moment M^C

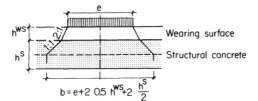

Wearing surface

Figure 5.49
Distribution of a wheel load

Structural concrete

e) Distribution of Wheel Loads

As shown in figure 5.49, wheel loads spread out in both horizontal directions at a 2:1 slope in the wearing surface and at a 1:1 slope in the structural concrete. The effective wheel load surface is measured at the middle surface of the deck slab.

f) Use of Influence Surfaces

Influence surfaces are an extension of influence lines to two-dimensional structural systems. The ordinate κ of the surface at an arbitrary point (x, y) gives the value of a sectional force or reaction at a specific location (x_0, y_0) when a unit load is applied at (x, y). Influence surfaces are represented two-dimensionally using contour lines joining points of equal influence. The parameter κ is dimensionless for moments and concentrated reactions, and has dimensions $[\mathrm{m}^{-1}]$ for shear forces.

Wheel loads are distributed over a finite area. The sectional forces they induce must therefore be evaluated as the product of the given load q and the volume V of the influence surface above the effective wheel footprint. The bending moment due to a load q [N/m²], uniformly distributed over an area A [m²] would thus be calculated using the following equation:

$$m = qV = q \int_A \kappa(x, y)\, dA \quad [\mathrm{Nm/m}] \tag{e}$$

Influence surfaces calculated from linear elastic plate theory have a singularity at the point (x_0, y_0); κ is thus infinite at the point where the sectional force or reaction is to be evaluated. The integral of equation (e) is nevertheless finite for all areas A. When the area of integration contains the point (x_0, y_0), calculations can be simplified by truncating the singularity. The value of κ is thus considered constant for all points (x, y) enclosed by a suitably chosen contour, κ_0. The error, equal to the truncated volume, decreases sharply as the value of κ_0 increases.

Published influence surfaces (Pucher 1977) normally contain contours of sufficient magnitude to enable truncation of the singularity with negligible error, provided the loaded area encloses the highest given contour. Attention must be paid to the coordinate system of the influence surface, which may be different from the system conventionally used in bridge design.

The volume V can be evaluated numerically by: (a) using an uneven number of equally spaced, parallel vertical planes, and summing the ordinates using Simpson's rule; (b) using horizontal planes cutting through the surface at the contour lines and summing using Simpson's rule; or (c) subdividing the loaded area into segments ΔA_i, determining the average ordinate of each segment $\bar{\kappa}_i$, and summing the products $\Delta A_i \bar{\kappa}_i$. Method (c) is not as exact as (a) or (b), but is most convenient and thus most commonly used. When concentrated loads are given instead of uniform loads, the summation $\Sigma \Delta A_i \bar{\kappa}_i$ is replaced by $\Sigma \Delta Q_i \bar{\kappa}_i$, where ΔQ_i is the portion of the concentrated load applied to area segment ΔA_i.

Example 5.4:
Use of influence surfaces

The moment m_x at midspan of an infinite one-way slab, fixed at both edges, is to be calculated for the single row of wheel loads given in figure 5.50. The effective footprint is 800 mm square. The wheel loads have been drawn to scale on the influence surface in figure 5.51. Due to the symmetrical arrangement of the loads, only three areas need to be considered. The calculation has been summarized in table 5.1. The value of m_x is 35.6 kN · m/m.

g) Moment Envelopes

Influence surfaces have been published for only a few selected locations, typically at the slab edges and at midspan. Moment envelopes for the entire slab can be estimated from the available influence surface moments using linear interpolation according to figures 5.52 through 5.54.

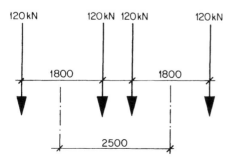

Figure 5.50
Wheel load arrangement of example 5.4
(dimensions in mm)

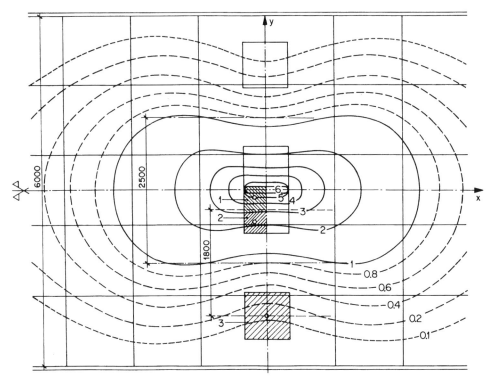

Figure 5.51
Influence surface of example 5.4 (adapted from Pucher 1977) (κ = contour/8π) (dimensions in mm)

Table 5.1
Calculation of m_x Using Influence Surfaces

Area	n	ΔQ_i (kN)	Contour$_i$	$\bar{\kappa}_i$ ($=$ Contour$_i$/8π)	$n \Delta Q_i \bar{\kappa}_i$ (kN · m/m)
1	4	30	4.9	0.195	23.4
2	4	30	2.3	0.092	11.0
3	2	120	0.13	0.005	1.2
Total $n \Delta Q_i \bar{\kappa}_i = m_x$					35.6

5.3.2 Webs

a) Web Thickness

Web thickness is governed primarily by detailing considerations and construction requirements. Webs should be wide enough to provide at least 100 mm clear spacing between tendon ducts in cast-in-place girders. This will ensure proper

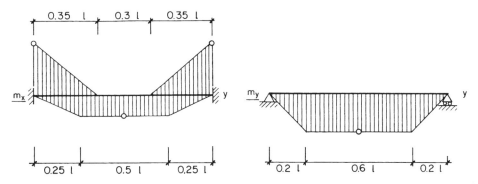

Figure 5.52
m_x and m_y moment envelopes due to concentrated loads for the interior deck slab span (circles indicate moments obtained from influence surfaces)

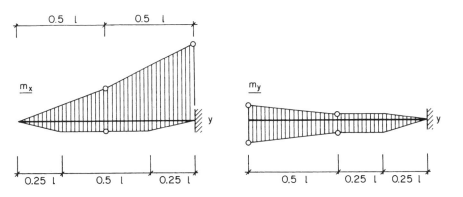

Figure 5.53
m_x and m_y moment envelopes due to concentrated loads for the deck slab cantilevers (circles indicate moments obtained from influence surfaces)

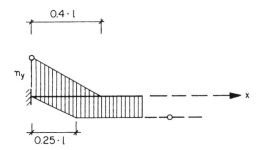

Figure 5.54
m_y moment envelope due to concentrated loads in the vicinity of a diaphragm (circles indicate moments obtained from influence surfaces, l = interior span length)

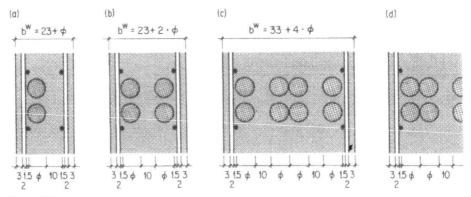

Figure 5.55
Minimum web width b^w (dimensions in cm, ϕ = tendon duct diameter): a, b, c, tendon arrangements that permit proper placement and vibration of concrete; d, arrangement to be avoided

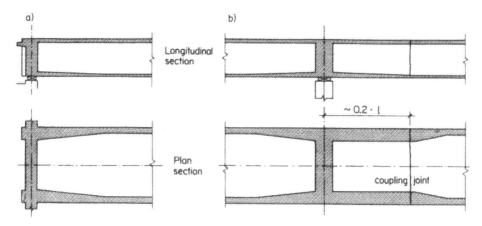

Figure 5.56
Increase in web width near supports: a, end supports; b, intermediate supports (l = girder span length)

placement and vibration of concrete (fig. 5.55). Local web thickening is often required to accommodate tendon anchors or couplers near abutments and supports (fig. 5.56).

In normal cases, web thickness is fully utilized for shear resistance only in the immediate vicinity of the supports. The minimum required web thickness b^w_{req} is obtained by setting the shear resistance of the stirrups $V_{R,s}$ equal to the shear resistance of the concrete $V_{R,c}$. (It is assumed that the given web reinforcement is fully utilized at the critical section):

$$V_{R,s} = \frac{A^w_s \, f_{sy}}{s} \, z \, (\cot\alpha + \cot\beta) \, \sin\beta \qquad \text{(equation (4.3))}$$

$$V_{R,c} = f_{c,red}\, b^w_{req}\, z\,(\cos\alpha + \cot\beta\,\sin\alpha)\,\sin\alpha \qquad\qquad \text{(equation (4.5))}$$

$$b^w_{req} = \gamma_R\,\gamma_S\,\frac{A^w_s\, f_{sy}}{s}\,\frac{\sin\beta}{f_{c,red}\,\sin^2\alpha} \qquad\qquad \text{(a)}$$

The notation of Section 4.3.3 is used throughout.

For constant $V_{R,s}$, an increase in the angle of inclination of the stirrups β will result in a decrease in required web thickness. Equations (4.3), (a) and (4.4) are used to formulate the following expression:

$$b^w_{req}(\beta) = b^w_{req}(90°)\,\frac{\cot\alpha}{(\cot\alpha + \cot\beta)\,\sin\beta}$$

which gives the required web thickness for arbitrary stirrup inclination, $b^w_{req}(\beta)$, as a function of the width required when vertical stirrups are used, $b^w_{req}(90°)$. When $\alpha = \beta = 45°$, the relation simplifies to

$$b^w_{req}(45°) = \frac{1}{\sqrt{2}}\,b^w_{req}(90°)$$

which implies a 30 percent reduction in web width.

b) Web Design for Shear and Transverse Bending

Webs must be designed to resist both longitudinal shear forces, V, and transverse bending moments, m. Simply summing the reinforcement required to resist each effect alone is not consistent with the actual behaviour of the web at ultimate limit state. The web reinforcement should rather be designed directly based on the combined effects of shear and transverse bending.

The minimum web width required to resist the factored shear force $\gamma_S\, V$ is obtained from equation (4.5):

$$b^w_{req} = \frac{\gamma_R\,\gamma_S\, V}{f_{c,red}\, z\,(\cos\alpha + \cot\beta\,\sin\alpha)\,\sin\alpha}$$

If b^w_{req} is less than the actual web width b^w, the excess capacity can be used to resist transverse bending. When the web is in a state of pure shear, the resultant concrete compressive force $V/z \cot\alpha$ is located at the web centreline. In the presence of transverse bending, this force shifts towards the flexural compression side of the web (Kaufmann and Menn 1976). In the context of the truss model used for longitudinal shear design, this phenomenon can be regarded as an inclination of the compression diagonal out of the plane of the web axis to maintain equilibrium.

The effect of shifting the resultant compressive force is illustrated schematically in figure 5.57. Figure 5.57a shows the equilibrium of a web segment in pure shear. When the compressive force is shifted to the left, equilibrium is maintained by

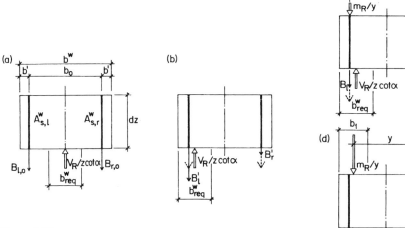

Figure 5.57
Equilibrium of a web segment: a, in pure shear; b, with resultant
compressive force shifted as far as possible to the left; c, with
small transverse bending moment; d, with large transverse
bending moment

changes in the stirrup forces (fig. 5.57 b). The superimposed moment m can then be
resisted by further changes in the stirrup tensions (fig. 5.57 c) or, when the flexural
compressive force m/y is greater than B_l', by an increase in the total concrete
compressive force (fig. 5.57 d). A simple superposition of the stresses due to shear
and the stresses due to transverse bending, without considering the shifting of the
resultant compressive force, would result in a different state of equilibrium.

Based on these considerations, a simplified model for the ultimate resistance (V_R,
m_R) of web sections subjected to shear and transverse bending can be formulated.
It is assumed that the inclination angle of the compression struts, α, and the
internal lever arm, z, have been given (fig. 5.58) and that the stirrups are vertical
($\beta = 90°$). Concrete tensile strength is neglected and the concrete compressive
stress distribution is modeled as a uniform stress block. Expressions for the stirrup
forces B_l and B_r required to equilibrate (V_R, m_R) are derived here for the two cases
shown in figures 5.57 c and 5.57 d:

1. Shear predominates (fig. 5.57 c):
    The resultant compressive force $V_R/z \cot \alpha$ is shifted as far as possible to the left.
    A net tensile force B_l remains in the left stirrup after the displacement of
    $V_R/z \cot \alpha$ and the addition of the force couple due to m. Equilibrium is achieved
    without additional concrete compressive forces. The equilibrium equations

$$\frac{V_R}{z \cot \alpha} - B_l - B_r = 0$$

$$B_r b_0 - \frac{V_R}{z \cot \alpha} \left(\frac{b_{req}^w}{2} - b' \right) - m_R = 0$$

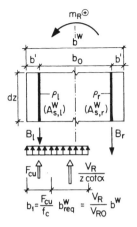

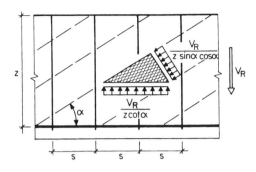

Figure 5.58
Symbols and sign convention

are solved to obtain the stirrup forces:

$$B_l = \frac{V_R}{z b_0 \cot\alpha}\left(b_0 - \frac{b^w_{req}}{2} + b'\right) - \frac{m_R}{b_0}$$

$$B_r = \frac{V_R}{z b_0 \cot\alpha}\left(\frac{b^w_{req}}{2} - b'\right) + \frac{m_R}{b_0}$$

2. Transverse bending predominates (fig. 5.57 d):
   In this case, the tensile force B_l vanishes, and the additional concrete compressive force F_{cu}, equal to the product of the design compressive strength f_c and the unknown width b_1, must be introduced to maintain equilibrium. The concrete compression due to shear must be shifted back towards the centreline of the web. The equilibrium equations are as follows:

$$\frac{V_R}{z\cot\alpha} + F_{cu} - B_r = 0$$

$$\frac{V_R}{z\cot\alpha}\left(\frac{b^w_{req} + b_1}{2}\right) + m_R - B_r\left(b^w - b' - \frac{b_1}{2}\right) = 0$$

The unknown stirrup force B_r is given by

$$B_r = \frac{m_R + \dfrac{V_R}{z\cot\alpha}\left(\dfrac{b^w_{req} + b_1}{2}\right)}{b^w - b' - \dfrac{b_1}{2}}$$

In both cases, the web reinforcement must be sufficient to resist B_r and, where applicable, B_l. The familiar equations are used: $A^w_{s,r} = s B_r / f_{sy}$ and $A^w_{s,l} = s B_l / f_{sy}$, where s denotes stirrup spacing.

Given web reinforcement $A_{s,r}^w$ and $A_{s,l}^w$, the transverse bending resistance m_R can be calculated as a function of shear resistance V_R using the relationships derived above. Alternatively, V_R can be calculated as a function of m_R. It is convenient to represent the relationships among $A_{s,r}^w$, $A_{s,l}^w$, V_R, and m_R graphically in the form of an interaction diagram.

The symbols and sign conventions shown in figure 5.58 are used in the generation of the diagrams. Shear resistance V_R is expressed as a dimensionless ratio V_R/V_{RO}, where V_{RO} is the maximum resistance of the concrete in pure shear:

$$V_{RO} = f_{c,red}\, b^w \cos\alpha \sin\alpha$$

Web reinforcement is described by the geometrical reinforcement ratios ϱ, ϱ_l, and ϱ_r, all of which are defined relative to the full web width:

$$\varrho = \frac{A_s^w}{b^w s}, \qquad \varrho_l = \frac{A_{s,l}^w}{b^w s}, \qquad \varrho_r = \frac{A_{s,r}^w}{b^w s}$$

where $A_s^w = A_{s,l}^w + A_{s,r}^w$. The ratio ϱ is expressed as a percentage of the ratio ϱ_0, defined by the following equation:

$$\varrho_0 = \frac{f_{c,red}}{f_{sy}} \sin^2\alpha \tag{b}$$

The parameter ϱ_0 corresponds to the total web reinforcement A_s^w such that $V_{R,s}$ is equal to V_{RO}. Equation (b) is obtained by substituting b^w for b_{req}^w in equation (a), assuming $\beta = 90°$.

Transverse bending resistance is expressed as the dimensionless ratio m_R/m_{RO}, where m_{RO} is the pure flexural resistance of the web with reinforcement ratio ϱ_0, assuming the stirrups are arranged symmetrically about the web centreline:

$$m_{RO} = \omega_0 (b_0 + b')^2 f_c \left(1 - \frac{\omega_0}{2}\right)$$

The ratio ω_0 is the mechanical reinforcement ratio relative to the statical depth $b_0 + b'$ of the section:

$$\omega_0 = \frac{\varrho_0}{2} \frac{b^w}{b_0 + b'} \frac{f_{sy}}{f_c} = \frac{f_{c,red}\sin^2\alpha}{2f_c} \frac{b^w}{b_0 + b'}$$

Interaction diagrams have been generated for the typical case defined by $f_{c,red}/f_c = 0.67$, $b'/b^w = 0.1$, and $\alpha = 45°$. They are given in figures 5.59 through 5.61 for $\varrho_r/\varrho_l = 1.0$, 1.5, and 2.0, respectively.

The design procedure consists first of calculating the reinforcement ratios ϱ_0 and ω_0 based on the chosen web width (b^w), reinforcement location (b'), the given

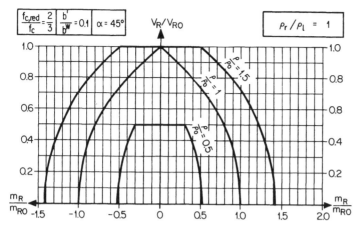

Figure 5.59
Shear-transverse bending interaction diagram for $\varrho_r/\varrho_l = 1.0$

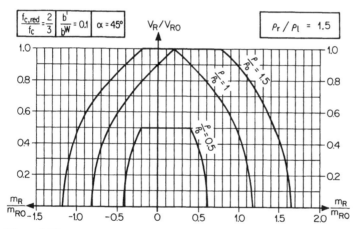

Figure 5.60
Shear-transverse bending interaction diagram for $\varrho_r/\varrho_l = 1.5$

material strengths (f_c, $f_{c,\mathrm{red}}$, and f_{sy}) and the inclination of the compression struts (α). The basic resistances V_{RO} and m_{RO} are then calculated. A point on the interaction diagram is obtained by dividing the required sectional resistance ($V_R = \gamma_R \gamma_s V$, $m_R = \gamma_R \gamma_s m$) by the basic resistances (V_R/V_{RO}, m_R/m_{RO}). The required reinforcement ratio ϱ is then read directly from the appropriate diagram. Linear interpolation can be used between the curves plotted on a given diagram and between any two diagrams.

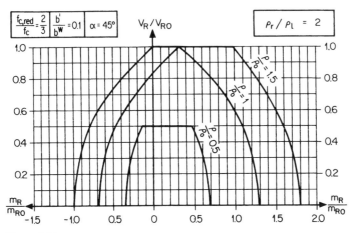

Figure 5.61
Shear-transverse bending interaction diagram for $\varrho_r/\varrho_l = 2.0$

5.3.3 Bottom Slab

The thickness of the bottom slab near the supports is determined from structural behaviour. Away from the supports, it is normally governed by detailing and construction requirements. Ductile behaviour is ensured when the axial resistance of the bottom slab, neglecting any contribution from the webs, is greater than the yield force of the girder reinforcement at the support. Conservatively chosen bottom slab dimensions in the support region will result in no significant increase in overall cost, since the moments produced by the additional weight will be small. Away from the supports, the bottom slab should be at least 0.15 m thick, to enable proper placement and vibration of the concrete, and to minimize differential shrinkage and temperature strains in the girder.

In regions of high shear, high diagonal compressive forces will be induced in the bottom slab near the slab-web junction. Haunching the bottom slab is recommended to reduce the accompanying diagonal compressive stresses. The required haunch thickness can be calculated from the state of equilibrium shown in figure 5.62. The diagonal compressive force per unit length, $\bar{D}$, is expressed as a function of shear stress at the slab-web interface, τ, and bottom-slab thickness t^{bs}:

$$\bar{D} = \frac{V}{\cos \alpha} = \frac{\tau t^{bs}}{\cos \alpha} \tag{a}$$

The diagonal compressive stress in the concrete is limited to $f_{c,\text{red}}$:

$$\frac{\bar{D}}{t^{bs} \sin \alpha} \leq f_{c,\text{red}} \tag{b}$$

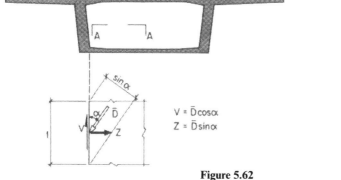

Section A-A

Figure 5.62
Force equilibrium at the bottom slab-web connection

Substitution of equation (a) into equation (b) yields the following expression for the maximum allowable shear stress at the slab-web interface, τ_{max}:

$$\tau_{max} = f_{c,red} \sin \alpha \cos \alpha \qquad\qquad\qquad\qquad (c)$$

Equation (c) is used to formulate an expression for the required bottom slab thickness in terms of the ultimate shear flow at the slab-web interface $v^* = \gamma_R \gamma_S v$:

$$t_{req}^{bs} = \frac{v^*}{\tau_{max}} = \frac{v^*}{f_{c,red} \sin \alpha \cos \alpha}$$

($f_{c,red}$ is a function of α_0, the inclination angle of the principal stresses, assuming homogeneous elastic behaviour:

$$f_{c,red} = f_c \left(1 - 0.3 \frac{\alpha_0}{45°} \right) \qquad\qquad\qquad \text{(equation (4.4))}$$

Since the normal stress in the transverse direction σ_y is very small, α_0 can be calculated from the ultimate longitudinal normal stress $\sigma_x^* = \gamma_R \gamma_S \sigma_x$ and the ultimate shear stress $\tau_{xy}^* = \gamma_R \gamma_S \tau_{xy}$, using the equation $\tan 2\alpha_0 = 2\tau_{xy}^*/\sigma_x^*$.)

The reinforcement necessary to resist the corresponding transverse tension Z is designed using the equation

$$A_s = \frac{s v^* \tan \alpha}{f_{sy}}$$

This reinforcement can be arranged according to the actual shear flow, decreasing away from the region of maximum shear. Proper anchorage of the bars into the webs is of utmost importance, since the stress in the reinforcement due to both shear and transverse bending will be highest at the slab-web interface.

To facilitate their construction, the haunches should be straight and their slope should be sufficiently flat to eliminate the need for top forms. Haunch thickness can be reduced by limiting the flexural compressive stress at ultimate limit state ($\sigma_x^* < f_c$), or by providing additional transverse reinforcement in the slab to reduce the inclination angle of the compression diagonals.

The design of the bottom slab for the combined effects of shear and transverse bending is carried out as described for the webs in Section 5.3.2.

5.3.4 Diaphragms

a) Overview

Diaphragms can be classified into several different types, according to their location in the superstructure: (1) diaphragms at abutments, (2) diaphragms at internal hinges, (3) diaphragms at piers, and (4) intermediate diaphragms.

The function of diaphragms at abutments, hinges, and piers is to transfer forces from the superstructure onto bearings or columns. Diaphragms at abutments and piers are normally proportioned rather generously, since their weight produces no moments in the superstructure.

The function of intermediate diaphragms is to stiffen the superstructure cross-section against in-plane deformation. They make possible a direct conversion of externally-applied torques into torsional sectional forces, which considerably reduces transverse bending in the section. They also contribute to the transverse distribution of loads in multiple-web cross-sections. Intermediate diaphragms are normally proportioned as thinly as possible, since they resist relatively small shear stresses and their weight produces additional moments in the superstructure.

Intermediate diaphragms are usually not necessary. Box cross-sections are sufficiently stiff to resist transverse bending without significant additional cost in both straight bridges and lightly curved bridges. Furthermore, when girder spacing is small, the deck slab can normally be relied on for adequate transverse load distribution in multiple-girder bridges. Only for curved girders with open cross-section will intermediate diaphragms be more practical than strengthening the deck slab and webs to resist transverse bending.

The use of diaphragms at abutments and piers can also be avoided by strengthening the cross-section and locating the bearings directly under the webs. In normal cases, however, it will be both practical and economical to use diaphragms at these locations.

Diaphragms are incompatible with certain mechanized construction techniques, for example the construction of double-T girders using a launching girder located between the webs. Under these circumstances, it may be preferable to cast the

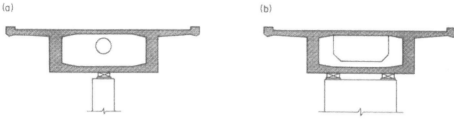

Figure 5.63
Diaphragm types: a, solid; b, frame

diaphragms after completion of the superstructure, or to eliminate them altogether. The former option is made possible by providing temporary openings in the deck slab and dowels extending from the deck slab and webs. The dowels can be bent into position after the launching girder and superstructure formwork have been advanced.

Diaphragms can be designed as solid walls (fig. 5.63 a) or as frames (fig. 5.63 b). Bending stresses predominate in frame diaphragms; shear stresses predominate in solid diaphragms. Frame diaphragms will thus require more reinforcement than their solid counterparts. Interior access and displacement of formwork in box girders is facilitated through the use of frame diaphragms. Solid diaphragms are normally provided with a small-diameter manhole. Interior formwork must therefore be removed through temporary openings in the deck slab when solid diaphragms are used.

b) Abutment Diaphragms

Abutment diaphragms generally extend over the entire width of the cross-section. They thus provide rigid support for the deck slab cantilevers and enable the proper anchorage of expansion joint hardware (figs. 5.64, 5.65). Their thickness normally lies in the 0.6 to 0.8 m range.

The bearings should be located at the intersection of the girder web and diaphragm axes. Vertical shear in the webs can thus be transferred directly to the bearings without bending or shear stresses in the diaphragm. Only horizontal shear and torsional moments from the superstructure need therefore be considered in the design of the diaphragm reinforcement.

Horizontal shear V_y results primarily from wind and earthquake loads. In continuous bridges, V_y is also produced by horizontal thermal gradients due to unequal exposure to the sun. Torsional moments T result from the antisymmetrical component of live load and differential settlement. In curved and skew bridges, additional torsional moments are produced by dead load and the symmetrical component of live load.

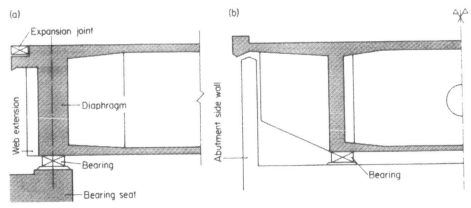

Figure 5.64
Abutment diaphragm: a, longitudinal section, b, cross-section

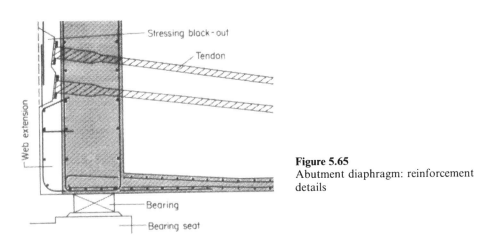

Figure 5.65
Abutment diaphragm: reinforcement details

Figures 5.66 through 5.69 show the reactions and shear flow due to V_y and T in open and closed sections. The diaphragm reinforcement can be calculated from these states of equilibrium using truss models. Solid diaphragms are normally reinforced with relatively closely spaced bars. The inclination of the compression diagonals is assumed equal to $45°$. Shear forces originating from the deck slab are directed to the lower edge of the diaphragm, from where the reinforcement transfers them to the fixed bearing. Shear forces that are applied outside the bearings produce bending in the diaphragm.

The vertical $(A_{s,v})$ and horizontal $(A_{s,h})$ reinforcement is designed at ultimate limit state using the truss model of figure 5.70:

$$A_{s,v} = A_{s,h} = s \, \frac{\gamma_R \gamma_S \, v}{f_{sy}} \tag{a}$$

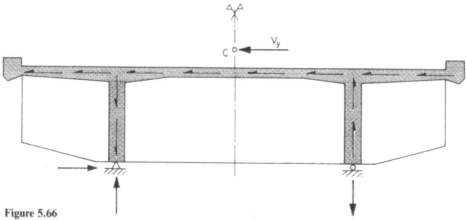

Figure 5.66
Double-T girder: shear flow and reactions due to V_y

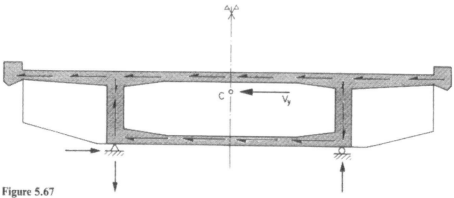

Figure 5.67
Box girder: shear flow and reactions due to V_y

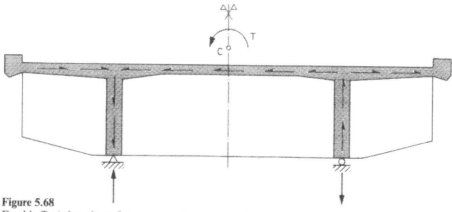

Figure 5.68
Double-T girder: shear flow and reactions due to T

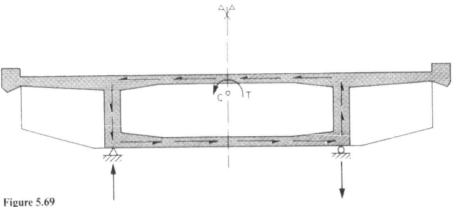

Figure 5.69
Box girder: shear flow and reactions due to T

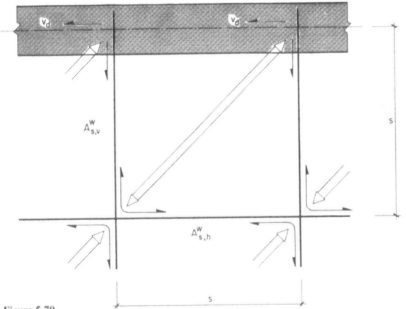

Figure 5.70
Truss model for diaphragm reinforcement

where s denotes bar spacing, and v denotes shear flow. The vertical and horizontal reinforcement must be properly anchored into the slabs and webs of the superstructure cross-section, since these bars are fully utilized at the diaphragm edges.

The diaphragm reinforcement for crack control should be designed not for the cracking load of the diaphragm, but rather for the cracking load of the superstructure. In this way, the redundant forces in the girder due to torsion and horizontal shear in the girder are considerably reduced after cracking. The minimum reinforcement in the diaphragm is thus designed for a superstructure shear flow $v = f_{ct} t$, where f_{ct} is concrete tensile strength and t is the super-structure slab or web thickness. Equation (a) is used; under service conditions, $\gamma_R = \gamma_S = 1.0$. Taking $f_{ct} = 2$ N/mm² and $t = 0.25$ m as typical values, a shear flow v of 500 kN/m is obtained. Assuming $s = 0.3$ m and $f_{ct} = 460$ N/mm², the reinforce-ment required for crack control is thus: $A_{s,v} = A_{s,h} = 330$ mm², or 16 mm diameter stirrups every 0.3 m.

c) Internal Hinge Diaphragms

Although they fulfill essentially identical functions, diaphragms at internal hinges are more thinly proportioned than abutment diaphragms. Their thickness normally lies in the 0.4 to 0.6 m range. Internal hinge diaphragms must be detailed to permit access to the bearings and to incorporate, when required, drainage for the expansion joint (fig. 5.71).

d) Pier Diaphragms

When the superstructure webs are not directly supported by the pier or bearings, the diaphragm must be designed to resist the vertical shear transferred from the superstructure in addition to transverse shear and torsion. The thickness of the diaphragm t^D depends on the type of connection between superstructure and pier:

1. *Direct support* (bearings directly under webs):

$$t^D_{\min} = 0.4 \text{ m} \quad \text{to} \quad 0.6 \text{ m},$$

depending on girder depth and span length

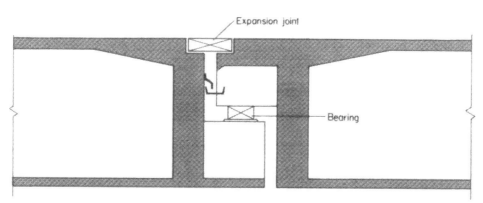

Figure 5.71
Internal hinge diaphragm

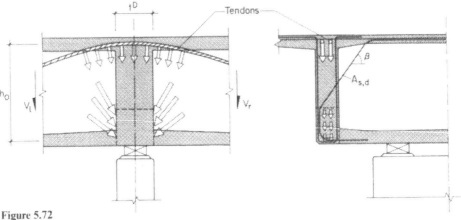

Figure 5.72
Pier diaphragm (indirect support)

2. *Indirect support* (bearings located away from webs):

$$t^D_{min} = 2b^w_{req}$$

3. Monolithic connection: t^D_{min} = column thickness

Figure 5.72 shows the forces acting in the vicinity of a diaphragm at an intermediate pier. It is assumed that, at ultimate limit state, the shear force originating from the girder webs is transferred to the lower third of the diaphragm. Reinforcement must be provided to transfer this force to the upper portion of the diaphragm above the webs. A combination of vertical stirrups, $A_{s,v}$ (fig. 5.73), and diagonal bars, $A_{s,d}$ is normally used for this purpose. Only the deviation force of the longitudinal prestressing tendons in the immediate vicinity of the diaphragm can be transferred directly to the bearings. This is accomplished by means of a

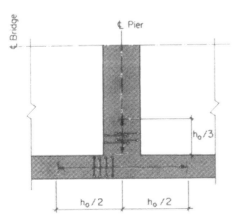

Figure 5.73
Plan section: arrangement of diaphragm reinforcement $A_{s,v}$ to transfer girder shear to the bearings

compression strut and the horizontal reinforcement in the upper portion of the diaphragm.

The factored shear force applied at the lower third of the diaphragm is

$$\gamma_s \, V(g+q) - 0.8Q(P)$$

where $V(g+q)$ is the girder shear force due dead and live load and $Q(P)$ is the deviation force of the tendon over the length h_0, centered on the pier axis. The factor 0.8 is used because the effect of $Q(P)$ is beneficial. The design of the reinforcement is based on the following inequality:

$$(A_{s,v} + A_{s,d} \sin\beta)\frac{f_{sy}}{\gamma_R} \geq \gamma_s \, V(g+q) - 0.8Q(P)$$

High bending and shear stresses result when the superstructure is supported indirectly. The amount of reinforcing steel required to resist these effects can be significantly reduced by partial prestressing. The most efficient tendon arrangement results when the anchors are located as low as possible in the diaphragm. Anchor height is normally limited by construction requirements. Since the diaphragm tendons must be stressed before removal of the falsework, the stressing anchors must be located to provide the necessary access (fig. 5.74). The sequence of longitudinal and transverse tendon stressing must be clearly stated in the construction documents. In normal cases, the diaphragm tendons are stressed before the longitudinal tendons.

The reinforcement in the web-diaphragm junction is closely spaced and must therefore be carefully detailed. Figure 5.75 shows a possible reinforcement arrangement.

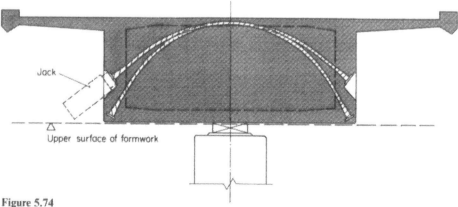

Figure 5.74
Prestressing layout for a pier diaphragm (superstructure indirectly supported)

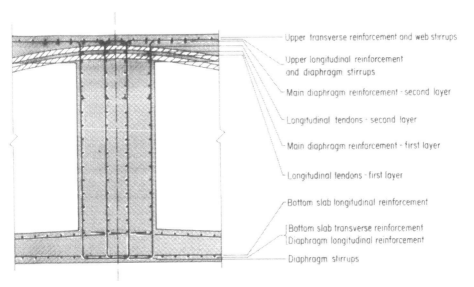

Upper transverse reinforcement and web stirrups

Upper longitudinal reinforcement
and diaphragm stirrups

Main diaphragm reinforcement - second layer

Longitudinal tendons - second layer

Main diaphragm reinforcement - first layer

Longitudinal tendons - first layer

Bottom slab longitudinal reinforcement

Bottom slab transverse reinforcement
Diaphragm longitudinal reinforcement

Diaphragm stirrups

Figure 5.75
Reinforcement arrangement at web-diaphragm connection

An unbroken flow of forces must be possible in monolithic superstructure-pier connections. The moment difference across the support is transferred to the pier through the internal forces in the diaphragm which must be in equilibrium. The vertical shear from the superstructure is converted to axial force in the pier, usually by a different flow of forces. The complete superposition of bending and axial force may therefore occur only at some distance below the girder soffit. The reinforcement at the top of the pier must therefore be designed for moment alone, without the beneficial effect of axial force. Additional discussion of the flow of forces at a monolithic girder/column connection can be found in Section 4.5.

e) Intermediate Diaphragms

Intermediate diaphragms, which complicate the interior formwork, should only be used when they result in substantial economic benefits. Their thickness is usually about 0.3 m. If intermediate diaphragms are required for transverse load distribution, the sectional forces in the diaphragm should be computed from a plane grid model of the superstructure.

Intermediate diaphragms that are used in curved bridges to convert the deviation forces $q = m_t/h_0 = M/rh_0$ into a closed shear flow are stressed in pure shear. High torques can be converted into torsional moments with relatively small shear stresses in the diaphragm. It is therefore recommended to provide interior diaphragms in curved bridges whenever transverse bending due to the introduction of torque requires special reinforcement.

References

Hetenyi, M. 1964. *Beams on Elastic Foundation*. 7th ed. Ann Arbor: University of Michigan Press.

Kaufmann, J., and C. Menn. 1976. *Versuche über Schub bei Querbiegung* (Investigations of shear and transverse bending). Institut für Baustatik und Konstruktion ETH Zürich, Bericht Nr. 7201-1. Basel and Stuttgart: Birkhäuser Verlag.

Pucher, A. 1977. *Einflußfelder elastischer Platten – Influence Surfaces of Elastic Plates*. 5th ed. Vienna and New York: Springer-Verlag.

Rüsch, H. 1972. *Stahlbeton – Spannbeton. Band 1: Werkstoffeigenschaften, Bemessungsverfahren* (Reinforced concrete – prestressed concrete. Vol. 1: Material properties, design procedures). Düsseldorf: Werner-Verlag.

6 Accessories

6.1 Bearings

6.1.1 General Remarks

Bearings are fabricated to higher standards of quality than most other bridge components and are normally furnished by firms that specialize in their manufacture. They restrict the movement of the structure in certain specified directions, while theoretically allowing it to displace and rotate freely in all others. The flow of internal forces is concentrated in the vicinity of bearings, resulting in high local stresses and tension transverse to the main direction of flow; special reinforcement must normally be provided to resist these effects. Bearings must be protected from moisture; direct contact with salt water must be prevented under all circumstances. Bearings must also inspected and maintained on a regular basis, and should thus be easily accessible. Their eventual replacement must be possible without undue cost and effort.

Bearings should be conservatively designed for the displacements and rotations which they must accommodate, since movements in excess of their capacity may have serious consequences for both bearing and structure. The possibility of extremely large displacements due to seismic or other extraordinary loads can be effectively dealt with by fitting expansion bearings with a positive stopping mechanism.

Movement in expansion bearings is always restricted by frictional forces which, although small, must nevertheless be considered in the design of structures and the detailing of bearing anchorages. The maximum coefficient of static friction is approximately 5 percent, while the minimum coefficient of dynamic friction is approximately 1 percent for expansion bearings with steel rollers or polytetrafluorethylene (PTFE)-coated sliding plates. In the calculation of ultimate loads, these nominal values should be factored by 0.8 for favourable action of the frictional forces and by 1.5 for unfavourable action.

A maximum of six independent reaction components and six independent displacements, or degrees of freedom, are possible at each bearing location (fig. 6.1). Reactions and degrees of freedom are complementary; each reaction

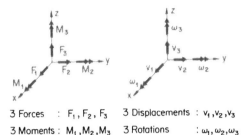

3 Forces   : F_1, F_2, F_3 3 Displacements : v_1, v_2, v_3
3 Moments : M_1, M_2, M_3 3 Rotations   : $\omega_1, \omega_2, \omega_3$

Figure 6.1
Independent reaction components and displacement degrees of freedom at a bearing

component corresponds to a restrained degree of freedom. The sum of the number
of independent reactions R and the number of independent degrees of freedom DF
is thus constant, i.e., $R + DF = 6$.

Bearings are normally classified according to their degrees of freedom. Small
rotations about the vertical axis are normally accommodated by all bearings, even
though they are not designed for such movement. The spectrum thus ranges from
the the simple rocker bearing, shown in figure 6.2, with two degrees of freedom, to
the confined elastomer or pot bearing, shown in figure 6.3, which can accommodate up to five degrees of freedom. The symbols given in figure 6.4 can be used to
represent the various bearing types on plans.

The materials commonly used for bearings include steel, concrete, and synthetic
materials such as neoprene and PTFE. Steel is well suited for simple rocker and
roller bearings. The combination of neoprene and steel has had considerable

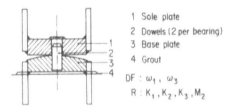

1 Sole plate
2 Dowels (2 per bearing)
3 Base plate
4 Grout

DF : ω_1, ω_3
R : K_1, K_2, K_3, M_2

Figure 6.2
Simple rocker bearing

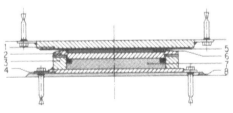

1 Sole plate   5 Stainless steel plate
2 Cover plate   6 PTFE sheet
3 Annular plate   7 Neoprene pad
4 Base plate   8 Grout

Figure 6.3
Pot bearing

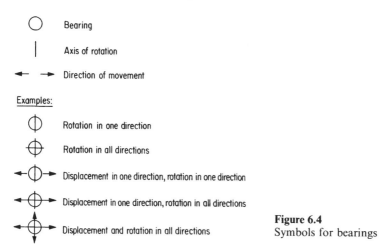

Figure 6.4
Symbols for bearings

success in recent years. The neoprene pad ensures a uniform distribution of the reaction, eliminating the need for thick steel plates. Although bearings made of neoprene alone are economical, they are normally only used for fixed bearings or for expansion bearings with small displacements. These bearings, which are very lightweight, must be installed with special care. Both all-steel and steel-neoprene bearings must be temporarily blocked before installation, in order to prevent any bearing movement during concreting operations.

Properly detailed and constructed concrete hinges have the advantage of high intrinsic durability. They can be designed to accommodate rotations of up to 0.015 radians (Leonhardt and Reimann, 1965). The proper functioning of the hinge requires that the compressive force be larger than a prescribed minimum value, and that the ratio of throat thickness to member thickness be less than a prescribed maximum value. Of particular importance is the design and detailing of the reinforcement for the transverse tension above and below the hinge. One disadvantage of concrete hinges is that superstructure and substructure are connected monolithically, thus preventing any adjustment of the bridge after its completion.

The highest standards of quality should be insisted on for the fabrication and installation of bridge bearings. The added cost, which is always small in comparison to the total construction cost, is always justified by lower maintenance costs and longer service life.

6.1.2 Structural Function of Bearings

Temperature change, shrinkage, creep, support displacements, and prestressing produce deformations in structures. When the structure is statically determinate, these deformations are not accompanied by sectional forces. When the structure is

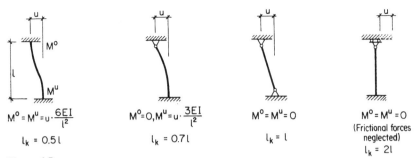

Figure 6.5
Redundant forces and effective lengths as a function of column end conditions

statically indeterminate, however, the deformations are restrained by statically redundant connections, which give rise to redundant sectional forces.

Bearings may be used to release one or more of the statically redundant connections between superstructure and substructure, thus reducing the degree of statical indeterminacy and the stresses arising due to redundant sectional forces. Any reduction in the statical indeterminacy of the system, however, is normally accompanied by a reduction in ultimate load. Bearings should therefore only be used when the saving due to the reduced redundant forces is significantly greater than the added expense resulting from the reduced ultimate load and the cost of the bearings themselves, including inspection, maintenance, and possible replacement costs.

These principles are illustrated in figure 6.5, in which the longitudinal displacements of a continuous superstructure are imposed onto relatively flexible columns. The redundant forces in the columns are a function of the degree of statical indeterminacy, i.e., of the column end conditions. Any reduction in the redundant forces due to more flexible end conditions is accompanied by an increase in the effective buckling length of the column, which will require additional expenditures to ensure adequate behaviour at ultimate limit state.

6.1.3 Superstructure Displacements

The displacements of superstructure and substructure must normally be considered in the calculation of the imposed displacements at the tops of columns and the required range of movement of expansion bearings.

Substructure displacements and rotations resulting from deformations of the foundation material can only be estimated from extensive geotechnical investigations. Long-term measurements are necessary when the soil movements occur over an extended period of time.

Bridge superstructure displacements due to temperature change, shrinkage, creep, and prestressing, however, can be determined more easily. The calculation depends on locating with reasonable accuracy the *neutral point* of the superstructure, defined as the point that remains fixed during the given deformation. The neutral point coincides with the point of direct connection of the superstructure to an immovable foundation, when such a connection exists. When the superstructure is connected to the foundations through flexible piers, however, the location of the neutral point must be calculated by considering the equilibrium of the superstructure under the imposed deformation, in which the sum of the forces due to pier displacements must equal zero. The location of the neutral point is thus a function of pier stiffness, which can only be roughly estimated due to the variability of foundation stiffness as well as cracking and creep in the columns. The calculation is therefore seldom exact for superstructures connected indirectly to their foundations.

The location of the neutral point can change during the course of construction. In the span-by-span construction of long bridges, for example, where construction begins at one abutment and proceeds towards the other, the expansion bearings can be temporarily blocked against horizontal displacement. These bearings are then released only when the construction reaches the other abutment, where the fixed bearings for the completed structure are located. Alternatively, the expansion bearings at the starting abutment can remain free to move; the location of the neutral point will then change at each phase of construction.

a) Longitudinal Displacements of the Superstructure

Longitudinal displacements in the superstructure are produced by the following actions:

1. Temperature Change:
   The effective temperature of the superstructure varies from $-20\,°C$ to $+30\,°C$ in temperate regions. Since average construction temperatures normally range between 10 and 20 °C, a temperature drop of 30 to 40 °C and a temperature increase of 10 to 20 °C are thus possible. Assuming a coefficient of thermal expansion of $10^{-5}/°C$, the design longitudinal strains due to temperature are

   $$\varepsilon_c(\Delta T) \cong -0.0004 \text{ to } +0.0002$$

2. Shrinkage:
   For a relative humidity of 60 to 80 percent, the long-term shrinkage strain in structural components of good-quality concrete, with average thickness 0.3 m, will be

   $$\varepsilon_{cs,\infty} \cong -0.00025$$

   The phenomenon of shrinkage in concrete is discussed at greater length in Section 3.1.1.

3. Prestressing:
   The strains due to prestressing consist of an elastic component and a plastic, or creep, component. The elastic strain at the centroid of the section is given by

   $$\varepsilon_{c,el}(P) = \frac{\sigma_{c,\infty}(P)}{E_c} = \frac{P}{A_c E_c}$$

   where P is the prestressing force, A_c is the area of the section and E_c is the modulus of elasticity at 28 days after the concrete has been cast. Assuming the average long-term creep coefficient $\phi_\infty \cong 2.5$, the plastic component of the total strain is

   $$\varepsilon_{cc}(P) = \phi_\infty\, \varepsilon_{c,el}(P) \cong 2.5\, \frac{P}{A_c E_c}$$

   The phenomenon of creep is likewise discussed in greater detail in Section 3.1.1.

   The elastic strain due to prestress is approximately -0.0001. This value therefore corresponds to a total long-term strain due to the combined effects of temperature, shrinkage and prestressing of $\varepsilon_{c,\infty} \cong -0.001$, or a shortening of 0.1 m in 100 m. Considerable variation from this result is, however, possible, depending on material properties, prestressing, and construction sequence.

4. Abutment displacements:
   Displacements are practically unavoidable when the abutment is founded on flexible material. The abutment can incline towards the bridge under the effect of earth pressure, or away from the bridge due to settlement of the foundation material.

5. Loads:
   Braking and acceleration loads must be considered when the superstructure is connected to the foundations by means of flexible piers. Because the resulting displacements depend to a large extent on column stiffness, their calculation is subject to considerable inaccuracy.

The computed superstructure and substructure displacements should be factored by roughly 1.4 to obtain the range of movement to be used in the design of the bearings. The resulting conservatively designed bearings will entail only an insignificant increase in total bridge cost. As stated in Section 6.1.1, a positive stopping mechanism to limit bearing displacement should always be provided, since bearing failures can be extremely costly to repair.

b) Influence of Girder Curvature

Changes in temperature and shrinkage produce a change in concrete volume, i.e. equal strains in the longitudinal, transverse, and vertical directions. Longitudinal prestressing, however, produces only a strain in the longitudinal direction. Curved

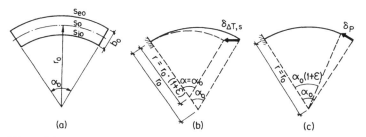

Figure 6.6
Curved girder: a, undeformed geometry; b, deformation due to temperature change and shrinkage; c, deformation due to longitudinal prestressing

girders thus exhibit qualitatively different displacements due to temperature and shrinkage on the one hand, and longitudinal prestressing on the other.

The following discussion relates to a curved girder fixed at one end. The symbols used are shown in figure 6.6a, where the subscript 0 is used to designate the geometrical quantities before deformation. In the undeformed structure,

$$s_0 = (s_{e0} + s_{i0})/2,$$

$$s_0 = r_0 \alpha_0$$

and

$$s_{e0} - s_{i0} = b_0 \alpha_0$$

It therefore follows that

$$r_0 = \frac{s_{e0} + s_{i0}}{2\alpha_0} = \frac{b_0}{2} \frac{s_{e0} + s_{i0}}{s_{e0} - s_{i0}}$$

Likewise, after deformation,

$$r = \frac{s_e + s_i}{2\alpha} = \frac{b}{2} \frac{s_e + s_i}{s_e - s_i} \tag{6.1}$$

and

$$\alpha = \frac{s}{r} \tag{6.2}$$

For the deformation due to temperature change and shrinkage, shown in figure 6.6b, $b = b_0(1 + \varepsilon)$, $s_e = s_{e0}(1 + \varepsilon)$ and $s_i = s_{i0}(1 + \varepsilon)$. Substituting these expressions into equations (6.1) and (6.2), the following equations are obtained:

$$r = \frac{b_0(1 + \varepsilon)}{2} \frac{s_{e0}(1 + \varepsilon) + s_{i0}(1 + \varepsilon)}{s_{e0}(1 + \varepsilon) - s_{i0}(1 + \varepsilon)} = r_0(1 + \varepsilon)$$

and

$$\alpha = \frac{s}{r} = \frac{s_0(1+\varepsilon)}{r_0(1+\varepsilon)} = \frac{s_0}{r_0} = \alpha_0$$

For the deformation due to prestressing (fig. 6.6c), $b = b_0$, $s_e = s_{e0}(1+\varepsilon)$ and $s_i = s_{i0}(1+\varepsilon)$. Thus

$$r = \frac{b_0}{2} \frac{s_{e0}(1+\varepsilon) + s_{i0}(1+\varepsilon)}{s_{e0}(1+\varepsilon) - s_{i0}(1+\varepsilon)} = r_0$$

and

$$\alpha = \frac{s}{r} = \frac{s_0(1+\varepsilon)}{r_0} = \alpha_0(1+\varepsilon)$$

Fixed bearings can be used at both ends of sharply curved bridges. Although temperature change, shrinkage, and prestressing will produce moments about the vertical axis in this arrangement, the resulting stresses will be low provided the curvature is high. This arrangement is particularly suited to sharply curved on-ramps and off-ramps, since complicated expansion joints can be eliminated at both ends of the bridge.

6.1.4 Bearing Layout

Bearings must be arranged to ensure a direct and positive transfer of the horizontal forces due to wind, earthquake, braking, acceleration, and centrifugal forces to the piers and abutments. Bearings provided primarily to resist vertical loads are installed level, so that no horizontal dead load forces are transferred from superstructure to substructure. When longitudinal forces are transferred directly to one of the abutments, the fixed bearings are best provided at the lower of the two ends of the bridge, so that the longitudinal drainage pipes in the structure can be connected to the abutment drainage system without expansion joints. Large horizontal forces should not be resisted by the bearings used for vertical loads, but rather by special bearings designed for this purpose only.

The difference in strain transverse to the bridge axis between superstructure soffit and bearing seat is normally very small. Since some transverse displacement can be accommodated by all bearings, the transverse strain difference can be neglected for bearing spacings of up to 10 m.

Bearings must always be arranged to permit easy access for inspection, maintenance, and replacement. A horizontal bearing seat and, under certain circumstances, projections in the superstructure will greatly simplify the installation of jacks.

Figures 6.7 and 6.8 show suggested bearing seat designs for superelevations and longitudinal grades, respectively. As a rule, projections are preferable to recesses when a horizontal surface is required.

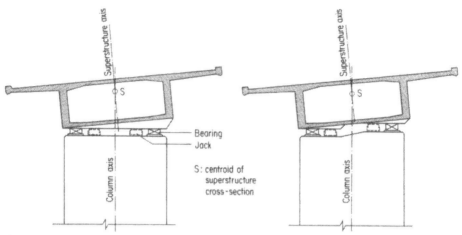

Figure 6.7
Bearing seat arrangements for superelevated bridges

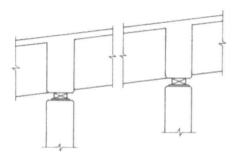

Figure 6.8
Bearing seat arrangement for bridges on
longitudinal grades

6.2 Expansion Joints

The design of expansion joints has many special difficulties. Expansion joints in
highway bridges are a relatively hard component relative to the much softer
asphalt wearing surface. As a result, asphalt is often rolled away from either side
of joints by the impact of passing wheels. This produces an uneven surface which
reduces rider comfort and subjects both the connection of waterproofing and the
anchorage of the joint to very high stresses. Large losses of wearing surface near
the joint can be especially dangerous if snow plows operate on the bridge;
extensive damage can result if the plow blade gets caught in the joint. Cracks in the
waterproofing membrane, which expose the joint anchorage to moisture, can
occur even when the asphalt remains reasonably intact. Whenever possible,
therefore, expansion joints for highway bridges are located only at the abutments,
where the deck slab can be easily thickened and reinforced to accommodate a
heavy-duty joint assembly; proper joint drainage can likewise be assured without
major difficulty. Expansion joints should be located away from stress concentrations due to reactions or prestressing anchors.

Expansion joints at internal hinges must normally, however, be located in a zone of introduction of force. The length of the overlapping hinge cantilevers, and hence the allowable joint displacement, are limited by the girder depth. The proper design and detailing of internal hinges is very difficult; for this reason, internal hinges should be avoided wherever possible.

Expansion joints are not normally required at abutments with fixed bearings, where the girder can be extended over the abutment back wall and backfilled directly.

The length of superstructure between expansion joints depends on whether the bridge carries highway or railway traffic. In the case of highway bridges, the proper detailing of expansion joints is difficult. Relatively large joint displacements can, however, be accommodated without undue problems. The opposite is true for railway bridges, where the proper detailing of a waterproof joint under the ballast presents no particular difficulties; large joint displacements, however, require relatively complicated and maintenance-prone rail expansion joints (figs. 6.9 and 6.10). The length between joints should therefore be smaller for railway bridges; the values given in table 6.1 can be used for preliminary design.

The actual expansion joint hardware is normally fabricated by specialty firms. Heavy-duty steel joints have performed better than lighter, less expensive joints. Because of the high dynamic stresses due to wheel impact, care must be taken in the design of corbels used to support the joint, as well as anchorages and the connection of the waterproofing membrane.

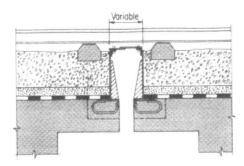

Figure 6.9
Railway bridge expansion joint: normal arrangement

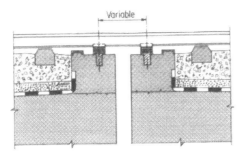

Figure 6.10
Railway bridge expansion joint: arrangement for large displacements

Table 6.1
Recommended Maximum Bridge Length Between Expansion Joints (m)

Location of Neutral Point	Highway Bridges	Railway Bridges
	Between abutment hinges	
At an Abutment	600	200
Away from Abutments	1000	300
	Between internal hinges[a]	
Away from Abutments	500	200

[a] These lengths apply only for bridges with superstructure depth ≥ 2.5 m.

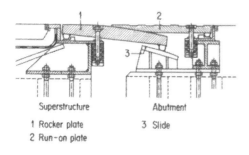

Superstructure   Abutment

1 Rocker plate   3 Slide
2 Run-on plate

Figure 6.11
Rocker-plate joint

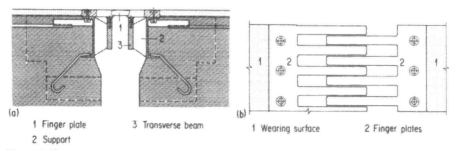

(a) 1 Finger plate   3 Transverse beam
    2 Support

(b) 1 Wearing surface   2 Finger plates

Figure 6.12
Finger joint: a, section; b, plan

Open expansion joints, shown in figures 6.11 and 6.12, must be provided with a drainage system that is accessible and easy to maintain. The possibility of water leakage must also be considered in the design of sealed expansion joints, shown in figure 6.13. A well detailed channel in the bearing seat is usually sufficient for the drainage of sealed expansion joints.

The direction of movement of the expansion joint must be compatible with that of the bearings. Wherever possible, the displacement of expansion bearings is set perpendicular to that of the expansion joint, thus facilitating joint design.

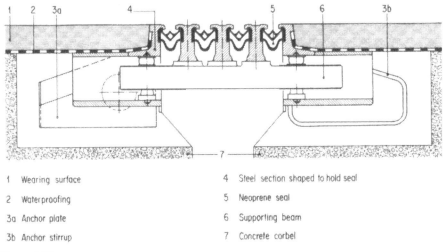

1	Wearing surface	4	Steel section shaped to hold seal
2	Waterproofing	5	Neoprene seal
3a	Anchor plate	6	Supporting beam
3b	Anchor stirrup	7	Concrete corbel

Figure 6.13
Modular sealed expansion joint

6.3 Drainage and Anchorage of Guardrails

6.3.1 Drainage

The drainage system should carry water off the bridge deck quickly and completely. Drainage is necessary not only for the safety of motorists, but also to allow the wearing surface to dry out as fast as possible. Water removed from the bridge deck should normally flow into a storm sewer system. Water that can discharge freely under the bridge can result not only in damage to objects underneath the bridge and to the bridge itself, but also in contamination of watercourses. Drainage inlets and pipes must be made of materials that resist abrasion and the effects of chlorides. Cast iron is normally chosen for drainage inlets, while a tough plastic, for instance high-density polyethylene (HDPE), is used for the pipes.

Regardless of the intensity of rainfall, drainage inlets should be located on both edges of the deck slab and spaced about 15 m apart along the length of the bridge, so that water can be carried away as quickly as possible. Providing inlets at the upper edge of a superelevated deck prevents the flow across the bridge deck of water from melting snow. Drainage inlets normally consist of three components: a basin, a frame, and a grate. The basin is provided with a flange which supports the frame and to which the waterproofing membrane can be attached. The height of the frame is detailed to match the depth of the wearing surface.

The drainage pipes should be inclined at at least a 2 percent slope. The calculation of the required pipe diameter is based on the intensity of precipitation. Transverse pipes normally require a thickening of the deck slab cantilevers. It is recommended that these pipes be placed inside ducts cast into the concrete to enable their replacement. The transverse pipes can, alternatively, be located in the open under the deck slab cantilevers if the visual appearance of the underside of the bridge is of secondary importance. For ease of maintenance and aesthetic reasons, longitudinal pipes should always be located inside box girders or between the webs of T-girders. Stronger pipes should be used for slab bridges, since the entire drainage system must be cast into concrete and thus cannot be replaced. Drainage pipes should be regularly maintained. For this purpose, clean-out vents should be provided at the edge of the deck at 50 to 70 m spacing, as well as at all intersections of transverse and longitudinal pipes.

6.3.2 Anchorage of Guardrails

Guardrail posts are usually attached to bridges using prefabricated, corrosion-resistant anchor bolts cast into the edge of the deck slab cantilevers using a template which matches the bolt holes in the base plate of the post. The diameter of the anchor bolts should be at least 16 mm; they should be capable of developing the ultimate resistance of the posts so that only the posts need be replaced after an accident. Guardrail posts that are grouted into holes in the deck slab have, in general, not performed well and are not recommended.

6.4 Waterproofing and Wearing Surfaces

The design of the covering layer of the deck slab, which normally consists of a waterproofing membrane and an asphalt concrete wearing surface, is of fundamental importance to the durability of bridges. The waterproofing membrane protects the surface of the deck slab concrete against freeze-thaw action and deicing chemicals. Deck waterproofing is almost always indispensible, since the deck slab concrete, worked by hand, rarely has the minimum surface impermeability necessary to guarantee resistance to freeze-thaw action in the presence of deicing chemicals and to limit the penetration of critical chloride concentrations (0.4 percent of cement by weight) to a depth of 20 mm.

The wearing surface is the outer protection layer of the deck slab; its role is to protect the relatively fragile waterproofing layer and reduce temperature variations in the concrete surface.

An effective waterproofing-wearing surface system for highway bridges can be detailed in one of the three following ways:

1. *Partially Bonded System* (fig. 6.14a). The actual waterproofing membrane consists of a bituminous layer, approximately 25 mm thick, which is applied in liquid form over a layer of asphalt-impregnated felt or glass fibres which separates the membrane from the deck slab. The separation layer must be vented at regular

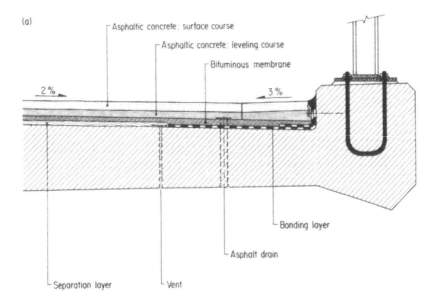

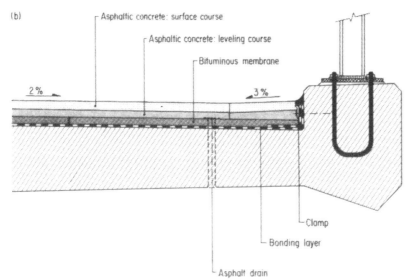

Figure 6.14
Arrangement of waterproofing and wearing surface at deck slab edge: a, partially bonded
waterproofing; b, fully bonded waterproofing

intervals through the deck concrete to prevent the buildup of vapour pressure underneath the membrane. The membrane is bonded to the deck slab only at the edges of the deck and at connections to expansion joints and drainage inlets. This is accomplished by reinforced bituminous material or elastomeric sheets, 0.5 to 1 m wide, applied with a hot adhesive. The concrete at these locations must be thoroughly cleaned by water blasting and allowed to dry before the edge waterproofing is applied. The bituminous membrane is extended over the bonded edge waterproofing to the edges of the deck slab. Bonded waterproofing should also be placed across the deck near the grout vents of prestressing tendons. The asphaltic concrete wearing surface should consist of a both a leveling course and a surface course, with a total thickness of at least 60 mm.

This system is relatively simple and economical. Salt water which penetrates through a leak in the unbonded bituminous membrane can, however, reach far beyond the location of initial penetration. The extent of the damage can be limited by partitioning the deck slab into several sections using transverse strips of bonded waterproofing. Any damage to the bituminous membrane can normally be detected by the presence of water at the vapour vents.

2. *Fully Bonded Waterproofing* (fig. 6.14 b). This system consists either of a bonded membrane, 1 to 3 mm thick, which is applied in liquid form to the deck concrete by spraying or screeding, or of a hot-glued elastomeric sheet. The concrete surface must first be prepared by waterblasting and allowed to dry out. Any blisters which may occur as the membrane is applied should be pierced and repaired. A few such small defects are not serious in this system, since water which penetrates the membrane is prevented from spreading to large areas of the deck. The wearing surface consists of a leveling course, preferably liquid-applied bitumin, and an asphaltic-concrete surface course, for a total thickness of about 80 mm. A fully-bonded system is relatively expensive and difficult to apply. When properly built, however, it is the most effective and practical waterproofing available.

3. *Concrete Wearing Surface*. A properly designed and constructed concrete wearing surface is extremely durable and provides reliable protection to the structural concrete underneath. An additional thickness of 80 to 100 mm is normally used. Since the spacing of cracks in the wearing surface will be approximately equal to its thickness, shrinkage will produce only microcracks of width less than 0.05 mm. Proper bond between the wearing surface and the structural slab is necessary to prevent the penetration of water to the structural concrete. For this reason, the deck slab must be waterblasted, maintained in a moist condition and coated with a thin layer of epoxy mortar before the wearing surface is cast using special machines. The bond between the structural concrete and the wearing surface is further enhanced through the use of dowels, which can be anchored into the hardened structural slab.

Concrete wearing surfaces must be sufficiently impermeable to resist freeze-thaw action in the presence of deicing chemicals and to restrict the penetration of critical chloride concentrations to no more than 50 mm below the surface. Impermeability results from the choice of aggregates, low water-cement ratio, air entrainment, proper placement, and prolonged curing. The edges of the wearing course must be treated with special care. Joints, which are sprayed with a waterproofing chemical and filled with epoxy, are normally provided between the wearing surface and curbs, expansion joint hardware and drainage inlets.

Reference

Leonhardt, F., and H. Reimann. 1965. *Betongelenke* (Concrete hinges). Deutscher Ausschuß für Stahlbeton, vol. 175. Berlin: Wilhelm Ernst & Sohn.

7 Design and Construction of Special Bridge Types

7.1 Overview

The most common types of prestressed concrete bridges are:

1. Girder bridges
2. Arch bridges
3. Frame bridges
4. Slab bridges
5. Cable-stayed bridges

The suitability of a particular type of bridge depends on many different factors, including topography, geotechnical conditions, height, clearance, and method of construction.

The most common type of concrete girder bridge has a prismatic, cast-in-place superstructure and is built on a straight alignment with no skew supports. This type of bridge will be referred to as *conventional*, even though it may be built using unconventional, mechanized construction procedures.

Conventional cast-in-place girder bridges are best suited for short and medium spans. Falsework accounts for a substantial portion of their total construction cost. The cost of conventional scaffolding supported directly from the ground is primarily a function of topography, geotechnical conditions, height, and clearance under the bridge. Intermediate supports can be eliminated when mechanized falsework is used; falsework costs are then independent of conditions below the bridge. Although mechanized systems are expensive, they can lead to lower construction costs for long bridges built in difficult terrain. When the alignment is straight or has constant curvature, the incremental launching method can be used.

Other types of girder bridges can be distinguished according to peculiarities of design and construction. The categories, which are not mutually exclusive, include:

1. Girder bridges with precast elements
2. Cantilever-constructed girder bridges
3. Skew girder bridges
4. Curved girder bridges

Precast beams are especially economical for long bridges with many short spans. Girders ranging from 25 m to 35 m in length can be easily erected by a self-propelled crane, provided they can be trucked directly to their proper location. Erection of girders in the 40 m to 50 m span range normally requires a launching girder, the high cost of which can be justified only for bridges of great length. The deck slab is at least partially constructed of cast-in-place concrete. The durability and ride quality of this type of bridge is substantially improved by making the individual spans continuous over the piers.

Precast segmental girder bridges are normally prestressed with external, un-bonded tendons. One of the principal advantages of this type of bridge is speed of construction.

Cast-in-place girders built using the balanced cantilever method are appropriate for long, high spans and difficult terrain. Falsework costs are independent of span length and bridge length. Construction time can be shortened relative to conventionally constructed bridges, since the girder can be built out from several piers simultaneously. The cantilever method can also be used for precast segmental spans up to 100 m in length. Because it requires expensive erection equipment, precast segmental cantilever construction is only economical for long bridges.

Skew bridges are used to achieve the shortest spans for crossings at angles of other than 90 degrees. They can be properly arranged into existing topography and result in the best clearance. Their construction, however, is always more difficult. In addition, the long, complicated abutments of short skew bridges can be expensive.

Arch bridges are economical only under special topographic and geotechnical conditions. In V-shaped valleys, for example, they may be preferable to girder bridges, which require costly tall piers of dubious aesthetic value. Arches can be built on conventional falsework or using cantilever construction; both methods are expensive. The high cost of the arch can be partially offset, however, by savings achieved in the approaches. Short spans and the use of mechanized falsework are recommended in this regard.

Frame bridges are normally built only for small spans and are particularly suitable for grade separations. Economy is difficult to achieve with long-span inclined-leg frames, which require expensive falsework; conventional girder bridges are usually preferable in spite of additional column height. In special situations, such as low-level river crossings, two-hinged frames may be feasible for spans up to 60 m.

Slab bridges can be constructed very simply and are economical for spans of up to roughly 30 m. This type of bridge is appropriate for grade separations where the available superstructure depth is limited. Solid slabs are particularly suitable for skew crossings, since reinforcement can be be arranged without difficulty in the directions of principal stress.

The feasible span range of concrete cable-stayed bridges extends beyond the current record of 440 m (see Chapter 1). The most economical arrangement has proven to consist of a slender girder suspended by two planes of closely spaced cables, which results in small bending moments in the girder. The ratio of span length to bridge width should be as large as possible, and should not be less than 12:1. The overall stability of the system and the resistance of the cables to fatigue and corrosion must be carefully investigated in the design. Cable-stayed bridges can be simply built using cantilever construction. The construction costs of cable-stayed bridges are primarily a function of span arrangement, tower height, and cable arrangement.

7.2 Conventional Cast-in-Place Girder Bridges

7.2.1 Conceptual Design

Girder bridges that are built non-segmentally should have constant depth over their entire length to reduce falsework and formwork costs. This type of bridge is economical for spans of up to roughly 80 m in length. An efficient use of materials and a simple layout of prestressing steel result from choosing span lengths to minimize the difference between the moment diagrams of any two adjacent spans.

Girder depth is determined by economic and aesthetic considerations and may also be influenced by clearance requirements. The ratio of span length to girder depth, l/h, should be chosen between 12 and 35.

The most economical girder depth corresponds to a ratio of span to depth of roughly 15. Greater depths often appear heavy, particularly when the bridge is low. Values of l/h ranging from 13 to 15 have nevertheless been successfully used for incrementally launched bridges, which require a greater internal lever arm to resist large stress reversals during launching. The deeper girder does not normally detract from the visual slenderness since most bridges of this type are high. The savings in falsework and formwork obtained by increasing l/h are normally outweighed by increases in the consumption of prestressing steel. The net increase in cost, however, is small in comparison with the improved appearance resulting from a more slender bridge profile. Raising l/h from 15 to 20, for example, increases total construction cost by roughly 5 percent. For this reason, the ratio of span to depth is often chosen between 17 and 22, which is slightly higher than the most economical value.

The increase in cost associated with values of l/h between 25 and 30 is, however, substantial and can only be justified when required for clearance under the bridge. Haunching one span is preferable to reducing the depth of the entire bridge when only one span is affected by the clearance requirement. Slender bridges, which are susceptible to large deformations and dynamic effects, should only be considered when spans are short.

Additional discussion of the effect of girder depth on visual appearance can be found in Section 2.3.

For bridges of up to 300 m in length, horizontal forces in the longitudinal direction originating from the superstructure (due to braking, acceleration, wind, and earthquake) should be transferred directly to a rigid foundation. This is best accomplished by providing fixed bearings at one of the abutments. Durability is enhanced by this arrangement, which requires only one expansion joint. Superstructure displacements due to creep, shrinkage, and temperature become too large to accommodate at one end of the bridge when total length exceeds 300 m. It is thus necessary to provide expansion bearings at both abutments and to transfer the longitudinal horizontal force from girder to foundation through flexible piers. Where there is no risk of foundation settlement, it is practical to connect several piers monolithically to the superstructure. Special care must be taken in detailing the reinforcement connecting the girder to stiff columns, where large moments are transferred. When foundation settlement is likely, the superstructure can be completely detached from the substructure using bearings at every pier. This arrangement enables the subsequent correction of unacceptably large settlements by jacking the superstructure.

7.2.2 Design of the Cross-Section

The form of the cross-section is primarily a function of the following factors:

1. Deck width and girder depth
2. Construction sequence
3. Arrangement of longitudinal prestressing
4. Use of space between the webs
5. Aesthetic considerations

Figures 7.1 through 7.9 show examples of typical cross-sections from bridges constructed in Europe since 1958. (Dimensions are given in m and cm.)

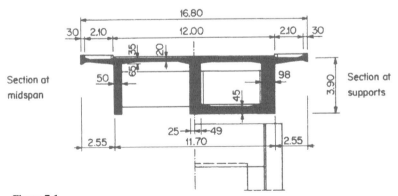

Figure 7.1
Weinland Bridge, Switzerland (built 1958): main span 88 m, span to depth ratio 22.6, span by span construction on conventional falsework

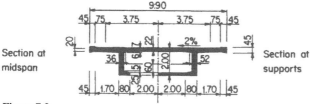

Figure 7.2
Quadinei Bridge, Switzerland (built 1967): main span 40 m, span to depth ratio 20.0, span by span construction on conventional falsework

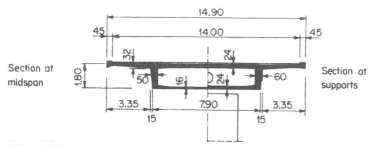

Figure 7.3
Pregorda Bridge, Switzerland (built 1973): main span 36 m, span to depth ratio 20.0, span by span construction on conventional falsework

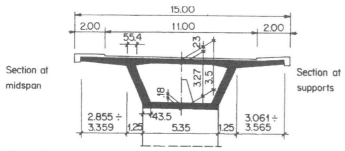

Figure 7.4
Krebsbachtal Bridge, Germany (built 1975): main span 45 m, span to depth ratio 12.9, incrementally launched with segment length 22.5 m

The two principal types of cross-section used for cast-in-place girders are hollow boxes and T-sections. Box girders are stiffer and stronger in flexure and torsion than T-girders of equal depth. (The strength of T-sections can, however, be improved through the use of a partial bottom slab at the supports, thus increasing the width of the compression block.) Box girders are also somewhat simpler to construct. In general, therefore, hollow boxes are more economical than T-sections. An additional advantage of box girders is that the danger of water freezing on the roadway surface is reduced by the heat retained inside the box.

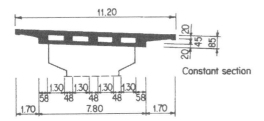

Figure 7.5
Grade separation, Zizers, Switzerland (built 1958): main span 32.5 m, span to depth ratio 38.2, built in 3 stages on conventional falsework (two frame systems, one suspended beam)

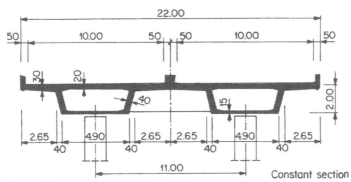

Figure 7.6
Grenzbrücke (Boundary Bridge), Basel, Switzerland (built 1980): main span 35.4 m, span to depth ratio 17.7, span by span construction on conventional falsework

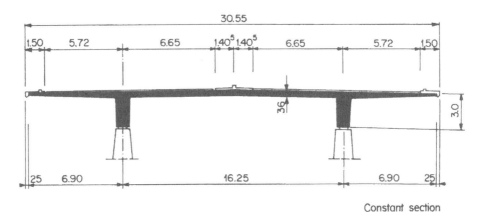

Figure 7.7
Vorland Bridge, Germany (built 1968): main span 39 m, span to depth ratio 13.0, span by span construction on launching girder

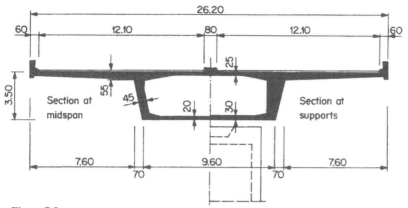

Figure 7.8
Felsenau Bridge approach spans, Berne, Switzerland (built 1975): main span 48 m, span to depth
ratio 13.7, span by span construction on conventional falsework

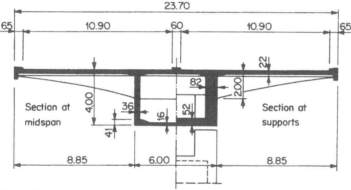

Figure 7.9
Lake Gruyère Viaduct, Switzerland (built 1978): main span 60.48 m, span to depth ratio 15.1,
span by span construction on launching girder

Access to and use of the area between the webs of T-girders requires scaffolding
suspended from the top slab. The cost of such a system is usually no less than the
cost of a bottom slab. Box sections are therefore preferable, for example, when a
large number of pipes must be installed between the webs.

In general, the number of webs should be as small as possible. Flexural strength
and stiffness in the longitudinal direction increase with increasing radius of
gyration ($\sqrt{(I/A)}$), which can be maximized for a given depth by minimizing the
total width of the webs. It is not possible, however, to design arbitrarily thin webs.
As discussed in Section 5.3.2, the minimum width of an individual web is
determined by the diameter of reinforcing bars and longitudinal prestressing
tendons, specified clear cover, and the required spacing between tendons. Total
web width must therefore be minimized by reducing the number of webs.

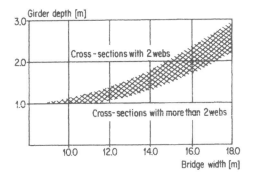

Figure 7.10
Estimation of number of webs according
to girder depth and bridge width

The minimum number of webs is normally governed not by longitudinal shear, but by transverse bending in the top slab. Webs that are too widely spaced result in transverse reinforcement costs that outweigh the associated savings in concrete, longitudinal reinforcement, and formwork. The feasibility of cross-sections with two webs can be assessed for particular combinations of bridge width and girder depth using figure 7.10.

The greatest ease of construction is achieved through the use of cross-sections with two webs. The entire reinforcement for both webs and, where applicable, the bottom slab can be placed without difficulty after the construction of the exterior forms. Two webs may nevertheless be insufficient in certain cases, for example when girder depth is limited and all tendons must be anchored at one section without thickening the webs.

Medium and long-span multiple-lane highway bridges can have either one wide cross-section with two webs or two narrower cross-sections. The ease of construction of a single wide section results in savings that far outweigh the increased transverse reinforcement required. Falsework costs can be reduced by casting the deck slab cantilevers after completion of the box using a form traveller. The use of single piers, which are more effective than twin piers in resisting lateral loads, is also made possible. A single wide cross-section appears narrower than twin sections of equivalent total width when viewed from below, resulting in a better appearance, particularly for high bridges. Wide sections appear slender even for relatively low ratios of span length to girder depth, due to the shadows cast by the wide cantilevers onto the webs. Furthermore, transparency is improved through the use of single piers.

The main disadvantage of single wide cross-sections is that the entire bridge must be closed during major repairs or demolition. Furthermore, single cross-sections are not recommended for low-level multiple-lane highway bridges, since the savings in falsework and pier costs are considerably smaller and an acceptable degree of visual slenderness cannot be achieved with economical values of l/h. Twin parallel box girders with continuous deck slab or double-T girders should be used instead.

7.2.3 Prestressing Concepts

a) Full Prestressing for Permanent Load

Prestressing designed according to this concept must be capable of preventing tensile stresses due to service dead load plus prestressing at all points along the girder after deduction of all losses. The prestressing force obtained in this way should be increased if deformations in the structure due to permanent load exceed allowable limits. Mild reinforcing steel is designed to ensure that the combined resistance of reinforcing and prestressing steel is adequate at ultimate limit state. The reinforcing steel provided must be greater than the minimum reinforcement required to control cracking due to restrained deformations and self-equilibrating stresses.

Moments in excess of the cracking moment will still be possible under the combined effects of permanent load and restrained deformations. The cracking moment will only be exceeded, however, by a small amount. The resulting cracks will be narrow and well distributed, provided the structure has been adequately reinforced for crack control.

b) Full Prestressing for Permanent Load only in the Support Region

Prestressing designed according to this concept need only prevent tensile stresses in the support regions of the girder due to dead load and prestressing, after deduction of all losses. This concept achieves greater economy by eliminating the restriction on permanent load stresses in the bottom slab, which is often better protected from exposure to deicing chemicals. Mild reinforcing steel is designed as outlined for Concept (a). A stabilized crack pattern can be produced when the cracking moment is exceeded by a significant amount under permanent load.

This concept normally leads to an economical use of prestressing steel and can easily be implemented by using tendons that overlap at the supports. As a result of the relatively weak prestressing in the midspan region, large deformations are possible for long, slender spans. This concept is therefore best suited to spans less of than 40 m.

7.2.4 Preliminary Design

An initial estimate of the reinforcing and prestressing steel in a continuous girder can be obtained from the behaviour of the system at ultimate limit state. Flexural resistance is computed using the equation

$$M_R = (A_s\, f_{sy} + A_P\, f_{Py})\, z$$

assuming the ratio of mild reinforcing steel in the top and bottom slabs is 0.6 percent and z is equal to $h - 0.75\,(t^{ts} + t^{bs})$. (The parameter h denotes total depth of

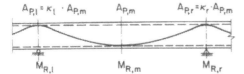

Figure 7.11
Notation for initial estimate of reinforcement in a continuous girder

the section, t^{ts} denotes thickness of the top slab, and t^{bs} denotes thickness of the bottom slab.) The area of prestressing steel at midspan, $A_{P,m}$ is obtained from the inequality

$$M_d^0 \leq \frac{1}{\gamma_R} \left(M_{R,m} + \frac{M_{R,l} + M_{R,r}}{2} \right)$$

where M_d^0 is the design moment due to dead load plus live load at midspan of a corresponding simply supported beam and $M_{R,m}$, $M_{R,l}$, and $M_{R,r}$ are the flexural resistances at midspan, left support, and right support, respectively (fig. 7.11). The areas of prestressing steel at the left and right supports, $A_{P,l}$ and $A_{P,r}$, are expressed as multiples of $A_{P,m}$:

$$A_{P,l} = \kappa_l A_{P,m}, \qquad A_{P,r} = \kappa_r A_{P,m}$$

The coefficients κ correspond to the proposed tendon arrangement.

The initial estimate of web width and any other cross-section dimensions that are a function of longitudinal prestressing should be corrected, if necessary, using the value of $A_{P,m}$ thus computed. The final design of the reinforcing and prestressing steel can then be carried out according to the chosen prestressing concept.

7.2.5 Tendon Layouts

a) Simply Supported Girders

Whenever possible, tendons in simply supported girders should extend from one end of the girder to the other. Provided both ends are accessible for jacking, it is preferable to alternate stressing and dead-end anchors (fig. 7.12). If the girder extends only a short distance past the supports, the tendons should be anchored in the lower portion of the section. This results in a direct flow of forces and good

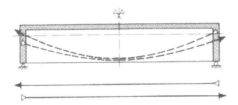

Figure 7.12
Recommended tendon layout for simply supported girders

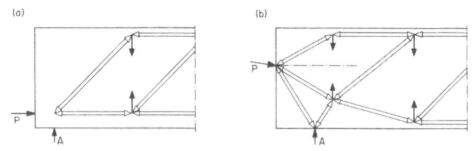

Figure 7.13
Flow of forces at girder ends (service conditions): a, tendons anchored in lower portion of section;
b, tendons anchored at centroid of section

protection against diagonal cracking immediately above the support (fig. 7.13 a).
Since the tendons are horizontal at the critical section for shear, increased shear
reinforcement will often be required. When the girder extends a sufficient distance
beyond the supports, the tendons can be anchored at the centroid of the section
(fig. 7.13 b). The inclined tendons then contribute to the shear resistance at the
critical section.

Varying the area of prestressing steel with bending moment does not reduce
construction cost, since any savings in prestressing steel are outweighed by
complications in detailing and construction. Tendons should therefore be
anchored away from the supports only if the girder ends are not accessible for
stressing. In such cases, stressing anchors can be located at the intersection of web
and top slab (fig. 7.14). The reinforcement of the deck slab cantilevers must be
interrupted at the anchorage block-outs and then spliced after the tendons have
been stressed. Alternatively, the stressing anchors can be located at the intersec-

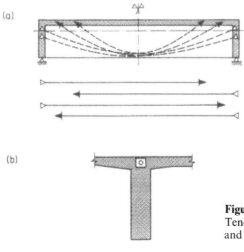

Figure 7.14
Tendons anchored at intersection of top slab
and webs: a, longitudinal section; b, cross-section

(a)

(b)

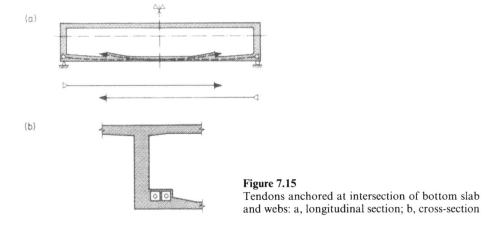

Figure 7.15
Tendons anchored at intersection of bottom slab
and webs: a, longitudinal section; b, cross-section

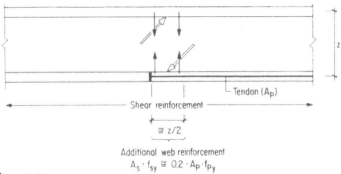

Shear reinforcement

$\cong z/2$

Additional web reinforcement
$A_s \cdot f_{sy} \cong 0.2 \cdot A_p \cdot f_{py}$

Tendon (A_p)

Figure 7.16
Additional web reinforcement near ends of tendons anchored at intersection of bottom slab and
webs

tion of web and bottom slab (fig. 7.15). It is recommended that additional web
reinforcement, equal to roughly 20 percent of the jacking force in the tendon, be
provided in front of the anchor (fig. 7.16).

b) Continuous Girders

The profile of the centroid of prestressing steel should be as close as possible to
parabolic in each span of a continuous girder. The deviation forces induced by the
curved tendons will then directly counteract the uniformly distributed dead load.
The percentage of dead load balanced by prestressing should be approximately
constant for all spans. In all practical cases, however, tendons are given a reverse
curvature at the supports. To balance dead load as closely as possible, the radius of
curvature of the tendons at the supports, r, should be minimized. Typical values of

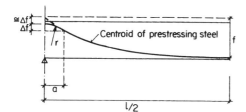

Figure 7.17
Tendon profile near intermediate supports of continuous girders

r range from 3 to 5 metres, depending on duct material and diameter. Expressions for the deviation forces were derived in Section 4.6.5:

$$q_P = \frac{8fP}{l^2} \qquad \text{in the span} \qquad \text{(equation 4.7)}$$

$$\bar{q}_P = -\frac{2\Delta f P}{a^2} \qquad \text{at the supports} \qquad \text{(equation 4.8)}$$

In both equations, P denotes prestressing force. The other parameters are defined in figure 7.17.

Estimates of a and Δf are obtained by considering the slope of the prestressing steel, α, at a distance a from the support. Assuming α is small and a is small compared to l, it follows that

$$\tan \alpha \cong \frac{4f}{l} \quad \text{and} \quad \tan \alpha = \sin \alpha = \frac{a}{r}$$

Therefore,

$$a = \frac{4r}{l/f} \tag{a}$$

Moreover,

$$\tan \alpha = \frac{2\Delta f}{a} = \frac{4f}{l}$$

and so

$$\Delta f = \frac{2a}{l/f} = \frac{8r}{(l/f)^2} \tag{b}$$

The value of f should be maximized by providing the largest possible tendon eccentricity at midspan and at the supports. The number and size of tendons at midspan and at the supports are required to calculate f exactly. As an initial estimate, f can be assumed equal to h, the total depth of the section.

The ideal tendon profile for haunched girders is also parabolic. Tendons can be laid out for an equivalent girder with straight centroidal axis (fig. 7.18).

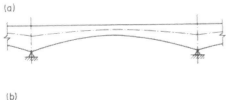

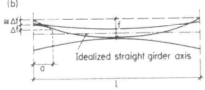

Figure 7.18
Tendon profile for haunched girders: a, real haunched girder; b, idealized straight girder

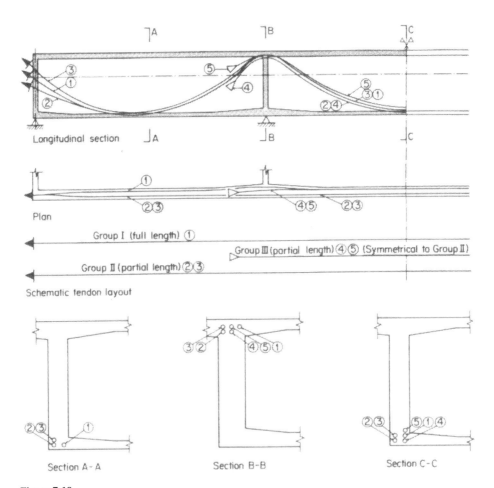

Figure 7.19
Tendon layout for three-span continuous girders: area of prestressing steel varied span by span

Two-span and three-span continuous girders are usually built on conventional falsework. Concrete is placed into all spans of the structure simultaneously without vertical construction joints. The prestressing can easily be varied from span to span according to the magnitude of the moments. It is therefore preferable not to carry all tendons from one end of the bridge to the other, but rather to locate the stressing anchors at the girder ends and the dead-end anchors near the intermediate supports (fig. 7.19). The webs must be locally thickened to accommodate the anchors. For long spans, additional tendons can be provided over the interior supports or at midspan in the top or bottom slabs. They should be anchored as close as possible to the webs to simplify the flow of forces.

Longer continuous girders are often built following a span-by-span construction sequence, for which it is common to stress only a portion of the tendons in a given span before the removal of falsework and formwork. Tendons are normally grouted only after falsework has been removed, several spans at a time. Unless a detailed calculation of the increase in stress at ultimate limit state is made for these unbonded tendons, flexural resistance must be calculated using the effective prestress, σ_P, and not the full yield stress, f_{P_y}. The flexural resistance of sections built by span-by-span construction can thus be significantly less during construction than the resistance of the completed structure; a verification of safety at ultimate limit state during construction is therefore indispensable. The full factor of safety, $\gamma = \gamma_S \gamma_R$, need not be developed during construction when prestressing forces and loadings are carefully controlled. Provided cracking and deformation behaviour are adequate, a reduced factor of safety of approximately 80 percent of γ can be used.

Several alternative tendon profiles and coupler arrangements suitable for span-by-span construction are illustrated in the following examples:

Example 7.1:
All tendons coupled at the same location

Figure 7.20 shows a possible prestressing layout in which all tendons are anchored, stressed, and coupled at each construction joint. The main advantage of this arrangement is that it requires no increase in web thickness, provided the anchors can be arranged in one vertical column. The tendons must, however, be spread wide apart in the vertical plane at the construction joint, which results in a large angular change and hence large friction losses. The introduction of a strong concentrated force at the construction joint requires local strengthening of the web reinforcement.

Example 7.2:
Only a portion of the tendons coupled at the same location

Figure 7.21 shows a tendon layout in which two thirds of the tendons are anchored, stressed, and coupled at each construction joint. The remaining tendons

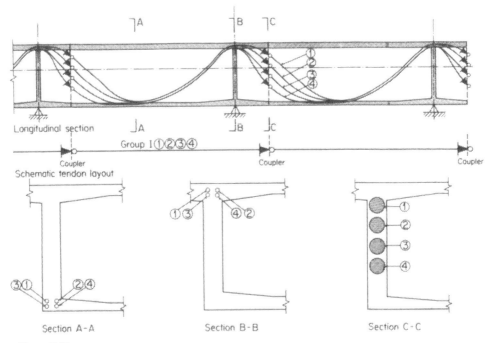

Figure 7.20
Tendon layout for continuous girders: all tendons coupled at the same location in each span

continue uninterrupted over the joint. They are placed into the forms and concreted in with segment i; the excess is kept on spools and placed into segment $i + 1$ after it has been formed. Storing tendons on spools in this way is usually more economical than using movable couplers.

The concentrated anchor force in this arrangement is smaller than in Example 7.1. It is normally necessary, however, to thicken the webs near the intermediate supports. The required web width is nevertheless smaller than if two vertical rows of anchors were used. Because some of the tendons will be unstressed when falsework is removed, safety at ultimate limit state, cracking, and deformations must be verified during construction.

Example 7.3:
Overlapping tendons at the intermediate supports

Figure 7.22 shows an arrangement in which couplers are replaced by overlapping tendons at the intermediate supports. Two alternate anchor arrangements are possible:

Alternate 1:
Stressing anchors are located at the face of the construction joint and dead-end anchors of the spliced tendons for the next span are cast into the webs behind the

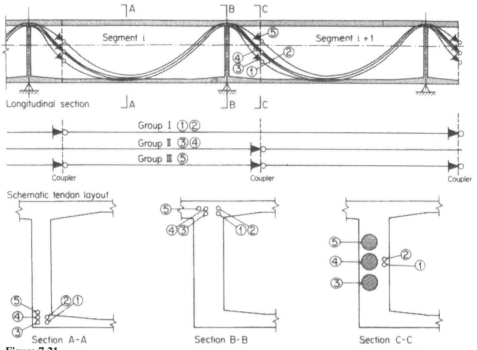

Figure 7.21
Tendon layout for continuous girders: combination of coupled and continuous tendons at construction joints

joint. Thickening of the webs can often be avoided in this way. All tendons for the next span must be cast into the support region of the previous span. The excess is stored on spools until the next span has been formed and reinforced.

Alternate 2:
Build-outs for the anchors are provided in the webs on the opposite side of the support from the construction joint. Only the ducts themselves must be cast into the concrete. The prestressing steel can thus be pushed or pulled into the ducts at any time. The stressing anchors remain accessible after construction, making possible a second stressing immediately before grouting.

In both alternate arrangements, all tendons for a given span are stressed before removal of falsework. Couplers are eliminated and, when the lap length is sufficient, the prestressing force is doubled in the support region. The stressing sequence can be freely chosen when Alternate 2 is used, enabling a substantial reduction of prestressing losses due to creep and shrinkage. The main disadvantage of Alternate 1 is that the tendons for the span that is not yet formed must be concreted into the previous span. The lapping of tendons can be complicated by the web build-outs required by Alternate 2.

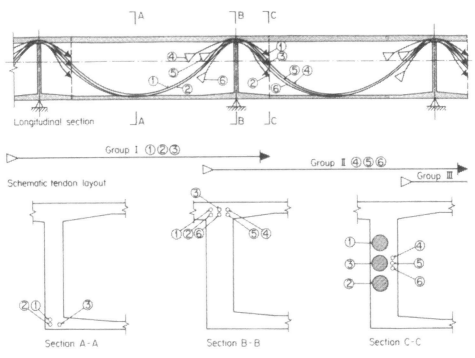

Figure 7.22
Tendon layout for continuous girders: tendons lapped at intermediate supports

Example 7.4:
Combinations

The systems presented in the preceding examples can be freely combined, as shown in figure 7.23. A simple, well-chosen tendon layout can significantly reduce losses due to friction and facilitate construction.

7.2.6 Incrementally Launched Bridges

The incremental launching method can be economical for the construction of long, multiple-span bridges when falsework costs are rendered prohibitive by bridge height or topographical conditions. The alignment must be straight or of constant curvature. All concreting is done behind one or both of the abutments, in a relatively small area that can be easily enclosed and protected from extremes of temperature and humidity. Quality comparable to factory-produced precast concrete can thus be achieved. Segments are cast against the previously constructed portion of the bridge. After hardening, they are prestressed for continuity with the existing portion and displaced away from the abutment, pushing the bridge along with them. The casting area is then prepared for the production of the next segment.

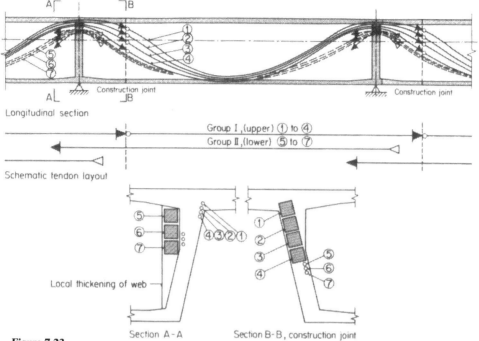

Figure 7.23
Tendon layout for the approach spans of the Felsenau Bridge, Berne, Switzerland

The casting area consists of steel beams to support the webs of the superstructure and mechanical formwork for the bottom slab and the deck slab cantilevers that can be rapidly swung into place. Steel plates are used to form the portion of the bottom slab that is in contact with the steel support beams; the contact surface between the beams and the plates is normally lubricated (fig. 7.24).

The launching apparatus consists of a horizontal jack, rigidly attached to the abutment, and a vertical jack which is pushed along a stainless steel surface by the horizontal jack (detail A of figure 7.24). The bottom of the vertical jack is coated with teflon to reduce friction against the stainless steel; its top surface is shaped to maximize friction between the jack and the underside of the bridge. A typical launching operation begins by lifting the bridge by several millimetres using the vertical jacks. It is then pushed forward by the horizontal jack to the full extent of its stroke. The vertical jacks are then unloaded and returned to their original position.

The prestressing of incrementally launched bridges normally consists of straight tendons in the top and bottom slabs and parabolically draped tendons in the webs. The former tendons are denoted A_P^{ts} and A_P^{bs}. They are used to cover the positive and negative moment peaks occurring at all points along the girder during construction (fig. 7.25). The parabolic tendons, located in the webs, are provided

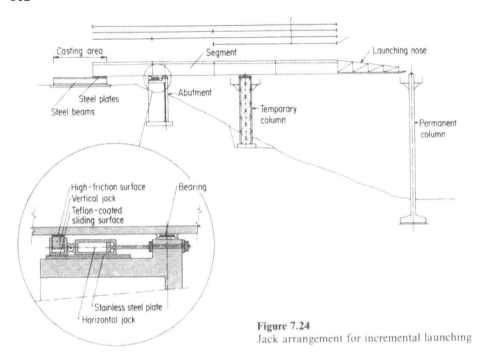

Figure 7.24
Jack arrangement for incremental launching

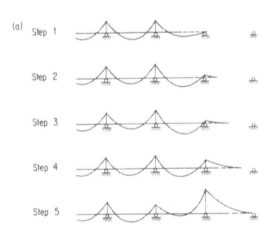

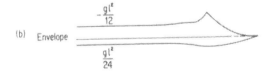

Figure 7.25
Incrementally launched bridges: a,
dead load moments during construction; b, envelope of dead load
moments

for additional live load resistance. Their area, A_P^w, is normally chosen between 1.0 and 1.2 times the total area of slab prestressing, $A_P^{ts} + A_P^{bs}$.

The slab tendons are coupled at the construction joints between segments. They can extend over two or three segments, depending on the arrangement of the abutment and the casting area. The web tendons are installed after completion of the girder and are normally lapped at the intermediate supports.

The construction sequence results in the characteristic moment history shown in figure 7.25 a, which yields the dead-load moment envelope of figure 7.25 b. Since the highest moments occur at the forward end of the girder, the slab prestressing must be increased in this area. A launching nose is normally used to reduce negative moments at the forward end of the girder.

7.3 Girder Bridges with Precast Elements

7.3.1 Conceptual Design

The principal advantage of precast components is ease of erection. Their use can substantially reduce construction time, enabling superstructures to be built at rates of up to 3000 m² per month. Shortened construction and elimination of falsework often result in low construction cost. Superstructures that use precast elements are often characterized, however, by thin cross-section components, large exposed surface of concrete, and many construction joints, all of which are disadvantages with regard to durability. The thin, finely reinforced precast elements must be therefore be manufactured and handled with care. Careful inspection is particularly important when precasting is done on the construction site, where well-trained and experienced staff are not always available.

The two most common types of precast components used in bridge superstructures are girders and cross-section segments.

1. Precast Cross-Section Segments. Bridge spans constructed by the *precast segmental method* are composed of many short precast segments which are prestressed together longitudinally after erection. Due to the high cost of casting and erection equipment, this type of construction is normally economical only for long bridges. The joints between segments are not normally crossed by mild reinforcing steel. They must therefore be prevented from opening by a residual compressive stress of 0.5 to 1.5 N/mm² under long-term dead load, long-term prestressing, live load, and temperature gradient. Economy and quality are enhanced by the use of external, unbonded tendons.

The shapes of precast segmental cross-sections are similar to the shapes used for cast-in-place girders (fig. 7.26). The webs of externally prestressed sections can be reduced to the minimum required for shear resistance or can be designed as

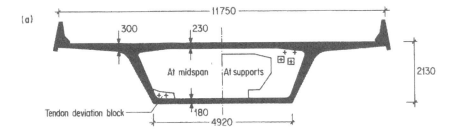

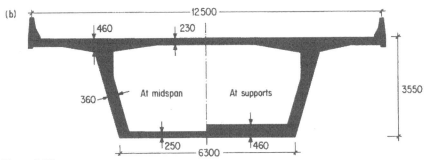

Figure 7.26
Cross-sections of precast segmental bridges (dimensions in mm): a, Long Key Bridge, U.S.A. (external, unbonded tendons); b, Kishwaukee River Bridge, U.S.A. (balanced cantilever construction)

concrete trusses, resulting in a significant reduction in dead load. When external prestressing is used, the vertical component of the tendon force at the deviation points does not act in the same plane as the opposing shear force in the web. This eccentricity must be taken into account in the design and reinforcement of the cross-section.

2. Precast Girder Elements. Precast girder elements can be used economically for spans of up to 30 m provided a minimum number of elements, typically 40, can be cast. Longer girders require more expensive casting and erection equipment and are therefore normally used only for long bridges.

Girder depth is determined using the same criteria as for cast-in-place bridges. The narrow deck slab cantilevers of precast girder bridges often have a negative impact on visual slenderness.

Simply supported systems have not provided adequate durability, appearance, or ride comfort. The neoprene seal in the expansion joint and the connection of the waterproofing membrane to the expansion joint hardware cannot be maintained watertight in service (fig. 7.27a). Water that penetrates at these locations endangers the ends of the girders, bearings, and tops of piers. Girder ends are often inaccessible for inspection and maintenance. The light expansion joint hardware

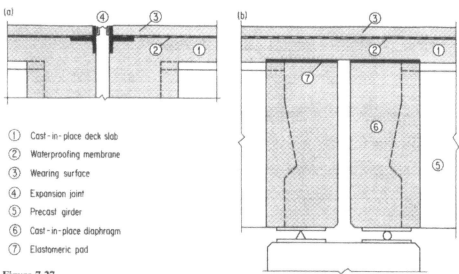

① Cast-in-place deck slab
② Waterproofing membrane
③ Wearing surface
④ Expansion joint
⑤ Precast girder
⑥ Cast-in-place diaphragm
⑦ Elastomeric pad

Figure 7.27
Details at supports of bridges with precast girder elements: a, simply supported spans with expansion joint; b, simply supported spans with linking slab

can rarely be anchored properly, resulting in damage to the surrounding concrete. Bearings must normally be located at girder ends and at the edge of columns, which results in a complicated flow of forces and the need for additional reinforcement.

The use of a continuous deck slab to link the girders over the supports is, at best, only a partial solution to the problem (fig. 7.27b). Although ride comfort is considerably improved, the waterproofing membrane over the relatively flexible linking slab is easily damaged and cannot be relied on to protect the slab, girder ends, bearings, and tops of piers from deicing salts. Linking the girders in this way results in no improvement in the bearing layout and complicates the jacking of the superstructure.

The best solution with regard to durability, appearance, and ride comfort is to transform the individual spans into a fully continuous system by using thick diaphragms and a continuous deck slab at the supports (fig. 7.28). The bridge can thus be supported on a small number of high-quality bearings. By eliminating bearings under each individual girder, the width and thickness of the piers can be substantially reduced. (Thickened or twin piers are nevertheless required at the intermediate expansion joints of long bridges.) Although temporary supports at the piers are needed for erection, their added cost can always be justified by the superior aesthetic qualities of the slender piers they make possible. Cracks in the deck slab over the supports can be prevented with a properly chosen prestressing concept. Tendons that are continuous over several spans should be coupled well below the deck slab surface to reduce the stress range in the couplers and to

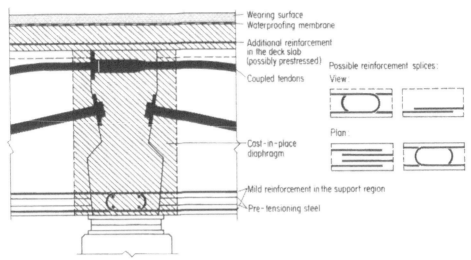

Figure 7.28
Details at supports of continuous bridges with precast girder elements

prevent tensile stresses in the lower fibres of the girder. The girder ends should be built out as shown in figure 7.28 to improve the transfer of shear from girders to diaphragm.

On curved alignments, precast girders are arranged along the chords of an arc, resulting in deck slab cantilevers of variable width. The axes of the piers can be set parallel to each other or radial to the roadway alignment (fig. 7.29). The former arrangement enables all girders of a given span to be of equal length. It is preferred for viaducts along mountain slopes, where the foundations must normally be aligned parallel to the gradient vector of the slope. Casting many girders of unequal lengths can usually be done without undue difficulty and cost. There is thus no penalty for aligning the piers radially. Tapered diaphragms must be used with this arrangement.

Sections 7.3.2 through 7.3.4 will be devoted to the use of precast girder elements.

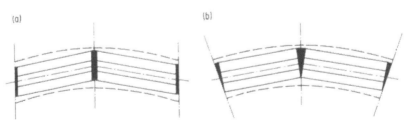

Figure 7.29
Arrangements of precast girders for bridges on curved alignments: a, axes of supports parallel; b, axes of supports radial

7.3.2 Design of the Cross-Section

The cross-sections of precast girders are determined by fabrication procedure, prestressing system, erection procedure, and construction of the deck slab. Girders are normally cast in steel forms that can be vibrated to consolidate the concrete. Pre-tensioning can be feasible, even for on-site precasting, provided a sufficiently large number of girders is cast. Pre-tensioning and post-tensioning can be combined. It is most common to stress the post-tensioning steel after erection of the girder or after the deck slab has been cast. The post-tensioning can be coupled at the intermediate supports to resist negative moments.

Although it would be theoretically possible to design a superstructure consisting entirely of precast girders, the heavy girders required would be expensive to transport and erect. Erection would be complicated by unequal girder deformations, which would have to be equalized somehow before the girders could be connected together transversely. Casting the deck slab in place after the girders are erected results in much lighter girders that are easier and more economical to transport and erect, and a system that is much less sensitive to unequal deformations in the girders.

Based on these considerations, the girder cross-section should have the following characteristics:

1. A compact bottom flange to facilitate consolidation of the concrete in the closed forms. It should be designed to limit compressive stresses during construction and detailed allow the proper distribution of pre-tensioning steel.
2. Webs varying in thickness between 180 mm and 220 mm for beams without post-tensioning and between 220 mm and 250 mm for beams with post-tensioning in the webs.
3. A wide, thin top flange to improve stability against overturning during erection and provide good bond with the cast-in-place deck slab. The top flange can be made wide without significant increase in formwork costs.

The web thickness at the girder ends is increased to roughly the width of the bottom flange to improve the transfer of force from precast girder to diaphragm and to increase the shear resistance of the girder near the supports (fig. 7.30).

The formwork for the deck slab can be supported from the bottom flange or suspended from the top flange of the girder (figs. 7.31 a and 7.31 b). Alternatively, precast concrete planks can be used (fig. 7.31 c). Pre-tensioning the planks substantially improves their strength and stiffness as formwork; the prestressing steel is fully effective as transverse deck slab reinforcement in the completed structure.

7.3.3 Prestressing Concepts

The prestressing concept should ensure adequate cracking behaviour under service conditions. This can normally be accomplished by eliminating tensile

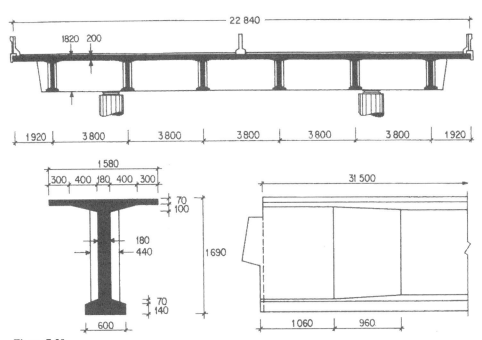

Figure 7.30
Superstructure cross-section and precast girder details of the Yverdon Viaduct, Switzerland
(dimensions in mm)

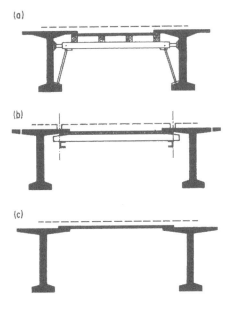

Figure 7.31
Formwork for deck slabs: a, supported from
bottom flange; b, suspended from top flange; c,
precast planks

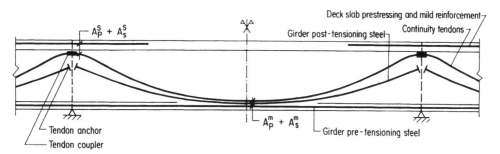

Figure 7.32
Main longitudinal reinforcement for bridges with precast girders

stresses due to dead load in the deck slab only, without restricting tensile stresses in the bottom fibres of the girder. Since the bottom flange of the girder must always be reinforced with well distributed prestressing and mild steel, cracks in the bottom fibres will be small and evenly spaced. This prestressing concept is usually implemented using continuity tendons, coupled at the supports, and additional tendons in the deck slab over the supports (fig. 7.32).

The first stage prestressing in the precast girders, i.e., the tendons that are stressed before the deck slab is cast, should be designed to minimize the redistribution of dead load moments. The procedure proposed in Section 7.3.4 can be used for this purpose. The simple beam moments due to the self-weight of the girders and the deck slab, effective during construction, will remain unchanged for the life of the bridge. The continuity tendons and the additional deck slab tendons need therefore only be designed to eliminate tensile stresses due to superimposed dead load. Since the positive moment at the supports due to prestressing will not be balanced by a negative moment due to dead load, reinforcement will be required to resist the tensile force at the bottom of the girders. These bars should be carefully spliced as shown in figure 7.28.

The calculation of stresses necessary for the design of prestressing is complicated by the effects of creep and shrinkage. A simplified calculation procedure is presented in Section 7.3.4.

7.3.4 Preliminary Design

a) Ultimate Limit State

It is assumed that the superstructure is constructed according to the following sequence:

1. Erect precast girders
2. Install continuity prestressing and reinforcing steel
3. Cast diaphragms at the supports and deck slab
4. Stress continuity prestressing

The precast girder cross-section is assumed to have a pronounced bottom flange, similar to the section shown in figure 7.30.

The self-weight of the precast girders and cast-in-place deck slab, g_0, is thus resisted by the simply supported girders and produces only positive moments. The moment at midspan is denoted $M^m(g_0)$; the corresponding design moment, denoted $M_d^m(g_0)$, is equal to the load factor γ_s times $M^m(g_0)$. Superimposed dead load, Δg, and live load, q, are resisted by the continuous system resulting from the addition of deck slab, diaphragms at the supports, and reinforcement for negative moment. The corresponding design moments at midspan and at the support are denoted $M_d^m(\Delta g, q)$ and $M_d^s(\Delta g, q)$, respectively.

The main reinforcement is shown schematically in figure 7.32. At midspan, it consists of:

1. First stage prestressing, installed and stressed before erection, $A_{P,0}^m$
2. Mild reinforcing steel in the bottom flange of the girder, A_s^m
3. Continuity prestressing stressed after the deck slab has been cast, $A_{P,C}^m$

(The total area of prestressing steel at midspan, $A_{P,0}^m + A_{P,C}^m$, is denoted A_P^m.) The reinforcement at the supports consists of:

1. Continuity prestressing in the girder and additional prestressing in the deck slab, A_P^s
2. Mild reinforcing steel in the deck slab, A_s^s

Behaviour at ultimate limit state must be investigated for two separate cases: the precast girders in the simply supported system during construction and the composite cross-section in the continuous system after completion of the bridge.

During construction, the following inequality must be satisfied:

$$M_d^m(g_0) \leqq \frac{1}{\gamma_R} M_{R,0}^m$$

where

$$M_{R,0}^m \cong (A_{P,0}^m f_{Py} + A_s^m f_{sy}) z_0$$

The internal lever arm, z_0, can be taken as the distance between the centroids of the top and bottom flanges of the precast girder (fig. 7.33).

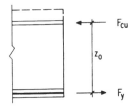

Figure 7.33
Flexural resistance at midspan during construction

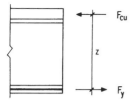

Figure 7.34
Flexural resistance at midspan in the completed structure

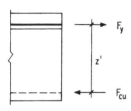

Figure 7.35
Flexural resistance at the supports in the completed structure

After completion of the structure, the following inequality must be satisfied:

$$M_d^m(g_0) + M_d^m(\Delta g, q) - M_d^s(\Delta g, q) \leq \frac{1}{\gamma_R}(M_R^m + M_R^s)$$

where

$$M_R^m \cong (A_P^m f_{Py} + A_s^m f_{sy}) z$$

and

$$M_R^s \cong (A_P^s f_{Py} + A_s^s f_{sy}) z'$$

The internal lever arm at midspan, z, is approximately equal to the distance between the centroid of the bottom flange and the centroid of the composite deck slab (fig. 7.34). At the supports, lever arm z' can be taken as the distance between the upper edge of the bottom flange and the centre of gravity of the reinforcement, taking into account the length and yield force of the continuity tendons, additional deck slab tendons, and mild reinforcing steel in the deck slab. As an initial approximation, z' can be assumed equal to the distance between the upper edge of the bottom flange and the centroid of the composite deck slab (fig. 7.35).

b) Service Conditions

The state of stress in the superstructure is influenced by: (1) self-equilibrating stresses due to differential creep and shrinkage in the precast girders and cast-in-place deck slab, and (2) the redistribution of moments due to the change from simply-supported spans to a continuous system. An exact calculation of stresses is thus analytically complex and, due to variability in the material properties, of doubtful accuracy. As stated in Section 7.3.3, moment redistribution can be eliminated by a suitable choice of first stage prestressing in the simply supported girders. In such a case, the primary effect of creep and shrinkage will be a redistribution of stress from the precast girders to the composite section (girders and deck slab), without changing the sectional forces.

Assuming no redistribution of moment, the long-term stresses due to dead load plus prestressing, σ_∞, can be approximated using the following expressions:

$$\sigma_\infty^{ds} \cong \tfrac{1}{2}\sigma_1^{ds} \qquad (\textit{deck slab}) \qquad\qquad\qquad (a)$$

$$\sigma_\infty^{pg} \cong \tfrac{1}{2}(\sigma_0 + \sigma_1^{pg}) \qquad (\textit{precast girder}) \qquad\qquad\qquad (b)$$

The subscript 0 denotes stresses due to dead load, g, plus prestressing, P, in the simply supported precast girders, assuming no participation of the deck slab; subscript 1 denotes stresses due to $g + P$ in an idealized simply supported superstructure, assuming girders and deck slab have been cast in place simultaneously. The long term stress σ_∞ is thus the average of stresses calculated from these two limiting cases.

When redistribution of moments is prevented, the additional strains due to creep and shrinkage in the top and bottom fibres of the precast girder at midspan will be equal:

$$\Delta\varepsilon_\infty^{tg} = \Delta\varepsilon_\infty^{bg} \qquad\qquad\qquad (c)$$

(The superscripts tg and bg denote top and $bottom$ fibres, respectively, of the precast girder.) By formulating expressions for $\Delta\varepsilon_\infty^{tg}$ and $\Delta\varepsilon_\infty^{bg}$ in terms of known stresses and strains, it will be possible to solve equation (c) for the required first stage prestressing force.

A general expression for ε_∞ was presented in Section 4.7.1:

$$\varepsilon_\infty = \frac{\sigma_0}{E_c}(1 + \phi_\infty) + \frac{\sigma_\infty - \sigma_0}{E_c}(1 + \mu\phi_\infty) + \varepsilon_{cs,\,\infty} \qquad (\text{equation 4.10})$$

It therefore follows that

$$\Delta\varepsilon_\infty = \frac{\sigma_\infty}{E_c}(1 + \mu\phi_\infty) - \frac{\sigma_0}{E_c}[1 - (1 - \mu)\phi_\infty] + \Delta\varepsilon_{cs,\,\infty}$$

or, assuming μ is 0.8,

$$\Delta\varepsilon_\infty = \frac{\sigma_\infty}{E_c}(1 + 0.8\phi_\infty) - \frac{\sigma_0}{E_c}(1 - 0.2\phi_\infty) + \Delta\varepsilon_{cs,\,\infty} \qquad\qquad\qquad (d)$$

where $\Delta\varepsilon_{cs,\,\infty}$ ist the residual shrinkage strain after continuity has been established. An expression for $\Delta\varepsilon_\infty^{tg}$ is formulated from the condition of compatibility of strain at the interface between girder and deck slab:

$$\Delta\varepsilon_\infty^{tg} = \Delta\varepsilon_\infty^{ds}$$

It follows from equation (d) that

$$\Delta\varepsilon_\infty^{tg} = \Delta\varepsilon_\infty^{ds} = \frac{1}{2}\frac{\sigma_1^{ds}}{E_c}(1 + 0.8\phi_\infty^{ds}) + \Delta\varepsilon_{cs,\,\infty}^{ds} \qquad\qquad\qquad (e)$$

where σ_∞ is obtained from equation (a). A similar expression for $\Delta\varepsilon_\infty^{bg}$ is formulated by substituting equation (b) into equation (d):

$$\Delta\varepsilon_\infty^{bg} = \frac{1}{2}\frac{\sigma_0^{bg} + \sigma_1^{bg}}{E_c}(1 + 0.8\,\phi_\infty^{pg}) - \frac{\sigma_0^{bg}}{E_c}(1 - 0.2\,\phi_\infty^{pg}) + \Delta\varepsilon_{cs,\,\infty}^{pg}$$

$$= \frac{1}{2}\frac{\sigma_1^{bg}}{E_c}(1 + 0.8\,\phi_\infty^{pg}) - \frac{1}{2}\frac{\sigma_0^{bg}}{E_c}(1 - 1.2\,\phi_\infty^{pg}) + \varepsilon_{cs,\,\infty}^{pg} \qquad (f)$$

where ϕ_∞^{pg} and $\Delta\varepsilon_{cs,\,\infty}^{pg}$ are residual creep and shrinkage coefficients, respectively, after creation of bond between girder and deck slab.

Equation (c) can now be solved for the first stage prestressing force required to prevent the redistribution of dead load moments. An additional margin of safety against positive moments at the supports due to superimposed dead load, continuity prestressing, and deck slab prestressing can be provided by satisfying the following inequality:

$$|\Delta\varepsilon_\infty^{tg}| > |\Delta\varepsilon_\infty^{bg}|$$

7.4 Cantilever-Constructed Girder Bridges

7.4.1 Conceptual Design

The balanced cantilever method is often appropriate for the construction of long spans when bridge height, topography, or geotechnical conditions render the use of conventional falsework uneconomical. Savings are achieved partly due to the construction method and partly due to the structural system. Previously cast portions of the superstructure can be immediately used to support the construction of new segments, making possible the use of short, economical form travellers. Formwork is adjustable and can be reused many times. The regular repetition of identical tasks for the construction of each of the segments substantially reduces the ratio of labour costs to material costs. The use of short segments makes possible an economical prestressing layout that closely matches the moment diagram.

The economical range of span lengths for cast-in-place cantilever construction begins at roughly 70 m and extends to beyond 250 m. For long spans, the ratio of falsework and formwork cost to total superstructure cost ranges between 25 and 35 percent, independent of bridge height and topography. This represents a considerable saving over conventionally constructed cast-in-place bridges, for which ratios of 40 percent are typical. For spans less than 70 m, the advantages of cantilever construction are outweighed by the costs of pier tables and erection of form travellers.

In classical cantilever construction, cantilevers are built out symmetrically from pier tables in segments ranging between 3 and 5 metres in length. The construction sequence for a given segment consists of the following steps:

1. Prestress the previously cast segment approximately two days after it has been cast
2. Advance form travellers to their new position
3. Anchor travellers to the newly prestressed girder end and adjust formwork
4. Place reinforcing steel, couple new tendon ducts with empty ducts in the completed portion, and pull prestressing steel into ducts
5. Cast the new segment

The completed cantilevers are linked together at midspan into a continuous frame system. They can be connected by a monolithic closure pour, a hinge, or a short suspended girder. Hinges at midspan are generally more economical than monolithic connections, since they prevent the redistribution of moments in the girder from the stronger support region to the weaker midspan region. Hinges have, however, significant disadvantages with regard to serviceability. Discontinuities in roadway grade at midspan due to creep deformations in the cantilevers are practically unavoidable when hinges are provided. The use of expansion joints at midspan not only creates special difficulties in detailing, but also reduces durability and ride comfort.

Hinges that permit longitudinal displacements of the superstructure must be provided with large, complicated bearings to transfer shear force due to live load (fig. 7.36). The height of the enclosure for the bearings must be carefully adjusted to ensure positive transfer of force and free rotation of the hinge, regardless of the horizontal position of the bearings. Provision must be made for the disassembly of the enclosure to replace the bearings.

A partial improvement over traditional hinges is the prestressed concrete hinge, which does not break the continuity of the deck slab (fig. 7.37). Its use may be appropriate when longitudinal expansion at midspan can or must be restrained. Since the long-term impermeability of the waterproofing membrane in the relatively flexible deck slab cannot be relied on, a corrosion-resistant cover plate should be provided at the axis of the hinge for additional protection. The underside of the membrane should be vented using small pipes cast into the slab to enable the rapid detection of defects in the waterproofing system. An accessible inspection chamber should be provided below the hinge.

Cantilever-constructed bridges are normally haunched to correspond to the cantilever moment diagram. In exceptional cases, however, the cantilever method can also be used to construct prismatic girders. Girder depth is then determined according to the same criteria as for cast-in-place girders. The redistribution of moments produced by the change from cantilever to the continuous system is larger for prismatic girders than for haunched girders. Prismatic girders will therefore require more prestressing and a more complicated tendon layout.

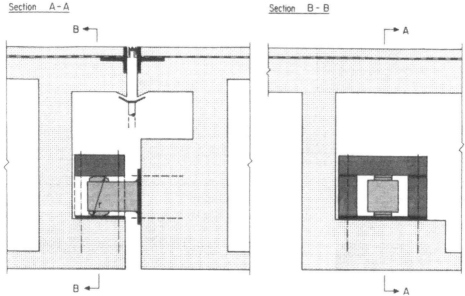

Figure 7.36
Details at midspan hinges permitting rotation and longitudinal expansion

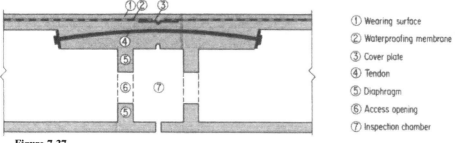

① Wearing surface
② Waterproofing membrane
③ Cover plate
④ Tendon
⑤ Diaphragm
⑥ Access opening
⑦ Inspection chamber

Figure 7.37
Details at midspan hinges permitting rotation only

7.4.2 Design of the Cross-Section

Box cross-sections, which are effective in resisting negative moments, are normally used for cantilever-constructed superstructures. As many section dimensions as possible should be kept constant to simplify the formwork.

a) Webs

The webs are most often vertical. Inclined webs complicate the placement of concrete and, in haunched girders, result in a bottom slab of variable width.

Figure 7.38
Girder cross-sections: a, normal; b, superelevated

Variable superelevations can be best accommodated by a rhombic deformation of the cross-section (fig. 7.38). The webs typically contain only a small number of tendons, and often none at all. Their thickness can therefore be determined on the basis of shear strength, provided additional thickness is not required to anchor the cantilever tendons or to place and vibrate the concrete. Vertical webs that do not contain tendons should not, however, be thinner than 0.35 m. As explained in Section 5.3.2, the use of inclined stirrups can substantially reduce the required web thickness. This will result in a considerable reduction in dead load in haunched girders. Inclined shear reinforcement, mild or prestressed, can be placed into the deep webs without difficulty.

b) Deck Slab

The form of the deck slab is chosen based on the considerations presented in Section 5.3.1. Its thickness must be sufficient to accommodate the cantilever tendons, which are typically spaced across the full width of the slab near the supports. Additional slab thickness may also be required beside the web stirrups for tendon anchors (fig. 7.39). The horizontal and vertical shear resistance of the slab is substantially reduced during construction by the large number of empty tendon ducts.

c) Bottom Slab

At the supports, the bottom slab must be sufficiently thick to resist the compressive force due to flexure and shear at ultimate limit state under maximum load. Its thickness should decrease linearly towards midspan to match the compressive force in the slab. This variation in thickness can be easily accommodated by the web formwork. The bottom slab should not be thinner than 0.16 m.

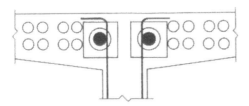

Figure 7.39
Anchorage of cantilever tendons at the intersection of deck slab and web

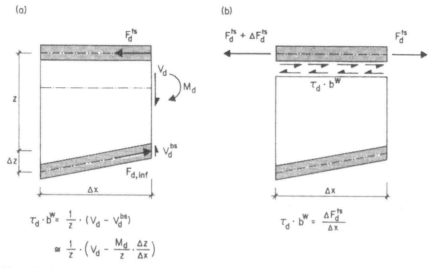

Figure 7.40
Shear resistance of haunched box girders; shear flow $\tau_d b^w$: a, as a function of the sectional forces; b, as a function of the tensile force in the top slab

d) Variation of Girder Depth

Based on aesthetic and economic considerations, the ratio of span to depth should be roughly 50 at midspan and 17 at the supports. The profile of the girder between these two points has a substantial influence on the internal forces due to flexure and shear. Girder shape should be chosen to achieve a suitable compromise between the thickness and reinforcement of the webs on the one hand, and the longitudinal reinforcement on the other.

The design shear force in a haunched girder, V_d, is resisted by shear flow in the webs, $\tau_d b^w$, and the vertical component of the compressive force in the bottom slab, V_d^{bs}. (The parameter τ_d denotes design shear stress in the web; b^w denotes web thickness.) For a given girder profile, the relationship between longitudinal reinforcement and shear reinforcement can be obtained by expressing the shear flow $\tau_d b^w$ as a function of the tensile force in the top slab, F_d^{ts}.

Such an expression is derived by considering the simplified girder segment of figure 7.40 a, in which the axes of girder and top slab are assumed horizontal. The shear flow is first expressed as a function of the sectional forces M_d and V_d:

$$\tau_d b^w = \frac{1}{z}(V_d - V_d^{bs}) = \frac{1}{z}\left(V_d - \frac{M_d}{z}\frac{\Delta z}{\Delta x}\right)$$

The sectional forces are now written in terms of F_d^{ts}:

$$M_d = F_d^{ts} z \quad \text{and} \quad V_d = \frac{\Delta M_d}{\Delta x}$$

The change in moment can be expressed approximately as

$$\Delta M_d = (M_d + \Delta M_d) - M_d = [(F_d^{ts} + \Delta F_d^{ts})(z + \Delta z)] - F_d^{ts} z$$
$$= F_d^{ts} \Delta z + \Delta F_d^{ts} z$$

(Higher-order terms have been neglected.) It therefore follows that

$$V_d = F_d^{ts} \frac{\Delta z}{\Delta x} - z \frac{\Delta F_d^{ts}}{\Delta x}$$

The shear flow can now be rewritten as

$$\tau_d b^w = \frac{1}{z}\left(F_d^{ts} \frac{\Delta z}{\Delta x} + z \frac{\Delta F_d^{ts}}{\Delta x} - \frac{F_d^{ts} z}{z} \frac{\Delta z}{\Delta x}\right) = \frac{\Delta F_d^{ts}}{\Delta x} \cong \frac{dF_d^{ts}}{dx} \qquad (a)$$

This relation is shown in figure 7.40 b.

The area of longitudinal steel required at a given section is proportional to F_d^{ts}. The total amount of longitudinal reinforcement is thus proportional to the integral

$$\int_0^{l/2} F_d^{ts}(x)\, dx \qquad (b)$$

where l denotes span length. The web width and web reinforcement required at a given section are proportional to $\tau_d b^w$, and hence to dF_d^{ts}/dx. The total consumption of shear reinforcement is thus proportional to the following integral:

$$\int_0^{l/2} \frac{dF_d^{ts}(x)}{dx} h(x)\, dx \qquad (c)$$

where h denotes girder depth.

The expressions derived above are now used to compare longitudinal reinforcement, web thickness, and web reinforcement for the two girders of figure 7.41 a. The profile drawn with solid lines varies more uniformly along the length of the girder than the dashed profile, which is flatter near midspan and steeper near the support.

The tensile force in the top slab, plotted in figure 7.41 b, represents the longitudinal reinforcement at a given section. The solid line is linear over most of the length of the girder. This implies that the girder corresponding to the solid line can be reinforced with constant increments of steel for segments of equal length. As can be observed by the areas under the two curves, less longitudinal reinforcement is required by the uniformly varying profile.

Figure 7.41 c represents the web thickness and web reinforcement required at a given section. It is apparent that web thickness and web reinforcement for the uniformly varying girder profile can be constant over a large region of the girder.

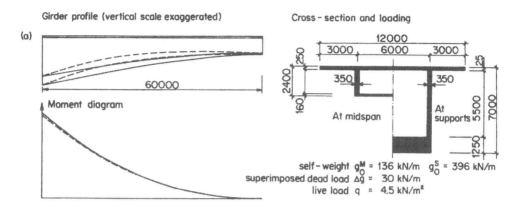

Girder profile (vertical scale exaggerated)

Cross – section and loading

(a)

60000

Moment diagram

12000
3000, 6000, 3000
250
2400
160
350
350
125
At midspan
At supports
5500
7000
1250

self-weight g_0^M = 136 kN/m g_0^S = 396 kN/m
superimposed dead load Δg = 30 kN/m
live load q = 4.5 kN/m²

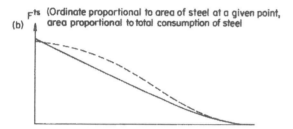

(b) F^{ts} (Ordinate proportional to area of steel at a given point, area proportional to total consumption of steel

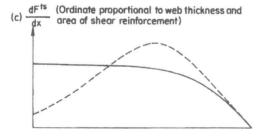

(c) $\dfrac{dF^{ts}}{dx}$ (Ordinate proportional to web thickness and area of shear reinforcement)

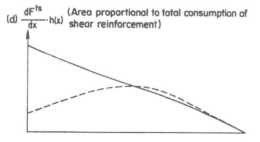

(d) $\dfrac{dF^{ts}}{dx} \cdot h(x)$ (Area proportional to total consumption of shear reinforcement)

Figure 7.41
Effect of girder profile on longitudinal reinforcement, web width, and shear reinforcement (dimensions in mm)

The minimum web thickness required for shear resistance can therefore be easily provided over a substantial portion of the span length. For the dashed, non-uniformly varying girder profile, the minimum thickness required for shear resistance varies considerably and is greatest away from the section of maximum shear force. To facilitate detailing and formwork, the web thickness corresponding to the peak of the curve is usually maintained for the entire length of the girder, resulting in an inefficient use of materials over much of the span.

The total consumption of shear reinforcement is proportional to the area under the curves of figure 7.41 d. It is evident that the uniformly varying girder profile requires more stirrups, since the maximum shear flow occurs in regions of greatest girder depth.

7.4.3 Prestressing Concept

The prestressing concept is chosen to make the ratio of prestressing moments to dead load moments as high as possible while still making efficient use of the prestressing steel. This will result in a prestress as close as possible to form-true, thus minimizing elastic and plastic deformations of the superstructure due to permanent load. For given a given tendon eccentricity and effective prestress, the ratio M_P/M_g is maximized by reducing the longitudinal mild steel to the minimum required for crack control, typically corresponding to a geometric reinforcement ratio between 0.6 and 0.8 percent. The area of prestressing steel, A_P, is then designed to ensure safety at ultimate limit state.

The ratio M_P/M_g near the supports is typically lower for cantilever-constructed bridges than for conventionally built bridges. This is a consequence of a low ratio of live load moment to dead load moment, which is typically equal to 0.2 for cantilever constructed bridges. The effect of M_q/M_g on the value of M_P/M_g can be investigated by considering the behaviour the section shown in figure 7.42. The following inequality must be satisfied for safety at ultimate limit state:

$$\gamma(M_g + M_q) \leqq (A_P f_{Py} + A_s f_{sy}) z$$

where γ denotes the global safety factor, assumed equal to 1.8. Efficient use of the steel is made when both sides of the inequality are equal:

$$\gamma(M_g + M_q) = (A_P f_{Py} + A_s f_{sy}) z \qquad\qquad (a)$$

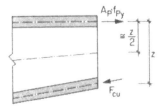

Figure 7.42
Internal forces due to flexure at ultimate limit state

Assuming $M_q = 0.2 M_g$, it follows that

$$\frac{1.2 \gamma M_g}{z} = A_P f_{Py} + A_s f_{sy}$$

The moment due to prestressing, M_P, is defined as

$$M_P = A_P \sigma_P \frac{z}{2} \tag{b}$$

Assuming σ_P and z are constant, the largest value of M_P is obtained by maximizing A_P, subject to the restriction of equation (a). It thus follows that A_s must be as small as possible. Setting A_s to zero, equation (a) is transformed into

$$\frac{1.2 \gamma M_g}{z f_{Py}} = A_P$$

Substituting this expression for A_P into equation (b) yields

$$M_P = 0.6 \gamma \frac{\sigma_P}{f_{Py}} M_g$$

Assuming $\sigma_P = 0.7 f_{Py}$, it follows that $M_P = 0.76 M_g$. The maximum moment due to prestressing consistent with efficient use of reinforcement is thus significantly less than the dead load moment.

Any difference between M_P and M_g will result in elastic and plastic deformations of the girder, which must be compensated by camber. When the structure is made continuous by a monolithic closure pour, these deflections will produce an undesirable redistribution of moments from the strong support region to the weaker midspan region. These effects should be prevented as much as possible by providing the largest possible ratio M_P/M_g, i.e., by minimizing the consumption of longitudinal mild reinforcing steel.

The axial component of stress due to prestressing, N_P/A_c, is substantially higher for cantilever-constructed bridges than for conventionally built bridges. In spite of the relatively large difference between M_P and M_g, therefore, the proposed prestressing concept normally results in full prestress for dead load and a substantial fraction of live load.

Minimizing the consumption of longitudinal mild reinforcing steel has the added advantage of reducing construction costs. Longitudinal mild steel cannot be used efficiently in cast-in-place segmental bridges since a large proportion of it will be taken up by lap splices at the construction joints between segments. In addition, bars projecting at the end of each segment complicate the installation and coupling of tendon ducts.

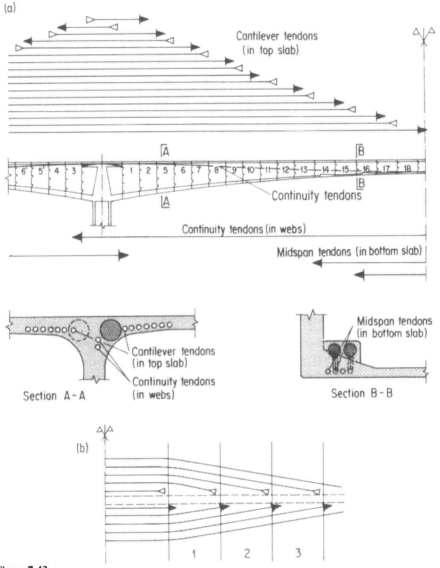

Figure 7.43
Prestressing arrangement in cantilever-constructed bridges: a, longitudinal section and schematic
layouts; b, plan: cantilever tendons

7.4.4 Tendon Layout

a) Cantilever Tendons

The cantilever tendons are the principal reinforcement of the structural system.
They are located in the deck slab and are anchored at the ends of the segments
(fig. 7.43). In this way, an efficient arrangement that closely balances the moment

diagram can be achieved. Anchors are preferably located at the intersection of deck slab and web. They can also be located lower down in the webs, however, provided the webs are thick enough to accommodate the anchor hardware. The small savings in the web reinforcement obtained in this way are usually outweighed by complications in detailing since the tendons must cross the plane of the stirrups to be anchored.

The top slab of each segment (other than the cantilever tip segments) must contain empty ducts for tendons that will be subsequently installed and stressed. These ducts must be spliced at the construction joints between segments. The arrangement of the tendons should therefore make possible the reuse of formwork at the segment ends by ensuring a regular pattern of holes for the tendon ducts. The clear distance between the ducts should not be less than 80 mm. The vertical shear resistance of the top slab is considerably reduced by the presence of the closely spaced ducts; splitting of the slab in the plane of the tendons must be prevented by stirrups between the ducts (fig. 7.44). Horizontal shear in the slab due to the spreading of the concentrated anchor forces must also be considered, especially during construction when the ducts are empty.

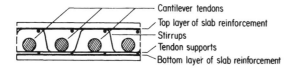

Cantilever tendons
Top layer of slab reinforcement
Stirrups
Tendon supports
Bottom layer of slab reinforcement

Figure 7.44
Top slab details

It is often practical to place the tendons directly onto the lower mat of slab reinforcement, thus eliminating the need for special tendon supports. The small reduction in tendon eccentricity is offset by greater ease in concreting. The arrangement of the tendons must not impair the flexural resistance of the slab by reducing the depth of the compression block.

Holes in the ducts or improperly made splices can lead to penetration of cement, which may hinder the placement of the prestressing steel. PVC pipes should therefore be placed inside the ducts during concreting. When these pipes are sufficiently stiff, it may be possible to eliminate tendon supports for short segments.

b) Midspan Tendons

The midspan tendons, located in the bottom slab near the webs, resist positive moments in the middle third of the span (fig. 7.43). Build-outs for the anchors should be located as near as possible to the webs and properly connected to them by reinforcement. The pull-out forces of the midspan tendons resulting from curvature of the girder soffit must be considered in the design of the bottom slab.

c) *Continuity Tendons*

Continuity tendons in the girder webs are sometimes used in special cases (fig. 7.43). These tendons can be used as reserve prestressing, designed on the basis of the *actual* deformations measured after closure. The ducts, which must be cast into the concrete before the definitive area of steel has been selected, should therefore be conservatively detailed. To limit their stressing length, continuity tendons are normally lapped over the pier table and anchored in the webs. Continuity tendons complicate the placement and vibration of concrete in the webs.

7.4.5 Preliminary Design and Special Design Considerations

a) *Estimate of the Main Reinforcement*

The thickness of the top slab and the arrangement of the intersection of top slab and webs cannot be detailed without knowing the number and size of prestressing tendons required. The estimates of A_P and A_s at the supports and at midspan presented in this section will normally be sufficient to finalize these details, and hence to establish the definitive cross-section dimensions at these two locations.

When hinges are provided at midspan, the reinforcement at the supports can be determined using the following inequality:

$$M_d^s(g + q) \leq \frac{1}{\gamma_R} (A_P^s f_{Py} + A_s^s f_{sy}) z$$

where $M_d^s(g + q)$ is the cantilever moment due to full dead load plus live load and z is assumed equal to the distance between the centroid of the deck slab and the centroid of the bottom slab. As stated in Section 7.4.3, A_s^s should correspond to a geometrical reinforcement ratio of 0.6 to 0.8 percent of the area of the top slab. The reinforcement in the midspan region will generally be very small.

When the cantilevers are monolithically connected at midspan, an envelope of design moments must be calculated at the supports and at midspan. The moment at the support consists of the following components:

1. $M^s(g_0)$:   cantilever moment due to dead load
2. ΔM: moment redistribution due to creep
3. $M^s(\Delta g, q)$:   continuous system moment due to superimposed dead load and live load
4. $M_{sP}(m, c)$:   redundant moment due to midspan and continuity prestressing

The reinforcement at the supports can therefore be determined from the inequality

$$M_d^s(g_0) + M_d^s(\Delta g, q) + \Delta M + M_{sP}(m, c) \leq \frac{1}{\gamma_R} (A_P^s f_{Py} + A_s^s f_{sy}) z \quad \text{(a)}$$

The moments ΔM and $M_{sP}(m, c)$ are positive and small relative to the negative moments $M^s(g_0)$ and $M^s(\Delta g, q)$. Since any error induced by neglecting ΔM and $M_{sP}(m, c)$ will be small and on the conservative side, inequality (a) can be rewritten as

$$M_d^s(g_0) + M_d^s(\Delta g, q) \leq \frac{1}{\gamma_R} (A_P^s f_{Py} + A_s^s f_{sy}) z$$

The moment at midspan consists of the following components:

1. ΔM: moment redistribution due to creep
2. $M^m(\Delta g, q)$:   continuous system moment due to superimposed dead load and live load
3. $M_{sP}(m, c)$:   redundant moment due to midspan and continuity prestressing

Since they increase $M^m(\Delta g, q)$, the moments ΔM and $M_{sP}(m, c)$ cannot be neglected at midspan. The value of ΔM can be estimated using the following expression, assuming the flexural stiffness at midspan is between 0.05 and 0.10 times the flexural stiffness at the supports:

$$-0.10 [M^s(g_0) + M_{0P}^s(ca)] \leq \Delta M \leq -0.15 [M^s(g_0) + M_{0P}^s(ca)]$$

where $M^s(g_0)$ and $M_{0P}^s(ca)$ are the moments at the support in the cantilever system due to dead load and cantilever tendons, respectively. When the prestressing concept proposed in Section 7.4.3 is used, the inequality simplifies to

$$-0.02 M^s(g_0) \leq \Delta M \leq -0.03 M^s(g_0)$$

The value of $M_{sP}(m, c)$ is a function of many different factors, including member stiffness, length of the midspan tendons, and layout of the continuity tendons. The following approximations for $M_{sP}(m, c)$ are valid for haunched girders:

$$M_{sP}(m, c) = M_{sP}(m) + M_{sP}(c)$$

where

$$M_{sP}(m) \cong -0.75 M_{0P}^m(m) \quad \text{and} \quad M_{sP}(c) \cong -0.5 M_{0P}^m(c)$$

and $M_{0P}^m(m)$ and $M_{0P}^m(c)$ are the primary system moments at midspan due to the midspan and continuity tendons, respectively.

The reinforcement required at midspan can now be calculated using the following inequality:

$$M_d^m(\Delta g, q) + \Delta M + M_{sP}(m, c) \leq \frac{1}{\gamma_R} (A_P^m f_{Py} + A_s^m f_{sy}) z$$

where A_P^m is the total area prestressing steel at midspan, including midspan and continuity tendons, A_s^m is the area of reinforcing steel in the bottom slab, and the internal lever arm, z, can be approximated by the distance between the centroids of the top and bottom slabs.

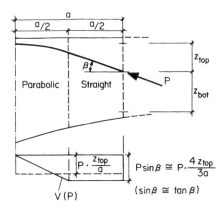

$P\sin\beta \cong P\cdot\dfrac{4\,z_{top}}{3a}$

$(\sin\beta \cong \tan\beta)$

Figure 7.45
Contribution of inclined longitudinal tendons
to shear resistance

b) Shear Resistance of Inclined Tendons Anchored in the Webs

Tendons that are carried down into and anchored in the webs can be considered in calculating the shear resistance of the cross-section. They are only effective in this regard, however, in the portion of the web located above the anchor. Stirrups must be provided in the web below the anchor to equilibrate the shear stresses in this portion of the web.

The vertical component of the force in the inclined tendon, $P\sin\beta$, acts opposite to the shear force at the section (fig. 7.45). The value of P at ultimate limit state is a function of the angle of inclination, β. If β is less than $45°$, P must be taken as the effective prestressing force. Higher values of P may be used only when β is sufficiently large.

The force $P\sin\beta$ is equilibrated by the shear flow τb^w, acting over the upper portion of the web:

$$P\sin\beta = \tau b^w z_{top}$$

This shear flow must be equilibrated in the lower portion of web with stirrups, assumed vertical:

$$\tau b^W z_{bot} = \frac{A_s^w f_{sy}}{s} z_{bot}\cot\alpha$$

where s is the stirrup spacing and α is the angle of inclination of the compression struts in the concrete. The area of steel required for the stirrups in the lower portion of the web is therefore given by

$$A_s^w = \frac{\tau b^w s \tan\alpha}{f_{sy}} = \frac{P\sin\beta s \tan\alpha}{z_{top} f_{sy}} \tag{b}$$

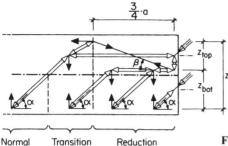

Normal   Transition   Reduction
zone   zone   zone

Figure 7.46
Truss model for development of shear resistance

Away from the zone of influence of the inclined web (denoted *Normal Zone* in figure 7.46), the shear flow τb^w must be equilibrated by stirrups extending over the full depth of the web. The required area of steel is given by

$$\frac{A_s^w f_{sy}}{s} z \cot \alpha = \tau b^W z$$

from which it follows that

$$A_s^w = \frac{\tau b^w s \tan \alpha}{f_{sy}} = \frac{P \sin \beta s \tan \alpha}{z_{top} f_{sy}}$$

which is identical to equation (b), derived for the stirrups in the region of influence of the inclined tendon. The area of web reinforcement corresponding to the contribution of the inclined tendons thus remains constant in the normal, transition, and reduction zones of the girder (fig. 7.46).

The upper end of the stirrups must be anchored above the tendon anchor in the reduction zone and at the top of the section in the normal zone. In the transition zone, the length of the stirrups varies linearly between these two extremes.

The total shear resistance of the section consists of the contributions of the inclined tendons, the corresponding partial-depth stirrups computed above, the vertical component of the compressive force in the bottom flange, V_d^{bs} (fig. 7.40), and any additional full-depth stirrups. The tensile force in the chord is determined directly from the truss model.

7.4.6 Calculation of Camber and Casting Elevations

Deformations of the falsework and of the structure itself must be compensated by *camber*, to ensure that the design profile of the bridge can be achieved after completion of the bridge. The prescribed camber is equal to the total deflection due to dead load plus a small portion of the deflection due to live load and temperature gradient. The plastic component of dead load deflection to be balanced is measured from time of construction to roughly one half of the design life of the bridge. A reliable calculation of girder deformations requires careful consideration of the modulus of elasticity of concrete, creep, shrinkage, relaxation of prestressing steel, and displacements of the foundations.

Large deflections occur in cantilever constructed bridges due to the peculiarities of the construction sequence and the prestressing concept. The following deformations must be considered:

1. Deflection of the traveller during concreting of the segments
2. Deformations of the cantilever system before closure due to the dead load of the cantilever segments, travellers, and closure segment
3. Deformations of the continuous system after closure due to midspan and continuity tendons, superimposed dead load, and possibly a small portion of the deformation due to live load and temperature gradient
4. Deflections of the girder induced by deformations of the columns and settlement of the foundations

Before casting, the formwork at the free end of segment $i+1$ must be set to a specific elevation, given by the sum of the deflections after casting and prestressing, Δ_{i+1}, and the form camber Δ^0_{i+1} (fig. 7.47).

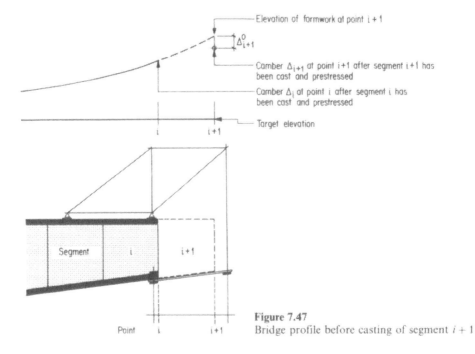

Elevation of formwork at point $i+1$

Δ^0_{i+1}

Camber Δ_{i+1} at point $i+1$ after segment $i+1$ has been cast and prestressed

Camber Δ_i at point i after segment i has been cast and prestressed

Target elevation

Segment $\quad i \quad\quad i+1$

Point $\quad i \quad\quad\quad i+1$

Figure 7.47
Bridge profile before casting of segment $i+1$

a) Form Camber Δ_{i+1}^0

Form camber, denoted Δ_{i+1}^0, compensates for the deflections at the tip of segment $i+1$ due to: (1) deformation of the traveller under the weight of segment $i+1$, and (2) deformation of the previously cast cantilever (segments 1 though i) due to the dead load and prestressing of segment $i+1$.

The deflections of the traveller can normally be calculated with good reliability and can also be checked on site at the beginning of the cantilever construction. Deformations of the traveller must be limited to prevent the webs from breaking away at the construction joint during casting of the deck slab. The formwork is normally attached to to the preceding segment, thus eliminating relative displacements of the two adjacent segments and reducing the load on the traveller by half (fig. 7.47). The deflections of the cantilever can be calculated using the method of virtual work with a unit load $\bar{Q} = 1$ at point $i+1$. The flexural stiffness of the uncracked section can be assumed.

b) Camber Δ_i

Camber Δ_i compensates for the deflections of point i. These deflections are accumulated beginning immediately after segment i has been prestressed up to approximately half the design life of the bridge. Three components are considered: (1) deflection of the cantilever *before closure*, δ_i^{bc}, (2) deflection of the cantilever due to the *closure* segment, δ_i^c, and (3) deflections of the continuous system *after closure* δ_i^{ac}.

1. Deflection of the Cantilever Before Closure δ_i^{bc}. This deflection is the sum of all deflections occurring from immediately after the prestressing of segment i to the completion of the final segment n and removal of the traveller (fig. 7.48). It can be calculated approximately using the following expression:

$$\delta_i^{bc(i,n)} \cong (1 + \Delta\phi)\,\delta_i^{bc(0,n)}(g_0, P) - (1 - \Delta\phi)\,\delta_i^{bc(0,i)}(g_0, P)$$
$$+ \Delta\phi\,\delta_i^{bc(n)}(T) - \delta_i^{bc(i)}(T)$$

where the following component deflections of point i are considered:

$\delta_i^{bc(0,n)}(g_0, P)$: due to g_0 and P applied over the entire cantilever length
$\delta_i^{bc(0,i)}(g_0, P)$: due to g_0 and P applied between points 0 and i
$\delta_i^{bc(n)}(T)$: due to weight of traveller, T, applied at point n
$\delta_i^{bc(i)}(T)$: due to T applied at point i

The partial creep coefficient $\Delta\phi$ is a function of the long-term creep coefficient, ϕ, and can be taken approximately as $0.1\,\phi$ for i in the first quarter of the cantilever, $0.05\,\phi$ for i in the third quarter of the cantilever, and 0 for i at the end of the cantilever. The prestressing force P is defined as the effective average force during the construction of segments i through n. Eventual foundation settlements and column deformations that occur during this time period must also be considered.

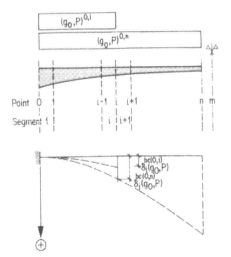

Figure 7.48
Elastic deflections during cantilever construction

2. Deflection of the Cantilever Due to the Closure Segment δ_i^c. These deflections of point i occur in the time interval that begins immediately the traveller has been removed from the tip of the completed cantilever and ends after the closure segment, denoted m, has been cast. They are caused by the following actions: (1) loss of prestress during this time interval, (2) creep during this time interval, and (3) self-weight of the closure segment.

3. Deflections of the Continuous System After Closure δ_i^{ac}. The deflections in this final phase consist of: (1) residual plastic deformations due to self-weight and cantilever prestressing, (2) residual loss of prestress in the cantilever tendons, (3) midspan and continuity prestressing, including all prestressing losses, (4) superimposed dead load, (5) possibly a small component of live load and temperature gradient, and (6) residual foundation settlement and column deformations occurring after closure. The calculation of items (2) through (6) is made for the final continuous system. Item (1) is calculated for the cantilevers, taking into account the redistribution of moments due to the change of system.

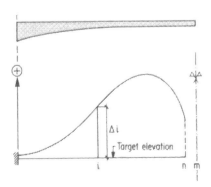

Figure 7.49
Camber curve Δ_i

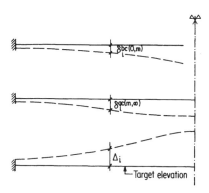

Figure 7.50
Deflections and camber for precast segmental cantilever construction

Camber of the entire girder can be reliably determined by interpolation between values of Δ_i calculated at four to six points. Resulting camber profile is shown schematically in figure 7.49.

The calculation of camber for precast segmental bridges is somewhat simpler. The desired camber profile, Δ_i, is the sum of the deflected shape of the cantilever at closure, $\delta_i^{bc(0,m)}$ and the deflected shape of the system in its final state $\delta_i^{ac(m,\infty)}$ (fig. 7.50). Camber Δ_i must be built into each segment during precasting. Cantilever construction with precast segments requires the precise alignment of pier segments. Adjustable bearings are helpful in this regard.

7.5 Skew Girder Bridges

7.5.1 Conceptual Design

The supports of skew grade separations and river crossings can often be aligned perpendicular to the longitudinal axis of the bridge. The use of skew supports, however, is normally preferable for the following reasons:

1. Abutments and piers can be properly integrated into the landscape
2. Span lengths are minimized
3. Piers of river crossings can be oriented parallel to the direction of flow and designed with an appropriate hydraulic profile, thus maximizing the channel cross-section and reducing scour

Skew bridges do, however, require long, complicated abutments with costly expansion joint details. In addition, the costs of superstructure falsework and formwork are generally higher than for non-skew bridges.

Skew girders tend to carry loads to the supports using the shortest path. The direction of principal stress due to bending and torsion in the horizontal plane of the top and bottom fibres lies between the girder axis and the normal to the support axes. The supports in the obtuse corners, moreover, are more heavily loaded than those in the acute angle corners.

The structural behaviour of skew bridges is a function of the ratio of torsional stiffness to bending stiffness. The edges of the cross-section (taken perpendicular to the longitudinal axis of the bridge) deflect differently; the cross-section twists as a result (fig. 7.51). Torsional moments proportional to torsional stiffness are required for compatibility; these moments are small in T-girders, which are relatively flexible in torsion.

The compatibility torsional moments induce a force couple and longitudinal bending moments at the ends of the girder. At the girder ends, rotations are only possible about a vector parallel to the support axis; rotations about the vector ω_n, normal to the support axis, are restrained by the end diaphragm (fig. 7.52). If the girder is also fixed against rotations about the vector ω_y, normal to the girder axis, total rotational fixity is produced, since ω_n and ω_y are linearly independent. Rotations and moments due to compatibility torsion are therefore prevented in girders that are restrained against flexural rotation and twist at both ends; their behaviour is not affected by the skew angle.

Flexural and torsional stiffness are reduced by the formation of cracks. Pure torsion in the absence of bending results in cracking over the full periphery of the section and thus in a sharp decrease in torsional stiffness. When coupled with bending, however, the decrease in torsional stiffness is less pronounced, since no shear cracks are formed in the flexural compression zone. The deformations of prestressed girders due to load and prestressing, and hence the torsional moments necessary for compatibility, are normally small. Under service conditions, therefore, the reduction in torsional stiffness due to cracking is not significantly greater than the reduction in flexural stiffness; the restraint at the ends of the girder does not fall off suddenly with a corresponding redistribution of moments. At ultimate limit state, the ratio of flexural stiffness to torsional stiffness is of little significance, provided torsional and flexural behaviour is ductile.

The end diaphragms are oriented along the support axis and are designed as panels, the behaviour of which is primarily in-plane. They must be extended out to the edge of the deck slab cantilevers, to provide a rigid support for the expansion joint and for the long deck overhang in the acute angle corner (fig. 7.53). All sharp corners should be broken off by surfaces of roughly 0.2 m in width.

The end diaphragms of box girders serve primarily to transform the moment due to the unequal vertical reactions, $X_1 = (A_2 - A_1) b_0'/2$, into a shear flow (fig. 7.54). The shear flow induces the end moment $M = X_1 \cos \alpha$ and the torsional moment $T = X_1 \sin \alpha$ in the girder. The end diaphragm is therefore stressed primarily in shear. The vertical reactions of open sections, which normally differ by only a small amount, produce bending and torsional moments in the girder webs. The end diaphragms are thus primarily stressed in bending (fig. 7.55).

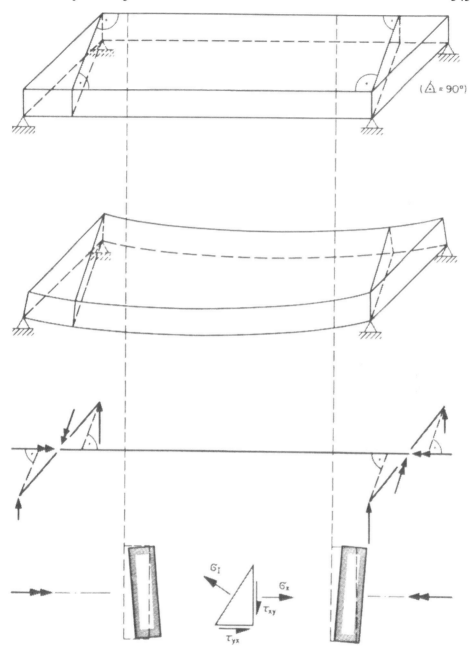

Stresses in the lower fibres of the girder

Figure 7.51
Structural behaviour of skew girder bridges

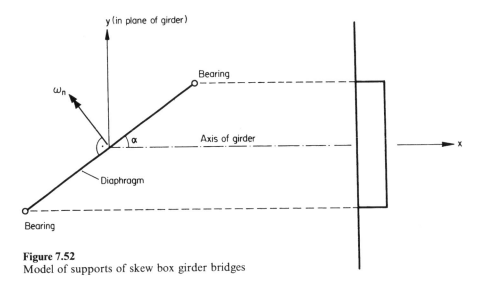

Figure 7.52
Model of supports of skew box girder bridges

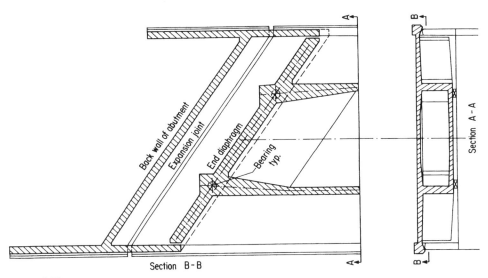

Figure 7.53
End diaphragms of skew box girder bridges

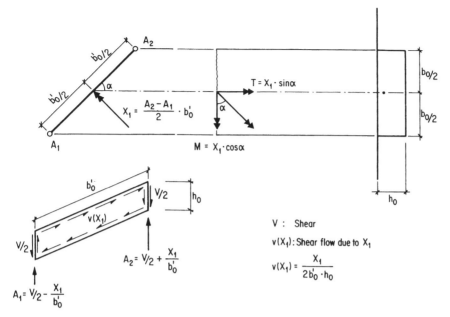

Figure 7.54
Shear flow in an end diaphragm of a skew box girder bridge

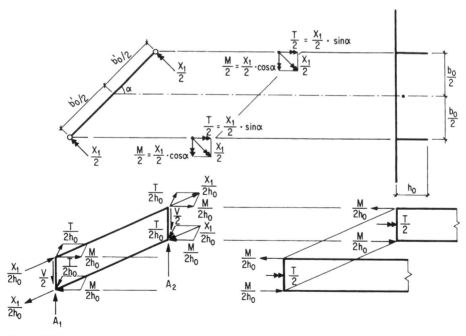

Figure 7.55
Bending in an end diaphragm of a skew double-T girder bridge

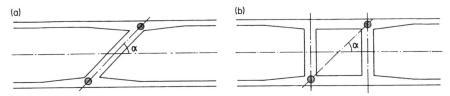

Figure 7.56
Intermediate diaphragms of skew bridges: a, single diaphragm parallel to axis of support; b, twin
diaphragms perpendicular to longitudinal axis of bridge

At intermediate supports, single skew diaphragms oriented along the axis of the
support or twin diaphragms oriented perpendicular to the longitudinal axis of the
bridge can be used (fig. 7.56). Whichever arrangement is chosen has little effect on
the structural behaviour of girders with open sections.

The supports of skew bridges are designed following the same principles as
outlined in Section 6.1 for bridges with supports at right angles to the bridge axis.
Whenever possible, fixed bearings should be provided at one of the abutments.
When skew bridges are supported flexibly in the longitudinal direction, the strong
axis of skew piers is not fully effective in resisting loads transverse to the
longitudinal axis of the bridge. Since the load cannot be directly transferred to the
abutment, a large portion of it is resisted by the weak axis of the pier (fig. 7.57).

For long bridges, the direction of movement of expansion bearings should always
be set parallel to the longitudinal axis of the bridge. When the system is sufficiently
flexible, the redundant forces induced by the component of displacement tangent
to expansion joints are small can be neglected. For short bridges, however, the
complications in joint detailing thus created can be avoided by setting the bearings
perpendicular to the axis of the joint. The resulting transverse displacement should
be accounted for in the design.

The axis of rotation of skew diaphragms (fig. 7.56a) is always identical to the
support axis. Deformations of the diaphragm are small and can be neglected in the
design of the bearings. When the diaphragms are oriented perpendicular to the
bridge axis (fig. 7.56b), the axis of rotation varies with girder loading. In such
cases, therefore, bearings that can rotate about two perpendicular axes should
always be provided.

It may be practical to support bridges with very pronounced skew on abutments
that are trapezoidal in plan (fig. 7.58). Although the effective length of the bridge
is increased somewhat by this solution, the abutment arrangement and detailing of
the expansion joints is considerably simplified. The abutments of twin parallel
bridges should be staggered according to the skew angle and designed as in the
normal case with the support axis perpendicular to the bridge axis (fig. 7.59).

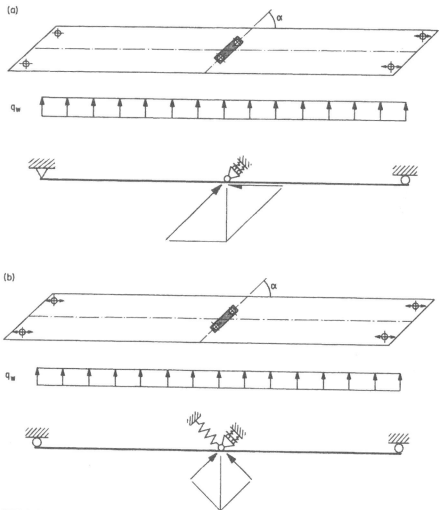

Figure 7.57
Models for the behaviour of skew bridges under transverse load: a, fixed supports; b, flexible supports

7.5.2 Calculation of the Sectional Forces

a) Single-Span Girders

Skew girder bridges can be modeled using beam elements, provided the ratio of span length to width of cross-section is greater than 1.5. The choice of model normally depends on the type of cross-section.

Girders with open section are modelled by plane grids. The longitudinal elements represent the webs and the transverse elements represent diaphragms and deck

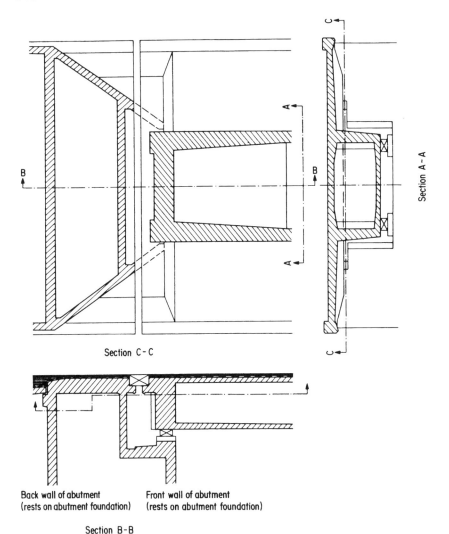

Section C-C

Section A-A

Section B-B

Back wall of abutment
(rests on abutment foundation)

Front wall of abutment
(rests on abutment foundation)

Figure 7.58
Trapezoidal abutment for single skew bridge

slab (fig. 7.60 a). This model yields sectional forces in the individual cross-section components and not the sectional forces of the entire cross-section. For preliminary design, the girder webs can be considered as independent of each other, i.e., not connected by transverse members.

Girders with closed section can be modelled as a single beam on skew supports (fig. 7.60 b). The diaphragms are assumed to be totally rigid. This model yields sectional forces of the entire section, which can then be apportioned to the individual webs. It should not be used for the design of diaphragms, for which local investigations are necessary.

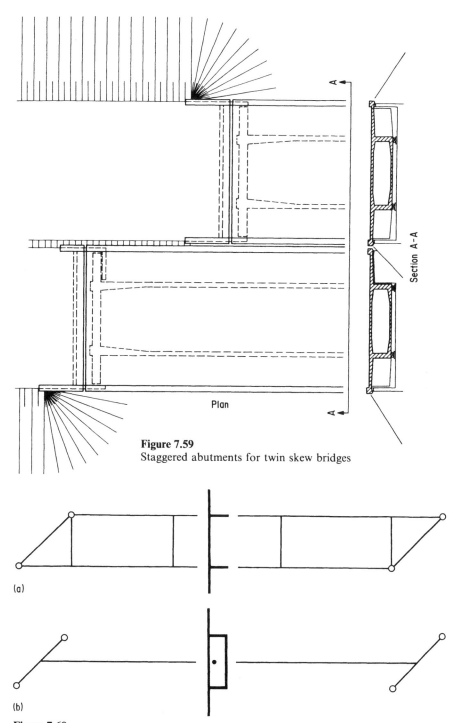

Figure 7.59
Staggered abutments for twin skew bridges

Figure 7.60
Models for single-span skew bridges: a, open section; b, closed section

Single-span skew bridges with closed cross-section can be most easily analyzed using the force method. The model is statically indeterminate to the first degree for vertical loading (four support reactions, three equilibrium equations). Releasing the moment at the connection of the girder to one of the diaphragms (point B in figure 7.61 a) yields the statically determinate primary system. The bending and torsional moments in the primary system due to the redundant force X_1 are shown in figure 7.61 b.

Example 7.5:
Single-span skew bridge with uniform load

The structural system and uniform load q are shown in figure 7.62 a. The skew angles at the supports are denoted α and β. The redundant force X_1 is obtained from the solution of the following compatibility condition:

$$\delta_1 = \delta_{10} + X_1\,\delta_{11} = 0$$

where

$$\delta_{ik} = \int \bar{M}\,\frac{M}{EI}\,dx + \int \bar{T}\,\frac{T}{GK}\,dx$$

It therefore follows that

$$X_1 = -\frac{ql^2}{8}\,\frac{\cot\alpha + \cot\beta}{\sin\beta\,(\cot^2\alpha + \cot\alpha\,\cot\beta + \cot^2\beta + 3\kappa)}$$

where $\kappa = EI/GK$. When $\alpha = \beta$, the expression simplifies to

$$X_1 = -\frac{ql^2}{8}\,\frac{2\cot\alpha}{3\sin\alpha\,(\cot^2\alpha + \kappa)}$$

The bending and torsional moments at the supports A and B are thus given by

$$M^A = M^B = X_1\,\cos\alpha$$
$$T^A = T^B = X_1\,\sin\alpha$$

The end restraint and the torsion in the girder thus increase with increasing skew angle, α, and decreasing values of κ.

When the skew angles at the support axes are not equal, the web span lengths will be unequal. The distribution of the sectional forces of the total cross-section to the girder webs can be done rigourously using the theory of shells, or using the approximate expressions of figure 7.63.

The moment X_1 is applied to the support diaphragm as a closed shear flow (fig. 7.64). The corresponding design shear flow, $v_d(X_1)$, is transferred from the

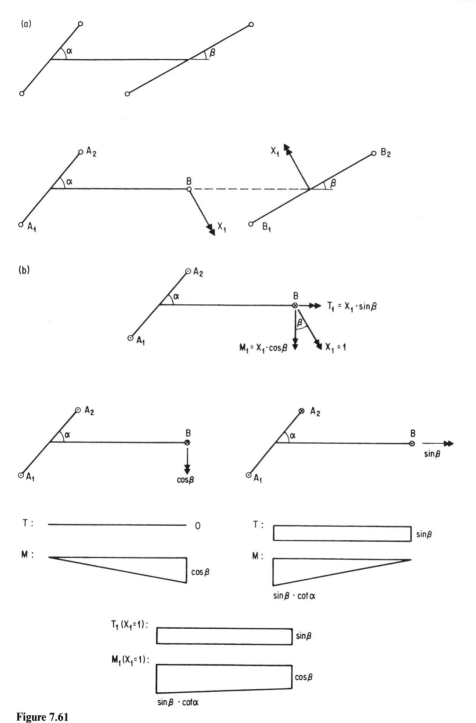

Figure 7.61

Analysis of a single-span skew bridge with closed cross-section: a, model and primary system; b, flexural and torsional moments due to $X_1 = 1$

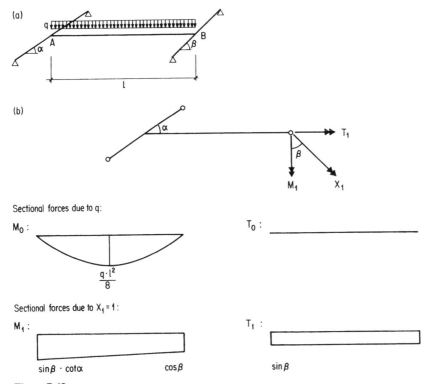

Figure 7.62
Example 7.5: a, model and loading; b, sectional forces in primary system

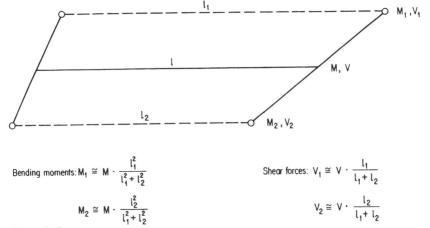

Figure 7.63
Distribution of total sectional forces to the girder webs

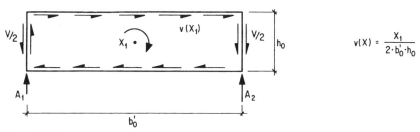

Figure 7.64
Shear flow in an end diaphragm

diaphragm to the supports A_1 and A_2 by an orthogonal layout of stirrups and longitudinal bars. This reinforcement must be properly anchored at the edges of the diaphragm. The design of the reinforcement is based in the following two inequalities:

$$v_d(X_1) \leqq \frac{1}{\gamma_R} \frac{A_s^W f_{sy}}{s^W} \qquad \text{(stirrups)}$$

$$v_d(X_1) \leqq \frac{1}{\gamma_R} \frac{A_s^L f_{sy}}{s^L} \qquad \text{(longitudinal reinforcement)}$$

where s^W and s^L denote the spacing of stirrups and longitudinal reinforcement, respectively.

In the top and bottom slabs, the shear flow $v(X_1)$ is in equilibrium with the slab tensile and compressive forces. These forces can be easily determined from the truss model corresponding to the given arrangement of reinforcement.

Example 7.6:
Truss models for the calculation of tensile and compressive forces in the top and bottom slabs due to the shear flow $v(X_1)$

The geometrical and statical relationships at the support are given in figure 7.65. The skew angle is $45°$. The bending and torsional moments at the ends of the girder are thus of equal magnitude and the moment X_1 is $M/\sqrt{2}$.

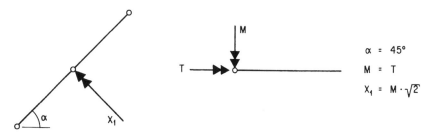

Figure 7.65
Example 7.6: geometry and sectional forces at the intersection of girder and end diaphragm

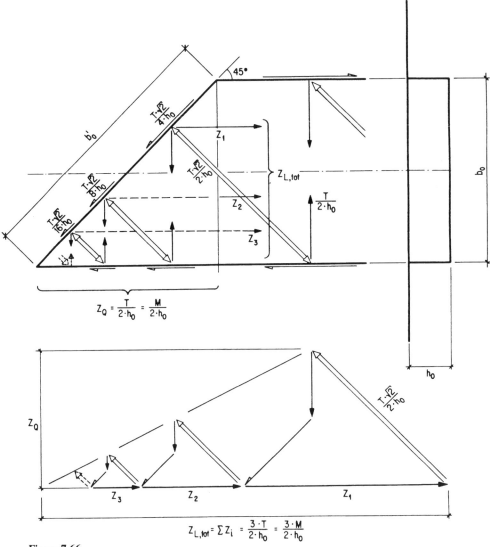

Figure 7.66
Example 7.6: truss model for the top slab

1. Top slab: It is assumed that the reinforcement is arranged orthogonally and the compression struts are inclined at 45° to the longitudinal axis of the girder. The longitudinal and transverse tensile forces can therefore be expressed as follows (fig. 7.66):

Longitudinal tension Z_L:

a) Due to bending: $Z_L(M) = \dfrac{M}{h_0}$

b) Due to torsion: $Z_L(T) = \dfrac{T}{2h_0 b_0} b_0 = \dfrac{T}{2h_0}$

c) Total: $Z_{L,\,tot} = \dfrac{3}{2}\dfrac{M}{h_0}$

Transverse tension Z_Q:

a) Due to torsion: $Z_Q(T) = \dfrac{T}{2h_0 b_0} b_0 = \dfrac{T}{2h_0} = \dfrac{M}{2h_0}$

The shear flow $v(X_1)$ at the connection of the slab to the diaphragm can be expressed as

$$v(X_1) = \frac{X_1}{2b'_0 h_0} = \frac{M\sqrt{2}}{2b_0\sqrt{2}h_0} = \frac{M}{2b_0 h_0}$$

The total shear force $V(X_1)$ along the interface is thus given by

$$V(X_1) = v(X_1)b'_0 = \frac{M}{2h_0}\sqrt{2}$$

The resulting state of equilibrium is shown in figure 7.67.

2. Bottom slab: For the given skew angle (45°) and orthogonal reinforcement arrangement, the longitudinal compressive force due to bending is greater than the longitudinal tensile force due to torsion. The longitudinal and transverse forces in the slab can be expressed as follows (fig. 7.68):

Longitudinal compression D_L:

a) Due to bending: $D_L(M) = \dfrac{M}{h_0}$

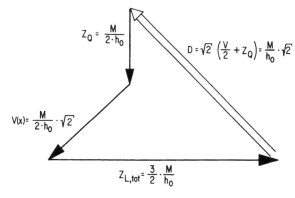

Figure 7.67
Example 7.6: equilibrium of forces at the connection of top slab to diaphragm

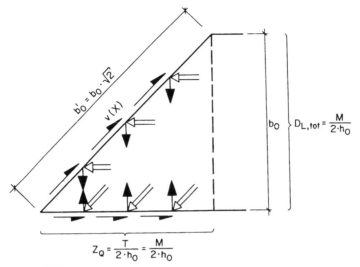

Figure 7.68
Example 7.6: truss model for the bottom slab

b) Due to torsion: $D_L(T) = -\dfrac{T}{2h_0 b_0} b_0 = -\dfrac{T}{2h_0}$

c) Total: $D_{L,\,tot} = \dfrac{M}{2h_0}$

Transverse tension due to torsion:

$$Z_Q(T) = \frac{T}{2h_0 b_0} b_0 = \frac{T}{2h_0} = \frac{M}{2h_0}$$

As in the top slab, therefore, the total shear force along the slab-diaphragm interface, $V(X_1)$, is $\sqrt{2}\,M/2h_0$. The resulting state of equilibrium is shown in figure 7.69.

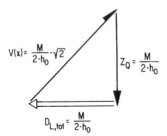

Figure 7.69
Example 7.6: equilibrium of forces at the connection of bottom slab to diaphragm

b) Continuous Girders

The arrangement of diaphragms at the intermediate supports has little influence on structural behaviour of girders with open cross-section. Plane grid models, by which the sectional forces in the girder webs, diaphragms, and deck slab can be directly calculated, are recommended.

The structural behaviour of girders with closed section, and hence the models used for analysis, vary with the arrangement of diaphragms. When skew diaphragms are used (fig. 7.56a), the girder is modeled as a single beam; the diaphragms are modeled as rigid elements. The elastic restraint of columns monolithically connected to diaphragms can be represented by a torsional spring (fig. 7.70). The bending moments and shear forces of the entire cross-section can be distributed equally to the webs, since the webs participate equally to the resistance of the total sectional forces.

The effect of skew diaphragms on the forces in the cross-section elements is small. Since the angle of flexural rotation of the girder at intermediate supports is normally small, compatibility torsional moments and the corresponding discontinuity in the bending moment diagram will be insignificant. For preliminary calculations, therefore, the skew at the supports can be neglected.

Diaphragm pairs perpendicular to the longitudinal axis of the bridge can be modelled as rigid members extending out from the girder (fig. 7.71). Their length is the distance between the girder axis and the bearing centreline. This model yields only the sectional forces of the entire cross-section; any difference in forces in the two webs is not considered. Grid models are therefore superior for girders with closed cross-section and perpendicular diaphragms. One half of the flexural and torsional stiffness of the total section is distributed to each web; the diaphragms are considered rigid in bending and perfectly flexible in torsion (fig. 7.72).

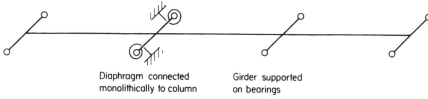

Diaphragm connected   Girder supported
monolithically to column   on bearings

Figure 7.70
Model for continuous skew girders with skew diaphragms

Figure 7.71
Single-beam model for continuous skew girders with diaphragms perpendicular to the longitudinal bridge axis

Figure 7.72
Plane grid model for continuous skew girders with diaphragms perpendicular to the longitudinal
bridge axis

At a given section cut parallel to the skew angle, moments and shear forces in the
webs are unequal. The moments at the supports and at midspan are somewhat
smaller than the corresponding forces in a system with skew diaphragms, since a
portion of the reaction at the perpendicular diaphragms is transferred to the other
web. At sections cut perpendicular to the bridge axis, the sectional forces
computed by the single beam model and the average forces computed by the grid
model agree well. The torsional moments computed by both models also agree
well, provided the torsional moments in the individual webs and the warping
torsional moment due to differential web bending are considered.

Example 7.7:
Models for continuous skew girders with closed cross-sections

The structural behaviour of a two-span continuous girder with two different
diaphragm arrangements is considered in this example. Sectional forces in a girder
with skew diaphragms are given in figure 7.73 a. Figures 7.73 b and 7.73 c give the
sectional forces in a girder with perpendicular diaphragms computed using a
single-beam model and a plane grid model, respectively.

The shear forces and torsional moments of the girder are transferred to the
diaphragms through shear flow in the slabs and webs. The resultant shear flow
applied to the diaphragms is calculated from the difference of the shear forces and
torsional moments on either side (fig. 7.74).

The shear flow due to the symmetrical components of shear force can be
calculated using the usual methods of structural mechanics. The shear flow due to
the antisymmetrical components is obtained from the superposition of an
antisymmetrical shear flow and a redundant St. Venant shear flow around the
cross-section. The shear flow in the webs must be equal to the vertical shear forces;
the resultant shear force in the top and bottom slabs must vanish. A closed shear
flow can be assumed for the torsional moment.

Figure 7.73
Example 7.7: models and sectional forces (dimensions in mm): a, skew diaphragms; b,
perpendicular diaphragms, single-beam model; c, perpendicular diaphragms, plane grid model

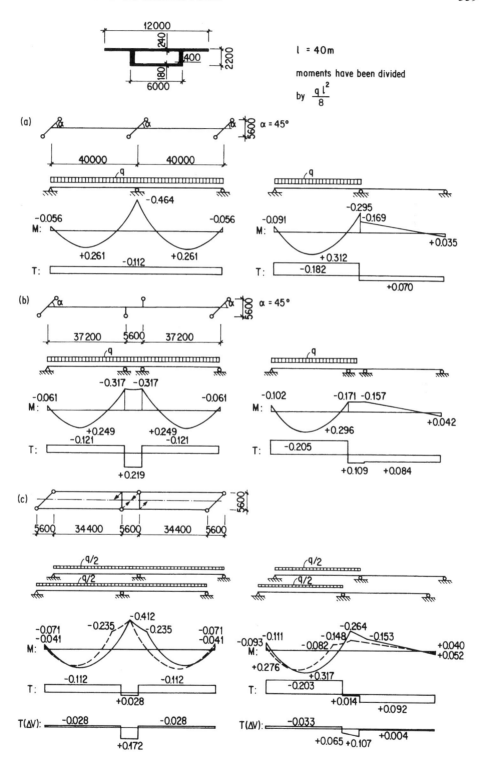

l = 40 m

moments have been divided

by $\dfrac{q\,l^2}{8}$

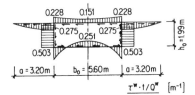

Figure 7.74

$\tau^w \cdot t / Q^w \quad [\text{m}^{-1}]$ Example 7.7: shear flow in perpendicular diaphragms

The calculation can be simplified by neglecting the shear flow in the top and bottom slabs due to vertical shear, since the resultant shear force in these two elements is zero. The stresses in the diaphragms are thus induced only by vertical shear in the webs and shear due to St. Venant shear flow due to torsion; a state of pure shear in the diaphragms results.

c) Frames

Single-span skew bridges can be designed as frames, in which case the girder is normally haunched. This results in a more slender and elegant apprearance as compared to a simply supported beam with constant depth. The abutment walls are strengthened with ribs. By decreasing the depth of the ribs from top to bottom, the connection of wall to foundation is effectively hinged. The redundant forces due to girder shortening are thereby significantly reduced (fig. 7.75).

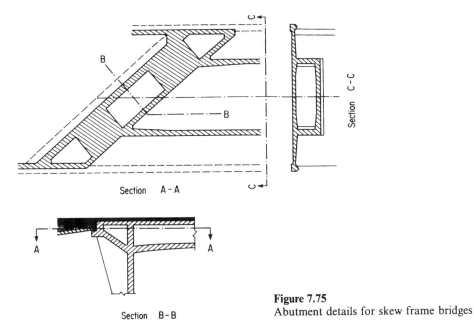

Section A - A

Section C - C

Section B - B

Figure 7.75
Abutment details for skew frame bridges

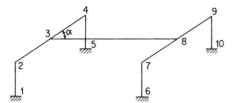

Figure 7.76
Model of a skew frame bridge

Statical models of skew frames are statically indeterminate to a high degree. For the model proposed in figure 7.76, the abutment walls are modeled by frames 1–2–4–5 and 6–7–9–10, respectively. It is assumed that beams 2–4 and 7–9 are rigid in bending and torsion; the columns 1–2, 4–5, 6–7, and 9–10 are assumed rigid in the plane of the frame. One half of the bending and torsional stiffness of each wall is distributed to each of the columns.

The detailing of frame corners requires special attention. The reinforcement connecting the abutment walls to the girder is laid out parallel to the prestressing tendons in the girder. A massive inclined beam at the connection of the abutment wall to the bottom slab must be provided to deviate these tensile forces (fig. 7.77).

Changes in the length of the girder due to prestressing, shrinkage, and temperature result in a rotation of the abutment walls (figure 7.78). The abutment walls should therefore be as flexible as possible so as not to restrict the free movement of the girder. The wing walls are therefore separated from the abutment wall using expansion joints or are designed as cantilevers. The tensile forces resulting from the restriction of the longitudinal girder movement must be considered in the calculation of stresses.

7.5.3 Prestressing Concepts and Tendon Layouts

The prestressing concept for skew girder bridges is essentially similar to that of cast-in-place girder bridges with non-skew supports. Full, form-true prestressing for dead load eliminates tensile stresses due to permanent load and effectively equalizes the reactions at the acute and obtuse corners. This results in small torsional moments in the girder.

Statical considerations favour locating the tendon anchorages as high as possible at the girder ends. The location of the anchors must, however, respect the clearance required for the expansion joints.

Example 7.8:
Moments due to prestressing in a single-span skew bridge

It is assumed that the skew angle, α, is 45° at both girder ends. The stiffness ratio EI/GK is 1.7. Two cable layouts are given (fig. 7.79):

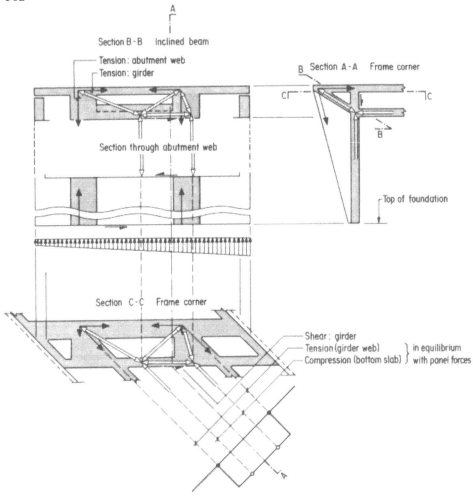

Figure 7.77
Truss model for the corner of a skew frame bridge

Layout 1:   Support moment $= 0.25\,Pf$,
            midspan moment $= -0.75\,Pf$

Layout 2:   Support moment $= 0.25\,Pf + 0.88\,P\Delta f$,
            midspan moment $= -0.75\,Pf - 0.12\,P\Delta f$

If the webs of a given span are of unequal length, the force and eccentricity of the prestressing are determined so that the deviation force is proportional to the share of the dead load carried by that web (fig. 7.80)

Tendons in continuous girders are always laid out so that the deviation force q_P balances the dead load, regardless of girder cross-section and diaphragm

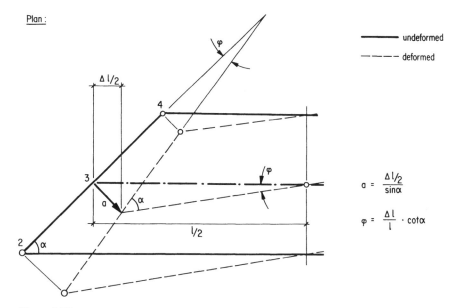

Figure 7.78
Abutment rotations due to longitudinal contraction of the girder

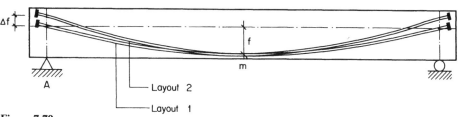

Figure 7.79
Example 7.8: cable layouts

arrangement. The portion of the load balanced should be transferred as directly as possible to rigid supports by providing sharp reverse curvature at these points (fig. 7.81)

Mild reinforcing steel is laid out orthogonally in the slabs, parallel and perpendicular to the longitudinal girder axis. This corresponds to the directions of the largest principal moments away from the immediate vicinity of diaphragms and supports.

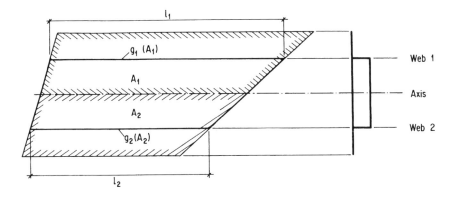

Prestressing force in web 1 : $P_1 \cong \dfrac{g_1 \cdot l_1^2}{8 \cdot f_1}$

Prestressing force in web 2 : $P_2 \cong \dfrac{g_2 \cdot l_2^2}{8 \cdot f_2}$

Figure 7.80
Prestressing force for webs of unequal length

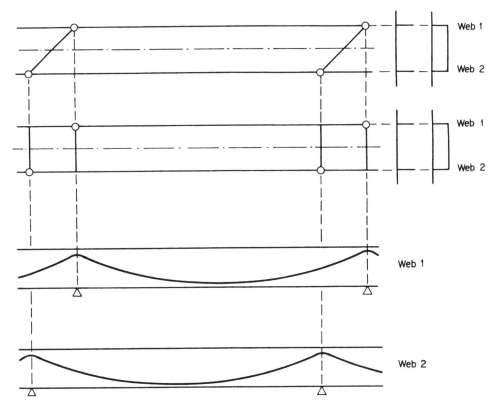

Figure 7.81
Tendon arrangement for continuous skew girders

7.6 Curved Girder Bridges

7.6.1 Conceptual Design

Modern highway construction frequently requires bridges that are located on curved alignments. Although such bridges can be designed by superimposing a curved deck slab onto straight girders, a superstructure that is truly curved normally results in simpler construction and an improved visual appearance.

Torsion in curved girders is induced by vertical loads, including those that are symmetrical about the longitudinal axis of the bridge. The flexural tensile and compressive forces acting on a curved girder element are statically equivalent to a radially directed couple of forces (see Section 7.6.2). The couple acts as a torque and must therefore be balanced by torsional sectional forces. In closed cross-sections, it is equilibrated by a closed shear flow; in open sections, it is equilibrated primarily by differential bending of the webs, referred to as *warping*. Either type of section must deform transversely in flexure to convert the torque into the torsional sectional forces. Although frame action is normally adequate to resist transverse bending in closed sections, girders with open sections require closely spaced diaphragms, which are expensive to build. Box cross-sections are therefore preferred for curved girders in most cases.

Bending moments may be redistributed at ultimate limit state only when the associated torques and torsional moments are redistributed in a corresponding manner. The absolute value of the torsional moments in a prismatic curved girder is thus increased when the bending moment diagram is shifted downwards (fig. 7.82). In a cantilever-constructed haunched girder, however, the absolute

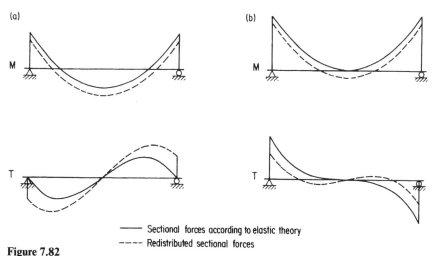

— Sectional forces according to elastic theory
---- Redistributed sectional forces

Figure 7.82
Effect of a redistribution of bending moments on the torsional moments in curved girders: a, continuous girder, constant depth; b, continuous cantilever-constructed girder, variable depth

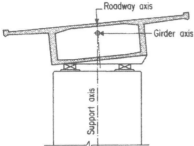

Figure 7.83
Axes of curved bridges

value of torsion decreases near the supports when the bending moments are redistributed in the same way.

Prestressing tendons and reinforcing steel must be restrained by special reinforcement against pull-out along the concave surfaces of the webs. The pull-out force, q_r, is equal to tensile force in the steel, Z, divided by the radius of curvature of the reinforcement, r. Due to inaccuracies in placing the reinforcement, the local radius of curvature of the steel is often smaller than the radius of curvature of the web. The detailing of curved prestressing tendons is discussed further in Section 4.6.4.

The difference in length between interior and exterior webs must be considered in detailing the reinforcement. The roadway axis, girder axis, and support axis are all different, and must be clearly distinguished on the drawings (fig. 7.83).

The equations of equilibrium for a curved girder subjected to vertical load,

$$\sum V_z = 0, \qquad \sum M_x = 0, \qquad \sum M_y = 0$$

can be satisfied for a two-span structure with three non-collinear vertical supports; the system is statically determinate and stable (fig. 7.84). This arrangement is not practical, however, since disproportionately large bending and torsional moments are produced over the entire length of the girder. The structure is thus sensitive to deformations. The ends of curved bridges should always be restrained against twist by at least two bearings at each abutment. To prevent uplift, it may be necessary to locate the bearings away from the webs by extending the end

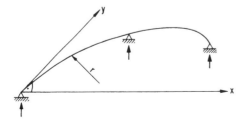

Figure 7.84
Statically determinate, continuous curved girder

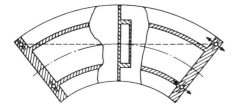

Figure 7.85
Bearing arrangement for a sharply curved, single-span bridge

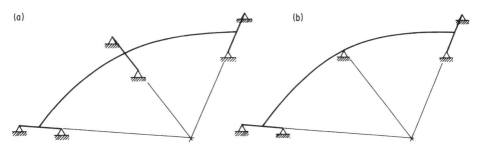

(a)   (b)

Figure 7.86
Recommended bearing arrangements for continuous curved girders: a, torsional restraint at all supports, 3 times statically indeterminate; b, torsional restraint at the abutments only, 2 times statically indeterminate

diaphragms (fig. 7.85). Stresses and deformations of continuous bridges are smaller when torsionally fixed intermediate supports are used (fig. 7.86). When clearance under the bridge is critical, for example at skew crossings, the narrower point supports may nevertheless be preferable.

As outlined in Section 6.1.3, the free end of a curved girder displaces in different directions due to temperature and due to prestressing (fig. 7.87). It is normally preferable to detail the expansion bearings at the free end to allow only tangential displacements, thus simplifying the detailing of the expansion joint which must displace in the same direction as the bearings (fig. 7.88). The associated redundant sectional forces in the girder and the columns due to change of temperature will be small. For the system shown in figure 7.88c, the redundant force in the column, X_1, can be solved from the following compatibility condition:

$$\delta_1 = \delta_{10} + X_1 \left(\delta_{11}^G + \delta_{11}^C \right) = 0$$

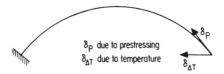

δ_P due to prestressing
$\delta_{\Delta T}$ due to temperature

Figure 7.87
Displacements at the free end of a curved girder (plan)

(a)

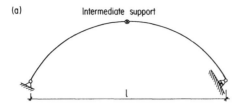

(b)

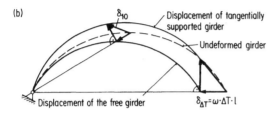

(c)

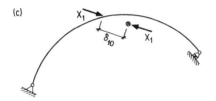

Figure 7.88
Restraint of thermal contraction of a curved girder: a, model; b, deflections; c, primary system and redundant forces

where δ_{11}^G and δ_{11}^C are the displacements of girder and column, respectively, when $X_1 = 1$. The redundant sectional forces in the girder and column can then be computed from the solution X_1.

Expansion bearings are not normally required at the abutments of sharply curved bridges, since changes in girder length produce only small axial forces and bending moments about the vertical axis. The corresponding stresses are so small that additional reinforcement is usually not required.

7.6.2 Analysis

The simplified method of analysis presented in this section is valid when the longitudinal axis of the bridge is a circular arc of constant radius. In general, the radius of curvature can vary along the length of the bridge. For design calculations, however, an average constant radius can be assumed for each span, thus making possible the use of the proposed simplified calculations. The notation and sign conventions are defined in figures 7.89 and 7.90.

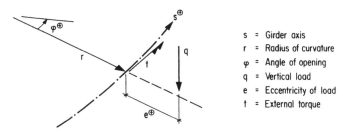

s = Girder axis
r = Radius of curvature
φ = Angle of opening
q = Vertical load
e = Eccentricity of load
t = External torque

Figure 7.89
Sign convention for geometry and loading

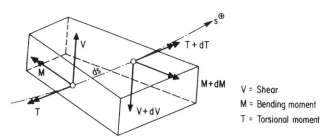

V = Shear
M = Bending moment
T = Torsional moment

Figure 7.90
Sign convention for sectional forces

The equations of equilibrium for the differential element of a curved girder shown in figure 7.91 are:

$$dV + qds = 0 \tag{a}$$

$$dT + Md\phi + (eq + t)\,ds = 0 \tag{b}$$

$$dM - Td\phi - Vds = 0 \tag{c}$$

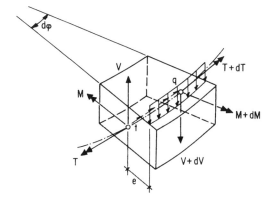

Figure 7.91
Differential element of a curved girder

Dividing each equation through by ds and substituting $1/r$ for $d\phi/ds$ yields the following three equations:

$$\frac{dV}{ds} = -q \tag{d}$$

$$\frac{dT}{ds} + \frac{M}{r} = -eq - t \tag{e}$$

$$\frac{dM}{ds} - \frac{T}{r} = V \tag{f}$$

By differentiating equation (f) with respect to s and then substituting equation (d) into equation (f), the following pair of differential equations is obtained:

$$\frac{d^2 M}{ds^2} = -\left(q - \frac{1}{r}\frac{dT}{ds}\right) \tag{g}$$

$$\frac{dT}{ds} = -\left(\frac{M}{r} + eq + t\right) = -m_t \tag{7.1}$$

where m_t denotes the total equivalent applied torque.

An approximate iterative solution to equations (g) and (e) can be developed by dividing equation (e) through by r:

$$-\frac{1}{r}\frac{dT}{ds} = \frac{M}{r^2} + \frac{eq}{r} + \frac{t}{r} \tag{h}$$

Expressing M as a linear function of ql^2, this equation can be rewritten as

$$-\frac{1}{r}\frac{dT}{ds} = \frac{1}{r^2}\frac{ql^2}{C} + \frac{eq}{r} + \frac{t}{r} = q\left(\frac{l^2}{r^2 C} + \frac{e}{r} + \frac{t}{rq}\right)$$

Assuming $l < r$, $e < r$, and $t/r \ll q$, it follows that the left-hand side of equation (h) is small relative to q. The term dT/rds can therefore be neglected in equation (g), which is transformed into

$$\frac{d^2 M_1}{ds^2} = -q \tag{i}$$

The bending moments in the curved girder are thus approximated by M_1, the moments in a straight girder of equal arc length. A first approximation of the torsional moments is obtained by substitution the solution of equation (i) into equation (e):

$$\frac{dT_1}{ds} = -\left(\frac{M_1}{r} + eq + t\right) \tag{j}$$

This is the equation of torsional equilibrium of a straight bar, which can thus be solved using the classical methods of structural mechanics. The term M_1/r is normally the dominant load in this equation.

The initial estimate M_1 can then be refined by substituting T_1 into equation (g):

$$\frac{d^2 M_2}{ds^2} = -\left(q - \frac{1}{r}\frac{dT_1}{ds}\right)$$

The iterative calculation converges rapidly.

This approximate method does not normally satisfy compatibility when used for statically indeterminate systems, for which an exact solution is obtained only when the ratio of flexural stiffness to torsional stiffness, EI/GK, is 0. The method does, however, satisfy equilibrium. The solution computed using the approximate method thus corresponds to a small redistribution of the sectional forces of the exact solution.

Example 7.9:
Simplified calculation of M and T in a curved girder

The girder is shown in figure 7.92a. Twist is restrained at both ends; flexural rotations are restrained at one end. A uniformly distributed load is applied over the entire length. It is assumed that $l/r = 0.2$ and $EI/GK = 1.0$. The results of the analysis are given in figure 7.92b and in table 7.1. The results of the approximate analysis agree closely with the exact solution.

(a)

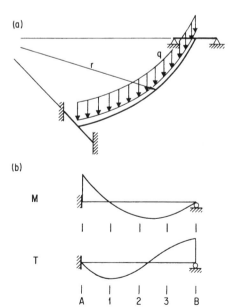

(b)

Figure 7.92
Example 7.9: a, structural system and loading; b, sectional forces

Table 7.1
Bending and Torsional Moments of Example 7.9 (times $ql^2/10^{-3}$)

Section	M_1	T_1	M	T
	1st Iteration		Exact Solution	
A	−125.00	0.00	−125.00	−0.02
1	0.00	2.87	− 0.28	2.86
2	62.50	0.78	62.46	1.05
3	62.50	−2.34	62.55	−2.34
B	0.00	−4.17	0.00	−4.16

7.6.3 Transformation of Torque into Torsional Sectional Forces

The vector sum of the bending moments M and $M + dM$ in a differential element of a curved girder is denoted $m_t\,ds$ (fig. 7.93). Because it is tangent to the longitudinal axis of the element, $m_t\,ds$ can be regarded as a torque. It is apparent from the force triangle shown in figure 7.93 that $m_t = M/r$. This expression can also be formulated by considering the *deviation forces* acting on the element, defined as the vector sum of the flexural compressive and tensile forces, respectively (fig. 7.94). The deviation forces are formulated as follows:

$$q_D\,ds = D\,d\phi \qquad q_z\,ds = Z\,d\phi$$

The internal lever arm is denoted z. Making the substitutions $D = M/z$, $Z = -M/z$, and $d\phi = ds/r$, the following equations are obtained:

$$q_D = \frac{D}{r} = \frac{M}{zr} \qquad q_z = -\frac{Z}{r} = -\frac{M}{zr}$$

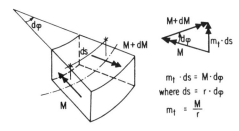

$$m_t \cdot ds = M \cdot d\phi$$
$$\text{where } ds = r \cdot d\phi$$
$$m_t = \frac{M}{r}$$

Figure 7.93
Equilibrium of bending moments and torque in a differential beam element

Figure 7.94
Deviation forces due to girder curvature

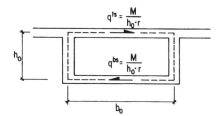

Figure 7.95
Deviation forces of flexural compression and tension in a box girder

The couple formed by q_D and q_Z is denoted m_t:

$$m_t \, ds = q_D \, z \, ds = -q_Z \, z \, ds$$

$$m_t = q_D \, z = -q_Z \, z = \frac{M}{r} \tag{a}$$

which is the same result as obtained from the model of figure 7.93.

In box cross-sections, the deviation forces are localized in the top and bottom slabs (fig. 7.95). Therefore,

$$q^{ts} = \frac{M}{h_o r} = \frac{m_t}{h_o} \quad \text{and} \quad q^{bs} = -\frac{M}{h_o r} = -\frac{m_t}{h_o} \tag{b}$$

When e and t are zero in equation (7.1), the moment $m_t \, ds$ will be in equilibrium with the torsional moment T:

$$\frac{dT}{ds} = -m_t = -\frac{M}{r} \tag{c}$$

It is assumed that torsion in box sections is resisted primarily by a closed shear flow v (fig. 7.96). The difference in shear flow dv across the element can therefore be expressed in terms of the difference in the torsional moment:

$$dv = \frac{dT}{2b_o h_o} \tag{d}$$

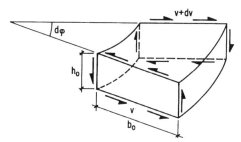

Figure 7.96
Shear flow in a curved girder element

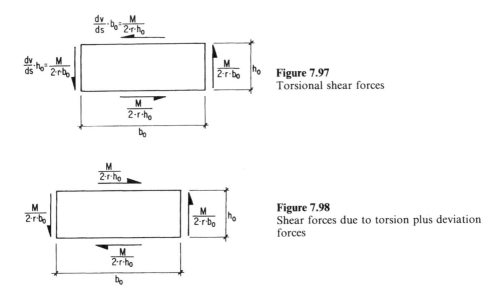

Figure 7.97
Torsional shear forces

Figure 7.98
Shear forces due to torsion plus deviation
forces

Combining equations (c) and (d), the following expression for dv/ds as a function of M is obtained:

$$\frac{dv}{ds} = \frac{dT}{ds}\frac{1}{2b_0 h_0} = -\frac{m_t}{2b_0 h_0} = -\frac{M}{2b_0 h_0 r}$$

The corresponding shear forces per unit length in the top slab, bottom slab, and webs are given in figure 7.97. Adding the deviation forces q^{ts} and q^{bs} (equations (b)) yields the state of equilibrium shown in figure 7.98.

The shear forces resulting from the introduction of torque produce transverse and longitudinal bending in the girder. This phenomenon was discussed in detail in Section 5.1.2. If the stresses induced by transverse bending are high, it may be preferable to provide intermediate diaphragms. The deviation forces q^{ts} and q^{bs} will then be transferred through longitudinal bending to the diaphragms, where they are introduced into the cross-section as a concentrated torque.

In curved girders with open cross-section, the torque m_t can be resolved into a statically equivalent couple of vertical forces applied in the plane of the webs (fig. 7.99 c):

$$\bar{q} = \frac{m_t}{b_0} = \frac{M(q)}{rb_0} \tag{7.2}$$

where q is the given load, applied symmetrically about the longitudinal axis of the girder. The longitudinal bending moment in each half of the girder can be assumed equal to half the moment due to q, $M(q)/2$, plus the moment due to $\bar{q}$, $M(\bar{q})$:

$$M_{\text{int}} = \frac{M(q)}{2} - M(\bar{q}) \qquad M_{\text{ext}} = \frac{M(q)}{2} + M(\bar{q})$$

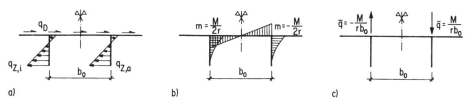

Figure 7.99
Warping torsion in a curved double-T section: a, deviation forces due to flexural tension and compression; b, approximate transverse bending moments due to the deviation forces; c, differential web loading due to the torque m_t

where "interior" and "exterior" are defined relative to the centre of curvature. When $M(q)$ is positive, $|M_{int}| < |M_{ext}|$; when $M(q)$ is negative, $|M_{int}| > |M_{ext}|$. (As a first approximation, it can be assumed that $M_{int} = M_{ext} = M(q)/2$.)

The deviation forces q_Z and q_D are calculated for the flexural tensile and compressive forces of each half of the girder. Assuming that $M(q)$ is positive, the tensile forces are given by the following equations:

$$Z_{int} = \frac{1}{z}\left(\frac{M(q)}{2} - M(\bar{q})\right) \qquad Z_{ext} = \frac{1}{z}\left(\frac{M(q)}{2} + M(\bar{q})\right)$$

where z is the internal lever arm. The corresponding deviation forces are

$$q_{Z,int} = \frac{Z_{int}}{r} \quad \text{and} \quad q_{Z,ext} = \frac{Z_{ext}}{r}$$

These forces induce transverse bending in the webs and in the top slab. At the intersection of webs and top slab, the transverse moments are given by the following equations:

$$m_{int} \cong q_{Z,int}\, z = \frac{1}{r}\left(\frac{M(q)}{2} - M(\bar{q})\right) \tag{7.3}$$

$$m_{ext} \cong q_{Z,ext}\, z = \frac{1}{r}\left(\frac{M(q)}{2} + M(\bar{q})\right) \tag{7.4}$$

Example 7.10:
Transverse bending moments in a curved bridge with open section

The single-span bridge is simply supported in flexure and torsionally fixed at both ends. The span, l, is 30 m and the radius of curvature, r, is 200 m. The distance between the centrelines of the webs, b_0, is 6.5 m. A uniform load of 150 kn/m, denoted g, is applied. The bending moment at midspan is calculated approximately for a straight girder:

$$M(g) = \frac{gl^2}{8} = \frac{(150)\,(30)^2}{8} = 16\,900 \text{ kN} \cdot \text{m}$$

The torque at midspan is calculated using equation (7.1):

$$m_t = \frac{M(g)}{r} = \frac{16\,900}{200} = 84.5 \text{ kN} \cdot \text{m/m}$$

and is resolved into the couple $\bar{q}$:

$$\bar{q} = \frac{m_t}{b_0} = \frac{84.5}{6.5} = 13.0 \text{ kN/m} \qquad \text{(equation 7.2)}$$

The load $\bar{q}$ is distributed parabolically along the length of the girder. The moment $M(\bar{q})$ can be approximated as follows:

$$M(\bar{q}) = \frac{\bar{q}l^2}{9.6} = \frac{(13.0)\,(30)^2}{9.6} = 1\,220 \text{ kN} \cdot \text{m}$$

The transverse moments at the intersection of the webs and top slab are calculated using equations (7.3) and (7.4):

$$m_{\text{int}} = \frac{1}{200} \left(\frac{16\,900}{2} - 1\,220 \right) = 36.2 \text{ kN} \cdot \text{m/m}$$

$$m_{\text{ext}} = \frac{1}{200} \left(\frac{16\,900}{2} + 1\,220 \right) = 48.4 \text{ kN} \cdot \text{m/m}$$

7.6.4 Prestressing

In statically determinate systems, the forces in the prestressing steel are in equilibrium with sectional forces in the concrete. The latter forces can therefore be obtained directly from the corresponding components of the former. Assuming the angle of inclination of the tendon is small, the components of prestressing force relative to the centroidal coordinate system of figure 7.100 are as follows:

$$P_x \cong P \qquad\qquad M_x = P_z\,a_y - P_y\,a_z$$

$$P_y = P_x\,\frac{da_y}{dx} \qquad M_y = P_x\,a_z$$

$$P_z = P_x\,\frac{da_z}{dx} \qquad M_z = -P_x\,a_y$$

The torsional moment is defined with respect to the shear centre:

$$T = M_x - P_z\,c_y + P_y\,c_z = P_z(a_y - c_y) - P_y(a_z - c_z)$$

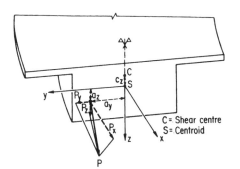

C = Shear centre
S = Centroid

Figure 7.100
Components of prestressing force

Each of these components corresponds to an equal and opposite sectional force in the concrete:

$$N_c = -P_x \qquad M_{c,x} = -P_x\left(a_y\frac{da_z}{dx} - a_z\frac{da_y}{dx}\right)$$

$$V_{c,y} = -P_x\frac{da_y}{dx} \qquad M_{c,y} = -P_x a_z$$

$$V_{c,z} = -P_x\frac{da_z}{dx} \qquad M_{c,z} = -P_x a_y$$

$$T_c = -P_x\left[(a_y - c_y)\frac{da_z}{dx} - (a_z - c_z)\frac{da_y}{dx}\right] \tag{7.5}$$

The concrete sectional forces produced by several tendons are obtained from the superposition of the force components of each individual tendon.

Curvature of the tendons, due to tendon profile or curvature of the girder itself, produces deviation forces normal to the longitudinal axis of the girder. These must be equilibrated in the concrete section, either by differential shear flow or by deviation forces of the normal stresses. Although the deviation force of the tendon can be assumed concentrated at one location, the equilibrating concrete stresses are typically distributed over the entire section. Transverse bending in the cross-section will thus be induced. Since the transverse bending moments are required for equilibrium, they must be considered at ultimate limit state.

The deviation force of the tendons due to *curvature* of the girder in the *x-y* plane is in equilibrium with the deviation forces of the normal stresses in the concrete due to N_c, $M_{c,y}$, and $M_{c,z}$:

$$q^{cg}_{P,y} = q^{cg}_{c,y}$$

where

$$q^{cg}_{P,y} = -\frac{P_x}{r} \quad \text{and} \quad q^{cg}_{c,y} = -\frac{1}{r}\sigma_x(N_c, M_{c,y}, M_{c,z})$$

The deviation forces of the tendons due to the *profile* of the *tendon* relative to the axis of the girder are given by the following expressions:

$$q_{P,y}^{pt} = \frac{dP_y}{dx} = P_x \frac{d^2 a_y}{dx^2} \quad \text{and} \quad q_{P,z}^{pt} = \frac{dP_z}{dx} = P_x \frac{d^2 a_z}{dx^2}$$

It is assumed that $q_{P,y}^{pt}$ and $q_{P,z}^{pt}$ are applied at the shear centre. The corresponding torque is therefore

$$m_{t,P} = -q_{P,y}^{pt} (a_z - c_z) + q_{P,z}^{pt} (a_y - c_y)$$
$$= P_x \left[-\frac{d^2 a_y}{dx^2} (a_z - c_z) + \frac{d^2 a_z}{dx^2} (a_y - c_y) \right]$$

These forces are equilibrated in the concrete section by the differential shear flow dv/dx due to $V_{c,y}$, $V_{c,z}$, and T_c. In a prismatic girder, the distribution of dv/dx in the cross-section is geometrically similar to the distribution of the shear flow, v. Typical shear flows in a box cross-section due to $V_{c,y}$, $V_{c,z}$, and T_c are shown in figure 7.101. For girders with variable cross-section, the shear flow on both sides of the girder element must be considered in the calculation of dv/dx.

The exact distribution of the differential shear flow is often not necessary for the calculation of the transverse bending moments. It is sufficient in most cases to compute shear *forces* in the top slab, bottom slab, and webs to equilibrate $q_{P,y}^{pt}$, $q_{P,z}^{pt}$, and $m_{t,P}$. Thus, shear forces in the top and bottom slab balance $q_{P,y}^{pt}$, shear forces in the webs balance $q_{P,z}^{pt}$, and $m_{t,P}$ is balanced by a closed shear flow around the box.

In statically indeterminate systems, the redundant moments due to prestressing can be easily determined using the approximate method presented in Section 7.6.2. Equation (7.1) is used to calculate the torque due to the redundant moment due to prestressing in the straight girder, $M_{sP,1}$:

$$m_{t,sP} = \frac{M_{sP,1}}{r}$$

(a)   (b)   (c)

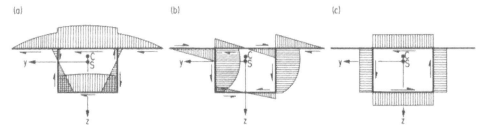

Figure 7.101
Shear flow τt in a box section: a, due to $V_{c,y}$; b, due to $V_{c,z}$; c, due to T_c

The redundant torsional moments due to prestressing, T_{sP}, are calculated from $m_{t,sP}$ and from the torsional moments due to prestressing in the statically determinate system, T_{0P}.

7.6.5 Prestressing Concept and Tendon Layout

The prestressing concept for curved bridges is normally based on the same considerations as for straight bridges. As a minimum, tensile stresses in the deck slab due to permanent load should be prevented. Torsion, which increases flexural tensile stresses, must be considered in the calculation of the required prestressing force. The tendons in curved bridges can be arranged as described in Section 7.2.5 for straight bridges. It is also possible, however, to arrange the tendons to enhance the behaviour of the structure not only in flexure and shear but also in torsion.

In box girders, tendons that counteract torsion can be arranged in the webs or in the top and bottom slabs. The required tendon profile can be chosen, for example, to balance some fraction of the torsional moments due to dead load. In a statically determinate system, the torsional moment induced by an individual tendon is given by the following expression:

$$T_c(P) = -P_x \left[(a_y - c_y) \frac{da_z}{dx} - (a_z - c_z) \frac{da_y}{dx} \right] \qquad \text{(equation (7.5))}$$

When the tendon is located in a web and $(a_y - c_y)$ is constant, the slope da_z/dx can be chosen to match the torsional moment diagram due to loads at each point along the girder. Similarly, when the tendon is located in the top or bottom slab and $(a_z - c_z)$ is constant, torsion can be balanced by an appropriate choice of da_y/dx.

In simply supported girders, it is possible to arrange the tendons to balance a given torsional moment diagram without altering the effect of the prestressing in bending. This is accomplished by locating the tendons in the exterior web above, and the tendons in the interior web below, the profile determined for flexural behaviour (fig. 7.102). As shown in figure 7.103, balancing torsion in this way increases transverse bending. Savings in torsional reinforcement are thus achieved at the cost of additional reinforcement for transverse bending.

Tendons in the top and bottom slabs can compensate torsion and transverse bending (fig. 7.104). In this arrangement, the savings in reinforcement for torsion and transverse bending are obtained at the cost of additional prestressing.

For these reasons, therefore, the balancing of torsion by prestressing is normally avoided. Regardless of the tendon arrangement, the savings in reinforcement are small and almost always outweighed by difficulties in construction.

In continuous girders, it is impossible to balance a given torsional moment diagram by adjusting the profile of tendons in the webs without reducing the

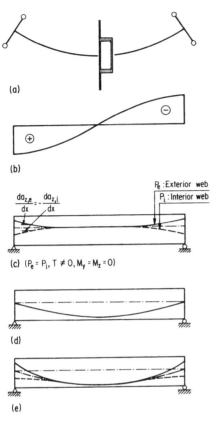

(a)

(b)

(c) $(P_e = P_i, T \neq 0, M_y = M_z = 0)$

$\dfrac{da_{z,e}}{dx} = -\dfrac{da_{z,i}}{dx}$

P_e :Exterior web
P_i :Interior web

(d)

(e)

Figure 7.102
Tendon arrangement in a simply supported girder to compensate bending
and torsion: a, model; b, torsional
moments due to dead load, T_g; c,
tendon arrangement to balance T_g; d,
tendon arrangement to balance bending moments due to dead load; e,
superposition of arrangements (c) and
(d) to balance torsion and bending

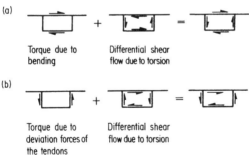

(a)

Torque due to
bending

Differential shear
flow due to torsion

(b)

Torque due to
deviation forces of
the tendons

Differential shear
flow due to torsion

Figure 7.103
Transverse forces in a girder element: a,
due to load; b, due to prestressing with
tendons in the webs arranged as in
figure 7.102

flexural effect of the prestressing (fig. 7.105). If torsion in continuous girders must
be compensated by prestressing, it is preferable to use additional tendons in the
top and bottom slabs (fig. 7.106).

In straight girders with non-skew supports, the effect of longitudinal prestressing
is essentially restricted to longitudinal structural behaviour. Prestressing contributes directly to the flexural resistance of cross-sections; the yield force of the

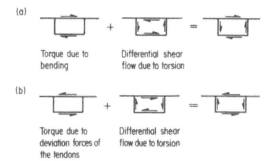

Figure 7.104
Transverse forces in a girder element: a, due to load; b, due to prestressing with tendons in the top and bottom slabs

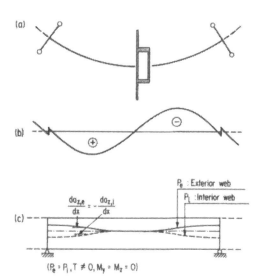

Figure 7.105
Theoretical tendon arrangement in a continuous girder to compensate bending and torsion: a, model; b, torsional moments due to dead load, T_g; c, tendon arrangement to balance T_g; d, tendon arrangement to balance bending moments due to dead load; e, superposition of arrangements (c) and (d) to balance torsion and bending

tendons is thus considered in the computation of M_R. The vertical component of prestressing, V_P, normally acts against the shear due to load. It can be added to the cross-section resistance, or, equivalently, to the design shear (see Section 4.3.3):

$$V_{d,\text{eff}} = V_d + V_P$$

Any transverse sectional forces induced by prestressing in straight girders are small and can be neglected.

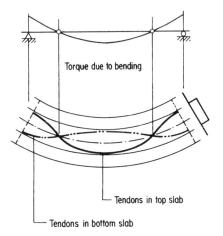

Torque due to bending

Tendons in top slab

Tendons in bottom slab

Figure 7.106
Arrangement of tendons in top and bottom slabs
of continuous girders to compensate torsion due
to load

In curved girders, the effect of longitudinal prestressing is more pervasive. As shown in figures 7.103 and 7.104, prestressing also induces torsion and transverse bending. The torsional moment induced by prestressing, T_P, can be added to the design torsion in a similar manner to longitudinal shear:

$$T_{d, \text{eff}} = T_d + T_P$$

Effective sectional forces are also used for the design of cross-section elements under the combined effects of shear and transverse bending:

$$m_{d, \text{eff}} = m_d + m_P$$

The components of the effective sectional forces due to prestressing (V_P, T_P, m_P) should be calculated using P_∞ when the design sectional force is decreased by prestressing and $1.2 P_\infty$ when it is increased by prestressing.

7.7 Arch Bridges

7.7.1 Conceptual Design

The cost of falsework and formwork for arch bridges is high in comparison to conventional cast-in-place girder bridges. As a result, arch bridges are economical only under a limited range of topographical and geotechnical conditions. Arches may be appropriate for crossings of rivers, canyons, or steep valleys, where a single long span is required for the main obstacle and several short spans can be used for the approaches. The economical range of reinforced concrete arch spans extends from 50 m to 200 m. The higher costs of arch bridges may sometimes be justified by their superior aesthetic qualities.

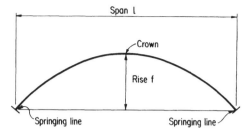

Figure 7.107
Terminology for arch bridges

The ratio of arch span to rise, l/f, should be chosen between $2:1$ and $10:1$ (for terminology, see figure 7.107). The sensitivity of arches to creep, shrinkage, temperature change, and support displacements increases with increasing values of l/f. Stresses and deformations due to these actions are normally small when the ratio of span to rise is less than $4:1$, regardless of the degree of statical indeterminacy of the arch. As l/f approaches $10:1$, it may be necessary to reduce or eliminate redundant moments due to restrained deformations by providing hinges at the springing lines and at the crown. The long-term deformations of flat, hinged arches, in particular the angle break at the crown hinge, will be large and must therefore be carefully checked. Excessive deformations are unavoidable when the span to rise ratio is greater than $10:1$, even when full rotational restraint is provided at the arch abutments. Reducing l/f below $2:1$ results in an awkward appearance and substantial increases in construction cost.

The most common arch bridge arrangement consists of a roadway girder supported by the arch from below (fig. 7.108 a). It is also possible, however, to suspend the roadway from twin arch ribs located above the roadway edges. The girder can be designed to resist the horizontal component of the arch reaction; this arrangement is called a *tied arch* (fig. 7.108 b). Suspending the roadway from the arch may be an appropriate solution for low-level crossings or when suitable foundation material for conventional arch abutments is not available. Economy and elegance are nevertheless difficult to achieve when this type of bridge is built of reinforced and prestressed concrete. Arch bridges with suspended roadway can be more successfully designed in structural steel.

The location of the arch abutments is largely a function of topography, geotechnical conditions, and construction method. Aesthetic considerations

(a)

(b)

Figure 7.108
Arch bridge types: a, conventional arch bridge with roadway supported above arch; b, tied arch with suspended roadway

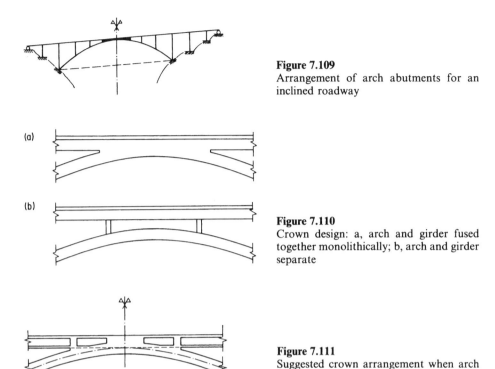

Figure 7.109
Arrangement of arch abutments for an inclined roadway

Figure 7.110
Crown design: a, arch and girder fused together monolithically; b, arch and girder separate

Figure 7.111
Suggested crown arrangement when arch and girder are fused together

require that the chord joining the springing lines be oriented parallel to the deck girder axis, as shown in figure 7.109. By locating a column directly above each of the arch abutments, two foundations can be eliminated and the arch reaction can be directed into the ground at a steeper angle.

The crown can be designed by fusing the arch and girder together monolithically (fig. 7.110a) or by separating the two with short columns (fig. 7.110b). The former arrangement enables a direct transfer of longitudinal horizontal forces from girder to arch and then to the arch abutments. A satisfactory visual treatment is obtained when girder soffit and arch are of equal width and when the arch axis is tangent to the extended girder soffit line (fig. 7.111). An uneven number of interior spans above the arch generally results in the best appearance; columns should thus not be located at midspan when the arch and girder are separated at the crown.

The arch, columns, and deck girder constitute a frame system. When the girder is divided into fewer than five interior spans, the arch should be simplified into an inclined-leg frame and the system designed following the guidelines presented in Section 7.8.

The moments in the frame system can be divided into two components: *fixed system* moments and *flexible system* moments. Fixed system moments are produced when vertical deformations of the arch are restrained and are thus equal to the continuous beam moments in the girder. Flexible system moments correspond to vertical displacements of the arch and are, in general, shared by arch and girder. Two idealized limiting cases are possible: *stiff arches*, which resist the entire flexible system moment with no participation of the girder, and *deck-stiffened arches* for which the entire flexible system moment is resisted by the girder.

Moments due to arch displacements can be further divided into dead load and live load components. To minimize bending in the arch due to dead load, the axis of the arch should be located along the dead load pressure line (see Section 7.7.4). Axial deformations in the arch due to dead load always displace the arch away from the pressure line, thus inducing moments due to dead load in the frame system. Far more important, however, are the frame system moments produced by arch displacements due to partial live load. Flexible system moments of either type are distributed to girder and arch according to their respective flexural stiffness. Bending in the arch need not be eliminated entirely. Since flexural capacity increases initially with axial force, moments of limited magnitude are actually resisted more efficiently by arches than by girders of equal depth, which do not benefit from "natural prestressing" at ultimate limit state.

Arch stiffness affects not only the behaviour of the completed structure but also its behaviour during construction. The arch can be made sufficiently stiff to carry the dead load of arch, columns, and girder without relying on girder stiffness for global stability. This enables falsework to be removed immediately after completion of the arch, or construction of the arch using the cantilever method. Although the falsework required for slender arches is lighter in comparison, it must normally be kept in place until completion of the girder.

Girder depth should be constant for the entire length of the bridge. In addition, the approach spans should not differ markedly in length from the girder spans above the arch. The girder moments in the arch region, the sum of a fixed system moment and a flexible system moment, will thus normally be greater than the moments in the approach spans. The girder must therefore be proportioned more generously than normal girder bridges; it is recommended that the ratio of span to depth be chosen between 12:1 and 15:1 in the approach spans. This enables the girder moments in the arch region to be resisted economically.

Stability of the frame is greatly enhanced by continuity in the deck girder. In addition, the detailing of properly functioning joints is difficult due to the relatively shallow girder. Expansion joints in the girder should therefore be provided only at the girder abutments, even when the approach spans are curved. Since the neutral point of the system is located at the crown of the arch, the girder supports must permit longitudinal displacements.

7.7.2 Design of the Cross-Section

The cross-section of the girder must be chosen in consideration of the interaction of girder and arch. For stiff or nearly stiff arches, bending moments in the girder will be a function of interior span length only. Double-T or solid slab sections can therefore be used, regardless of arch span length. When girder stiffness has a significant influence on the behaviour of the frame system, positive and negative moments in the girder will be of approximately equal magnitude over the entire length of the arch. Box cross-sections are therefore chosen for the girder of long-span arches; double-T sections can be used for short-span arches.

The cross-section of the arch is primarily a function of arch span length and the ratio of arch stiffness to girder stiffness. Deck-stiffened arches can be built as thin slabs, the thickness of which is normally controlled by the buckling resistance of the arch between columns. This criterion can be satisfied by providing a relatively large number of closely spaced columns. Slabs, twin ribs, or hollow boxes can be used for arches that must resist some or all of the flexible system moment. The choice of section depends primarily on span length.

Slab arches appear slender and elegant but are relatively expensive. The ratio of flexural resistance to material cost is small, particularly for wide slabs. Falsework for slab arches must be capable of supporting the entire dead load of arch, columns, and girder. Slab cross-sections are therefore normally used only for narrow bridges and for spans less than 120 m.

Twin ribs are economical for spans of up to roughly 150 m. A more favourable ratio of flexural resistance to material cost is obtained with rectangular sections oriented vertically. The solid, narrow cross-section reduces the cost of top forms, required when the outer face of the arch is steeper than 20 degrees. The considerable bending stiffness of ribbed arches helps to reduce falsework costs.

Box sections, which can be used for spans in excess of 200 m, are the most practical choice for long-span arches. The high ratio of flexural resistance to material consumption helps to offset the high cost of interior formwork. Hollow-box arches are normally sufficiently stable to enable removal of falsework before completion of columns and girder. Their light weight, moreover, makes them suitable for construction using the cantilever method.

7.7.3 Prestressing Concept and Tendon Layout

The approach spans, which resist only fixed-system moments, can be prestressed according to the prestressing concepts presented in Section 7.2.3 for conventional cast-in-place girder bridges. The tendons are preferably arranged in the webs. A different concept and tendon arrangement are required for the girder spans in the arch region, which must resist positive and negative moments of roughly equal magnitude at all locations. Tension reinforcement is therefore required at the top and at the bottom of the section. An efficient use of prestressing results when the prestressing steel and the minimum reinforcing steel are just sufficient to resist the ultimate moments due to dead load plus live load at the locations of lowest stress. The cross-section resistance is increased as required at other points by the addition of mild reinforcing steel.

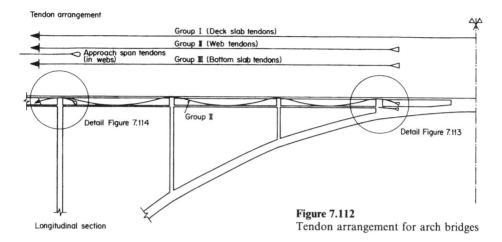

Figure 7.112
Tendon arrangement for arch bridges

It is recommended that the entire girder be fully prestressed for dead load. The prevention of permanent cracks is desirable not only for durability, but also because the resulting increase in stiffness improves the global stability of the frame system.

The tendon arrangement shown in figure 7.112 is recommended for hollow-box girders. Web tendons, designed to balance the dead load of the girder in the fixed system, are provided over the entire length of the bridge. They are lapped at the columns above the arch abutments and are terminated just before the crown. The interior spans above the arch are often progressively shortened towards the crown. By stressing the web tendons at the columns above the arch abutments, the prestressing force decreases towards the crown to match the reduction in span length. The full eccentricity of the tendons can thus be utilized for the entire length of the girder.

Tendons in the top and bottom slabs are provided only in the arch region and should be placed as close to the webs as possible. The bottom slab tendons terminate just before the crown (fig. 7.113). Top slab tendons that are continuous over the crown can help to balance the eccentrically applied arch force at this location.

The stressing anchors of all tendons in the arch region are located at the face of a buttress formed by widening the webs (fig. 7.114). A temporary opening in the deck slab must be provided to allow access for the jack. Overlapping the tendons is beneficial since negative moments in the girder are normally highest at this location.

7.7.4 Preliminary Design

The axis of the arch should be located as close as possible to the pressure line due to dead load. Since the weight of the girder and columns is applied as concentrated

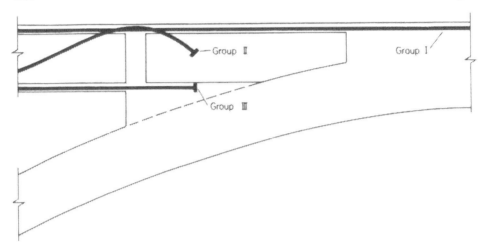

Figure 7.113
Tendon arrangement near crown

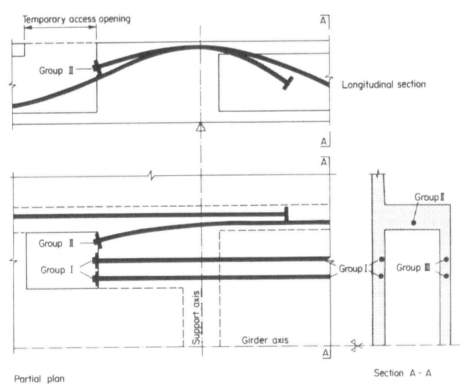

Partial plan

Section A - A

Figure 7.114
Tendon arrangement over the arch abutments

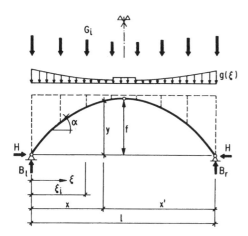

Figure 7.115
Pressure line due to dead load: notation

loads at the base of each column, the slope of the pressure line will be discontinuous at these locations. These small angle breaks should not be smoothened out for aesthetic reasons to achieve an arch of constant curvature; the true pressure line is always the most visually convincing form.

The pressure line is calculated assuming the arch is hinged at the springing lines and at the crown (fig. 7.115). The dead loads, G_i and g, span length l, and rise f are given. The moments produced when G_i and g are applied to a simply supported beam of length l are denoted $M_0(x)$. The horizontal reaction component, H, is obtained from moment equilibrium of one half of the three-hinged arch:

$$H = \frac{M_0(l/2)}{f}$$

The moments in the three-hinged arch, $M(x)$, are therefore given by the following equation

$$M(x) = M_0(x) - Hy$$

The arch ordinates, y, are obtained from the condition $M(x) = 0$, for all x. It therefore follows that

$$y = \frac{M_0(x)}{H} = \frac{M_0(x)}{M_0(l/2)} f$$

The total weight of arch, columns, and girder can be approximated as a uniformly distributed load, denoted $\bar{g}$. The corresponding value of the horizontal reaction is given by the equation

$$H(\bar{g}) = \frac{\bar{g}l^2}{8f}$$

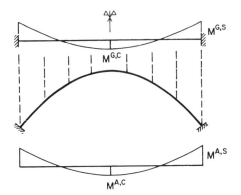

Figure 7.116
Moments induced by vertical deflection of
the arch due to dead load

The axial force in the arch, $N(\bar{g})$, induces axial deformations which produce a
vertical deflection at the crown, δ^C. Assuming $N(\bar{g}) = H(\bar{g})\cos\alpha$, the crown
deflection can be calculated using the following approximation

$$\delta^C \cong \frac{H(\bar{g})}{EA^{A,C}} \frac{l(1 + 3(f/l)^2)}{4f/l}$$

where $A^{A,C}$ denotes the cross-sectional area of the arch at the crown. When both
girder and arch are fixed at the ends of the arch span as shown in figure 7.116, the
moments induced by δ^C can be expressed as follows:

$$M^{G,C} = -\frac{1}{2} M^{G,S} \cong \frac{16\,EI^G}{l^2} \delta^C$$

$$M^{A,C} = -\frac{1}{2} M^{A,S} \cong \frac{16\,EI^A}{l^2} \delta^C$$

The moments thus produced in the girder must be superimposed with the fixed
system moments due to dead load.

Redundant moments due to dead load are also produced when the arch is built by
cantilever construction. The arch deforms due to creep after closure, inducing a
state of stress approximating that of an identical arch cast on conventional
falsework.

Moments in the flexible system due to dead and live load can be calculated in a
relatively straightforward manner for two limiting cases: (1) a *deck-stiffened arch*
with fixed-end girder (fig. 7.117a) and (2) a *stiff arch* with girder of negligible
flexural stiffness (fig. 7.117b). For equal geometry of the nodal points, the frame
moments in both systems will be identical. For the redundant forces shown in
figure 7.117c, the flexibility coefficients δ_{ik} can be formulated as follows:

$$\delta_{ik} = \int_0^l M_i \frac{M_k}{EI^G} d\xi \qquad\qquad \text{(deck-stiffened arch)}$$

$$\delta_{ik} = \int_0^l M_i \frac{M_k}{EI^A} ds = \int_0^l M_i \frac{M_k}{EI^{A,C}} d\xi \qquad \text{(stiff arch)}$$

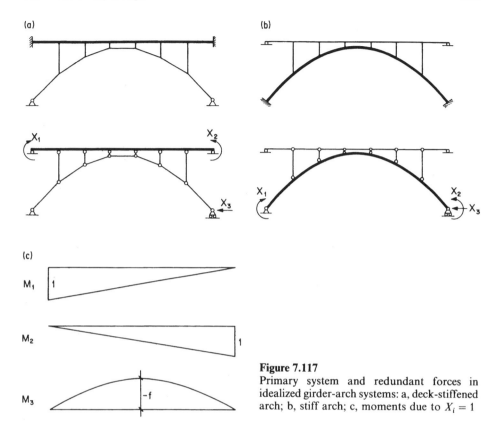

Figure 7.117
Primary system and redundant forces in idealized girder-arch systems: a, deck-stiffened arch; b, stiff arch; c, moments due to $X_i = 1$

where s denotes arc length along the arch axis and ξ is the horizontal coordinate. It is assumed that the flexural stiffness of the arch obeys the following equation:

$$EI^A \cos \alpha = EI^{A,C}$$

where $EI^{A,C}$ is the stiffness of the arch at the crown. Moment diagrams for the two common partial live load cases of figure 7.118 are given in figure 7.119 for a stiff arch; the moments in the girder of the corresponding deck-stiffened arch are identical.

In general, moments in the flexible system are shared by girder and arch. They can be calculated approximately by distributing the moments M obtained from either a deck-stiffened arch or a stiff arch to girder and arch according to their respective stiffnesses:

$$M^G = M \frac{I^G}{I^G + I^{A,C}} \qquad \text{(girder)} \tag{a}$$

$$M^A = M \frac{I^{A,C}}{I^G + I^{A,C}} \qquad \text{(arch)} \tag{b}$$

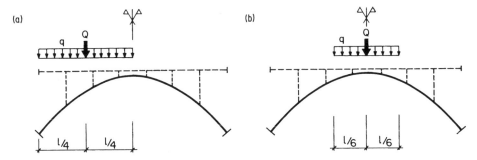

Figure 7.118
Partial live load arrangements: a, over the left half of the span; b, over the middle third of the span

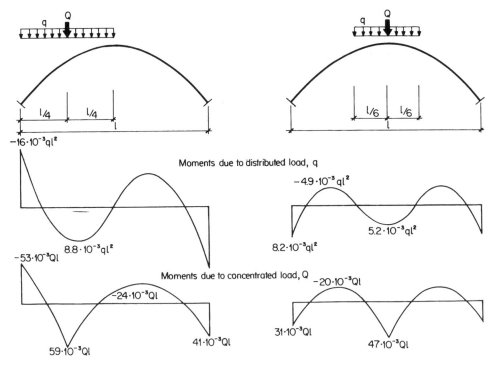

Figure 7.119
Frame system moments in a stiff arch due to partial live load

Influence lines are normally required for the calculation of the envelope of live load moments in the flexible system. For preliminary design, however, maximum moments at the springing lines, quarterpoint, and crown can be estimated from the two partial live load cases shown in figure 7.118.

Girders that are not supported over the springing lines have a longer "frame span" than the arch. The flexible system moments should in such cases be computed taking l as the arch span. The moment diagrams of figure 7.119 remain valid. The distribution of the flexible system moment to girder and arch according to equations (a) and (b) is valid at the quarterpoint and crown. At the springing lines, the flexible system moment will be resisted essentially by the arch alone, assuming $EI^G \cong EI^{A,C}$.

The total moment in the flexible system, M, must be computed using second-order methods. Moment M can be expressed as the sum of a first-order component, M_0, obtained from figure 7.119 and a second-order component M_1, equal to the axial force in the arch times the second-order deflection of the arch due to M_0. The following simplified method for calculating M_1 is valid for fixed-end arches with prestressed girder.

The critical load case for buckling normally consists of dead load plus live load applied to one half of the arch span. The deflections of the arch due to this load, and hence its buckled shape, are shown in figure 7.120. Provided $EI^A \cos\alpha = EI^{A,C}$ for all values of α, the calculation of M_1 can be carried out for an idealized straight member, shown in figure 7.121.

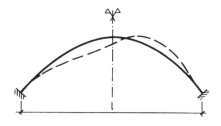

Figure 7.120
Antisymmetrical buckling shape

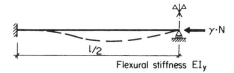

Flexural stiffness EI_y

Figure 7.121
Model for the calculation of second-order moments in the frame system

The second-order deformations, δ, are computed from the first-order deflections due to live load. The effective flexural stiffness of the idealized straight member at ultimate limit state, EI_y, must be used. Provided the arch profile follows the dead load pressure line and the deck girder is prestressed, EI_y can be approximated as

$$EI_y \cong \tfrac{1}{4}(EI^{A,C} + EI^G)$$

The factor 1/4 is used to account for cracking and material plastification at ultimate limit state.

The first-order deflection at ultimate limit state can be expressed approximately as

$$\delta_0 = \frac{\gamma M_{0,\max} \left(\frac{l}{2}\right)^2}{13\,EI_y}$$

where $M_{0,\max}$ is the maximum moment in the flexible system due to live load applied to one half of the span, obtained from figure 7.119. The total safety factor γ is the product of the load factor γ_S and the resistance factor γ_R.

The second-order deflection is obtained from the equation

$$\delta = \delta_0 \frac{1}{1 - \dfrac{\gamma N}{N_E}}$$

where the critical load of the arch, N_E, is given by

$$N_E = \frac{\pi^2 EI_y}{\left(0.7\,\dfrac{l}{2}\right)^2}$$

and N denotes the axial force in the arch due to dead load and live load.

The second-order moment M_1 can thus be computed from the following equation:

$$M_1 = 0.73\,\delta\gamma N$$

The factor 0.73 accounts for the end conditions of the straight beam model (fig. 7.122). The total moment M (equal to $M_0 + M_1$) must be distributed to the girder and the arch according equations (a) and (b).

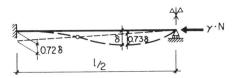

Figure 7.122
Buckling shape of the straight girder model

7.8 Frame Bridges

7.8.1 Conceptual Design

Frame bridges can be regarded as simplified arch structures. The most common types of frame bridges are single-span frames with straight or inclined legs (figs. 7.123 a, b, c) and multiple-span, inclined-leg frames (figs. 7.123 d, e).

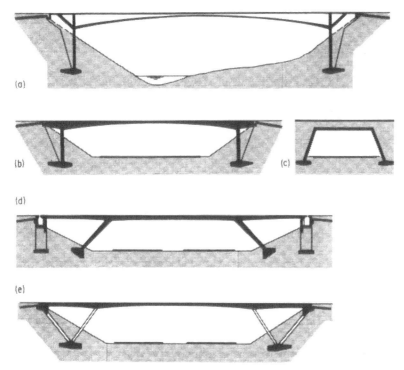

Figure 7.123
Frame bridges: a, two-hinged frame with box cross-section; b, two-hinged frame with slab cross-section; b, two-hinged frame with slab cross-section; c, trapezoidal frame for small underpasses; d, inclined-leg frame bridge; e, inclined-leg frame bridge with V-struts

Single-span frame bridges are appropriate when the span is long compared to the height of the bridge. The horizontal reaction at the base of the short, stiff columns produces positive and negative moments of similar magnitude in the girder. This type of structure is usually easier and less expensive to maintain than simply supported bridges, since expansion joints and bearings can be eliminated for spans of up to approximately 50 m. Because they use materials more efficiently, single-span frame bridges also appear lighter and more slender than simply supported bridges, especially when the girder is haunched. Single-span frames can be built without special problems for spans of up to approximately 70 m. For longer spans, it is difficult to design foundations capable of resisting the horizontal reaction.

Single-span frame bridges are normally designed as two-hinged systems. Hinge action can be achieved by tapering the columns towards the base; actual hinges are detailed only in exceptional cases. Because they are usually required to retain soil on either side of the bridge, the columns are designed as vertical or slightly inclined walls. Ribs can be provided on the backfilled face of the columns to increase their internal lever arm. Long-span structures may require a heavily reinforced, inclined diaphragm at the intersection of girder and column to help deviate tensile and compressive forces around the corner of the frame (fig. 7.123 a). A truss model can be used to design the diaphragm reinforcement.

The ratio of girder span to depth should be roughly 17:1 at the supports and 50:1 at midspan. A box section can be used for the entire length of the girder for spans greater than 50 m. The depth near midspan is normally insufficient for a hollow box when the span is less than 50 m. For spans in the 25 m to 50 m range, a double-T section can be provided over the middle three quarters of the bridge; a bottom slab is added near the ends for additional negative moment resistance. Solid slabs are most practical for spans less than 25 m.

Solid slabs can be used for girder and columns when the span length is less than 15 m (fig. 7.123 c). These small frames are normally covered with subgrade material to increase the homogeneity of the roadway.

Expansion joints must be used for span lengths greater than 50 m. Figure 7.124 shows a suggested detail in which the frame columns are located inside the abutment chamber. Expansion joints are not necessary, however, for spans less than 50 m. In such cases, backfill is retained by the columns; the approach slab is placed onto a neoprene strip on the end diaphragm (cf. fig. 7.125). Damage to the wearing surface at the ends of the bridge can be prevented by cutting a groove in the asphalt at the probable crack location and filling it with bituminous material.

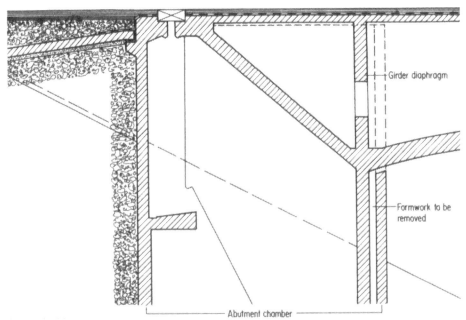

Figure 7.124
Detail at ends of long-span two-hinged frame bridges

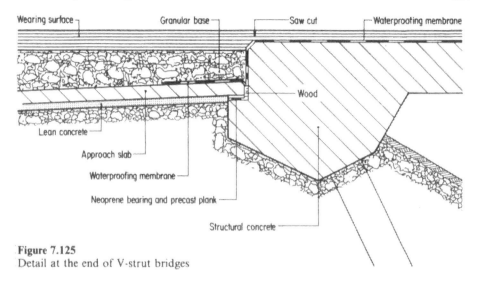

Figure 7.125
Detail at the end of V-strut bridges

The details shown in figure 7.125 require considerably less maintenance than an expansion joint.

Grade separations in the 30 to 35 m span range are often designed with V-struts. The truss action of the struts results in strong fixity of the girder. The columns, which are slender relative to the girder, can be connected monolithically to both girder and foundation; the bending moments in the columns will be insignificant. Tension columns that are covered with fill should be protected against vertical loads induced by soil settlement by a semicircular shell located above the columns. The girder ends should be backfilled directly without expansion joints, thus eliminating the need for abutments (fig. 7.125).

Inclined-leg frames can be an economical alternative to arches for small and medium spans. They are particularly appropriate when the use of inclined legs results in balanced, economical span lengths (fig. 7.126). The geometry of multiple-span inclined-leg frames should correspond to the pressure line due to dead load (fig. 7.127). This will eliminate girder deflections at the columns due to dead load and will result in a dead-load moment diagram in the girder equal to that of a continuous beam. The line connecting the foundations of the inclined

Figure 7.126
Suitable topography for inclined-leg frame bridges

(a)

(b)

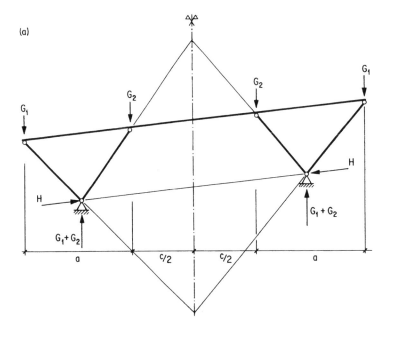

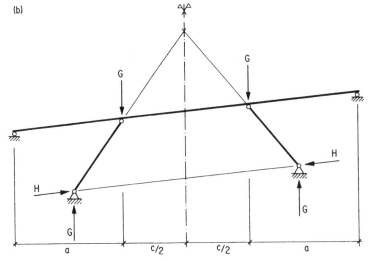

Figure 7.127
Pressure line for inclined-leg frame bridges (G denotes girder reaction plus the weight of the upper
portion of the column): a, skew-symmetrical V-strut bridge; b, skew-symmetrical inclined-leg
bridge; c, asymmetrical inclined-leg bridge

(c)

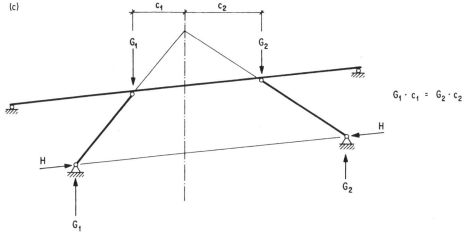

$$G_1 \cdot c_1 = G_2 \cdot c_2$$

Figure 7.127 c

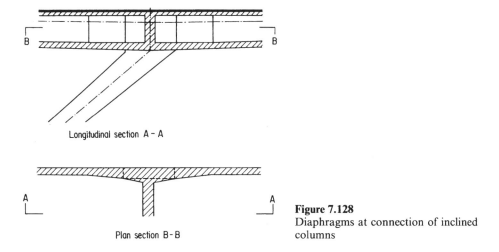

Longitudinal section A - A

Plan section B - B

Figure 7.128
Diaphragms at connection of inclined columns

columns or the points of intersection of the columns and the finished ground line should, for aesthetic reasons, be oriented parallel to the roadway grade.

Vertical diaphragms are used at the connection of the inclined piers to the girder. The width of the diaphragms should be less than the column as it connects to the girder (fig. 7.128).

7.8.2 Prestressing Concepts and Tendon Layouts

Long-span frame bridges are sensitive to deformations. The horizontal reaction at the base of the columns can be considerably reduced by deformations due to shrinkage, creep, and foundation displacements. This can lead to substantial redistributions of moments. It is therefore recommended to provide a concentric state of stress at midspan under permanent load, taking into account the loss of prestress and the reduction of the horizontal reaction. When frame bridges are used for grade separations, cracking in the underside of the girder should be prevented by full prestressing for dead load.

Slab frames with spans of less than 10 m, used for grade separations, need not be prestressed in the direction of the span. The behaviour of these structures is not impaired by deformations nor by flexural cracks. Adequate prestressing perpendicular to the span is of considerably greater importance for long grade separations, to prevent full-penetration cracks due to shrinkage and temperature.

The tendons in single-span frames usually extend over the entire girder length to achieve a strong prestress at midspan. The deviation forces due to prestressing should balance the dead load, regardless of haunching. The effective curvature of tendons in a haunched girder can be calculated as shown in figure 7.129. The curved axis of the girder is taken as horizontal and the upper surface of the girder is drawn curved, maintaining the distance between axis and surface.

The columns of single-span frames are prestressed only for long girder spans, since the wide webs of the columns can usually be properly reinforced with mild reinforcing steel. The lap splice of the column reinforcement and the girder prestressing should be generously proportioned. Adequate transverse reinforcement must be provided at the location of the splice.

The tendons in multiple-span inclined-leg frames can be arranged as for a conventional continuous beam, provided the form of the structure corresponds to the pressure line due to dead load.

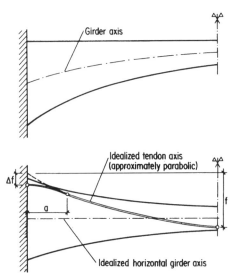

Figure 7.129
Idealized tendon layout for haunched girders

7.9 Slab Bridges

7.9.1 Conceptual Design

The solid slab is the least efficient of the cross-sections commonly used in bridge construction. Slab sections can achieve flexural stiffness and resistance only at the cost of high dead load. Since the thickness of solid slabs is usually no greater than 0.8 m, their economical range of spans is limited. Slab bridges are nevertheless a practical solution for grade separations, in particular those with a complicated layout, thanks to their minimal depth, two-dimensional isotropic structural behaviour, and simple construction procedure.

The geometry of skew slabs can be described by the following parameters (fig. 7.130): span length parallel to bridge axis, l; skew span length perpendicular to the axis of the supports, l_0; width, b; and skew angle, ϕ, measured between the longitudinal axis of the bridge and the axis of the supports. Economical values of l for a simply supported solid slab, depth 0.8 m, will vary between 20 m and 40 m, depending on ϕ and the ratio l_0/b (fig. 7.131). Span lengths can be increased by 10 to 20 percent when the slab is continuous over several supports or incorporated into a rigid frame.

The dead load bending moments in slab bridges are generally about five times greater than the cracking moment of the slab. A complete crack pattern will therefore be produced over a large area of a non-prestressed slab due to dead load

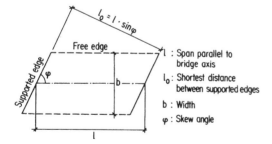

Figure 7.130
Geometry of skew slab bridges

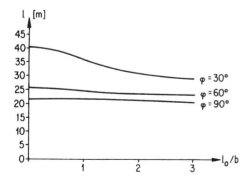

Figure 7.131
Maximum economical span length, l, for a simply supported slab of thickness 0.8 m

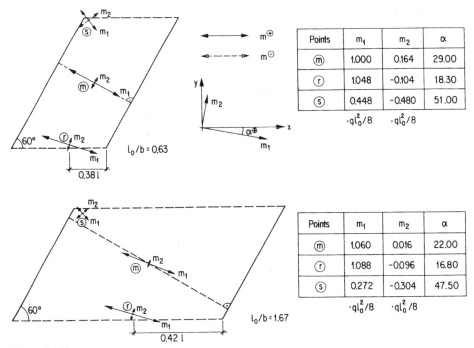

Figure 7.132
Principal bending moments m_1 and m_2 due to uniform load in a simply supported skew slab

alone. In addition, non-prestressed slabs are extremely sensitive to deformations, especially at the free edges. It is therefore recommended that slab bridges be prestressed and that their deformations due to permanent load be carefully checked.

In skew slabs, the direction of the principal bending moments due to dead load deviates only slightly from lines drawn perpendicular to the support axes (fig. 7.132). When the width of the slab, b, is large relative to l_0, the sectional forces in its middle region will be identical to the sectional forces in a slab of infinite width. A special investigation will thus only be required at the free edges.

Expansion joints and bearings should be avoided whenever possible. Single-span slabs, for example, can be connected monolithically to the abutment walls to form a frame system. Provided the slab is prestressed, the negative moments due to dead load plus prestressing at the slab-wall connection will be small under service conditions. The thickness of the abutment walls can thus be smaller than the slab depth (figs. 7.133 and 7.134). The abutment wall may be strengthened with perpendicular ribs that taper from the slab-wall connection down to the foundation.

Bearings will be required at at least one end of long continuous slab bridges. They should allow rotation in all directions and should be spaced at a distance of 3 to 5

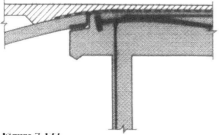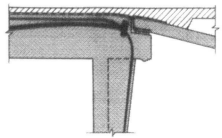

Figure 7.133
Details at intersection of slab and abutment wall

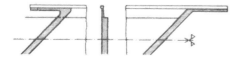

Figure 7.134
Plan: connection of abutment wall to slab

times the slab thickness. A single unindirectional bearing, normally located on the longitudinal axis of the bridge, determines the direction of movement. All other bearings are free to displace in all directions. The resistance of the slab to punching shear must be checked at the bearings, in particular at the corners. The danger of punching is best eliminated by locally thickening the edge of the slab rather than by reducing the spacing of the bearings.

The design of abutment wingwalls poses a difficult aesthetic problem for grade separations. Wingwalls that are parallel to the roadway above should not appear too tall relative to slab depth. Smaller, better proportioned walls can be achieved by lengthening the span (fig. 7.135). This will often, however, result in a substantial increase in cost. Wingwalls that are parallel to the roadway crossing underneath can create the impression of a tunnel. A better appearance is obtained when the walls are funnel-shaped or when they are split into upper and lower portions, as shown in figure 7.136. Short-span grade separations that are designed as trapezoidal frames can be incorporated particularly well into embankments (fig. 7.137).

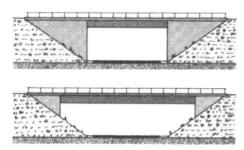

Figure 7.135
Visual effect of wing walls

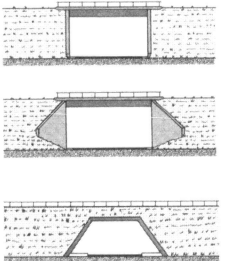

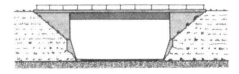

Figure 7.136
Possible wing wall arrangements

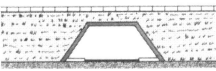

Figure 7.137
Trapezoidal underpass

7.9.2 Design of the Cross-Section

Solid cross-sections are normally used. Solid slabs are particularly easy to construct and, due to their two-dimensional isotropic behaviour, are appropriate for complicated plan layouts with large variations in the directions of the principal bending moments. The slab thickness should, for economic reasons, be limited to roughly 0.8 m.

Although hollow-core slabs made with stay-in-place forms are somewhat lighter, they are more difficult to construct and are less durable than their solid counterparts. The internal formwork must be anchored against uplift during concreting and all voids must be provided with vents and drainage openings. These added construction costs usually outweigh any saving in concrete and reinforcement for short span bridges.

If a slab thickness of 0.8 m is not sufficient for the given span length, a multiple-cell cross-section should be used (fig. 7.138); the depth should be increased to at least 1 m. Precast, prestressed concrete planks, 50 mm thick, can be used to form the deck slab. The prestressing in the planks can later serve as the main reinforcement of the deck slab in the completed structure. A large number of oval-shaped manholes must be provided in the bottom slab for access to each of the cells.

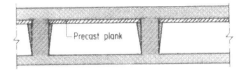

Figure 7.138
Cross-section of a multiple-cell slab bridge

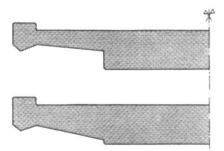

Figure 7.139
Details at free edges of slab bridges

The free edges of non-skew bridges can be designed as cantilevers, as shown in figure 7.139. Cantilevers should be avoided, however, for the free edges of skew bridges which are highly stressed and thus sensitive to deformations. Any weakening of the cross-section or additional loading from the cantilevers is therefore undesirable.

7.9.3 Prestressing Concept

It is neither necessary nor practical to eliminate tensile stresses due to dead load at every point on the slab. It is sufficient to prestress the structures to obtain a balanced state of stress under permanent load. The cracking moment should be exceeded only locally and deformations should be kept small, particularly along the free edges. This can be achieved by balancing between 60 to 80 percent of dead load with the deviation forces due to prestressing over the entire slab. Construction is facilitated when the prestressing steel is arranged in the smallest number of different directions, using parallel tendons of equal length.

7.9.4 Design

Sectional forces can be calculated directly from the differential equation of an elastic plate

$$\frac{\partial^4 w}{\partial x^4} + 2 \frac{\partial^4 w}{\partial x^2 \partial y^2} + \frac{\partial^4 w}{\partial y^4} = \frac{q}{D} \qquad \text{(equation 5.21)} \quad \text{(a)}$$

using equations (5.18), (5.19), and (5.20) of Section 5.3.1. Reinforcement designed purely on the basis of the elastic sectional forces, however, is needlessly complicated. A simpler, more rational arrangement of reinforcing and prestressing steel can be obtained by taking advantage of the capacity of slabs to redistribute moments. Significant reduction of peak stresses due to cracking can be assumed even under service conditions.

Design, detailing, and construction are considerably simplified by concentrating the main reinforcement into a small number of bands, each composed of parallel

bars or tendons of equal length. The arrangement will satisfy global equilibrium
and ensure adequate behaviour under service conditions when the design
moments are computed as the sum of the moments obtained from: (1) the elastic
solution of equation (a), and (2) a solution of the homogeneous equation

$$\frac{\partial^4 w}{\partial x^4} + 2 \frac{\partial^4 w}{\partial x^2 \partial y^2} + \frac{\partial^4 w}{\partial y^4} = 0 \qquad\qquad\text{(b)}$$

for an appropriately chosen structural system. The second, self-equilibrating
component can be regarded as a redistribution of a portion of the moments of
component (1) onto the main reinforcement bands. The load of magnitude zero is
decomposed into the following two loads:

1. $\bar{q}$, applied to an idealized slab in which the reinforcement bands are assumed to
   behave as independent beams
2. $-\bar{q}$, applied to the original slab

Equilibrium is unchanged since the net load is zero. Both systems satisfy the
statical boundary conditions of the given system. The sectional forces due $\bar{q}$ are
beam sectional forces and are thus easily computed.

The redundant moments due to prestressing, m_{sP}, which must be combined with
the moments due to dead load and live load at ultimate limit state, are a
particularly good choice for the self-equilibrating moments. They are defined by
the following equation

$$m_{sP} = m_P - m_{0P}$$

where m_P is the total moment due to prestressing and m_{0P} is the prestressing
moment in the statically determinate system:

$$m_{0P} = -\frac{Pe}{\Delta b}$$

The parameter Δb is the tributary width of one tendon and is equal to the lesser of
the tendon spacing or twice the slab thickness (fig. 7.140). The moments m_P are
obtained when the deviation force q_P is applied to the original slab; moments
$-m_{0P}$ are obtained when $-q_P$ is applied to the idealized slab in which the
prestressing bands are individual statically determinate beams. The redundant

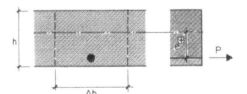

Figure 7.140
Model for the calculation of the moment
due to prestressing in slabs

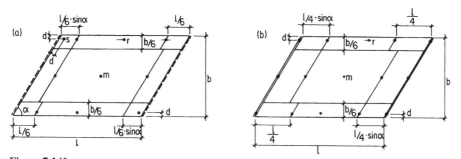

Figure 7.141
Points at which reinforcement must be designed (d denotes slab thickness): a, slab simply supported at two opposite edges; b, slab fixed at two opposite edges

moments m_{sP} are therefore self-equilibrating and can be readily combined with the moments due to dead load and live load. The load factor γ_P can be freely chosen between 0.8 and 1.4.

The elastic sectional forces due to the given loads are normally calculated using the finite-element method. The self-weight of cantilevers (fig. 7.139) is considered as an edge loading, (v_{edge}, m_{edge}). The sectional forces due to live load are obtained from several critical positions of the load.

The reinforcement is designed for the factored moments due to dead load, live load, and the redundant moment due to prestressing. The design is based on the principles outlined in Sections 4.3.5 and 4.4.2. The calculations need only be performed at the points indicated in figure 7.141; the reinforcement thus computed can be then used for the corresponding region of the slab.

Example 7.10:
Rectangular slab: Calculation of design moments due to dead load

The simply supported slab shown in figure 7.142 is loaded by its self-weight, g. Prestressing tendons are located at either edge of the slab in bands of width b_0,

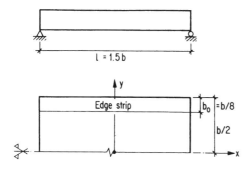

Figure 7.142
Example 7.10: structural system

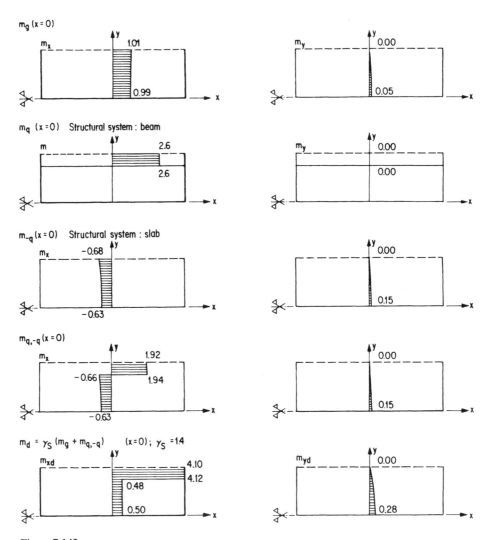

Figure 7.143
Example 7.10: moments m_x and m_y (divided by $gl^2/8$) at the centre of the slab

equal to $b/8$. The design moment in the x-direction will be concentrated along these bands by adding self-equilibrating moments due to the sum of: (1) $\bar{q}$ on the idealized system consisting of independent edge beams, and (2) $-\bar{q}$ on the original system. The total load due to $\bar{q}$ is 65 percent of the total dead load:

$$\bar{q} = 0.65g \frac{b}{2b_0}$$

The resulting design moments are shown in figure 7.143.

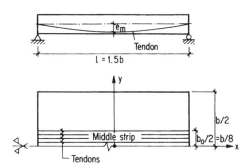

Figure 7.144
Example 7.11: structural system

Example 7.11:
Rectangular slab: Calculation of design moments due to dead load, live load, and prestressing

The slab shown in figure 7.144 is loaded by dead load, g, and a uniformly distributed live load, q, equal to $0.25g$. The deviation force due to prestressing, q_P, is chosen to give a total load equal to 80 percent of total dead load:

$$q_P = 0.8g \frac{b}{b_0}$$

The load factor for prestressing, γ_P, taken as 1.4. The design moments are shown in figure 7.145.

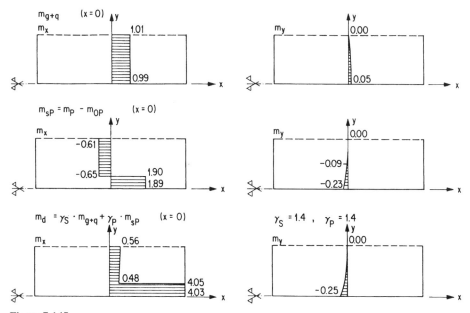

Figure 7.145
Example 7.11: moments m_x and m_y (divided by $(g+q)l^2/8$) at the centre of the slab

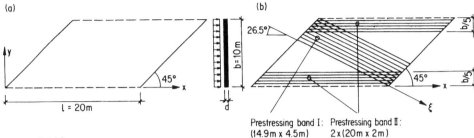

Figure 7.146
Example 7.12: a, structural system and loading; b, tendon arrangement

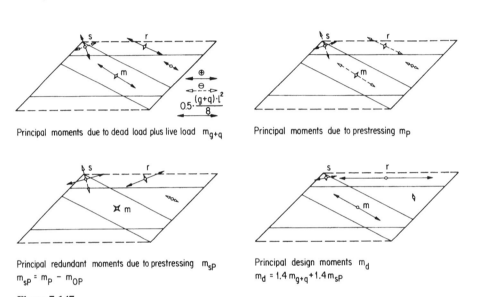

Figure 7.147
Example 7.12: principal bending moments at points m, r, and s of a simply supported slab (see figure 7.141)

Example 7.12:
Skew prestressed slab: Calculation of the design moments in the principal directions

The slab shown in figure 7.146 is loaded by dead load, $g = 10$ kN/m², and a uniformly distributed live load, $q = 2.5$ kN/m². The deviation force due to prestressing, q_P, is taken as $-0.8\,(g+q)$ or -10 kN/m². The factor γ_P is 1.4. The resulting design moments are shown in figure 7.147.

7.9.5 Reinforcement Layout

It is difficult to optimize of the total tonnage of reinforcement in skew slab bridges. Each position of the load yields different principal directions for the bending moments at all points in the slab. The theoretically most efficient reinforcement layout would therefore be complicated and expensive. It is more economical to arrange the reinforcement into regular patterns that are easy to fabricate and place. The savings in labour costs will always outweigh the added costs resulting from the higher consumption of steel.

It is not necessary to cover the entire surface of the slab with prestressing tendons. The sectional forces in the portions of the slab that are not crossed by tendons will be transferred to the prestressing steel by the redundant moments due to prestressing. It is preferable to provide bands of parallel tendons which, wherever possible, are oriented in the direction of principal stress. Parallel tendons must always be provided along the free edges, which are particularly sensitive to deformations. The resistance to punching shear at point supports can be enhanced by concentrating bands of prestressing tendons at these locations.

The eccentricity of the tendons need not be proportional to the theoretical moment diagram due to dead load. It is preferable to provide a parabolic cable profile, which produces distributed deviation forces that are roughly uniform. The prestressing will therefore balance a portion of dead load and the moment diagrams due to dead load and prestressing will be roughly proportional to each other.

Figures 7.148 and 7.149 illustrate some possible arrangements of prestressing in skew slab bridges.

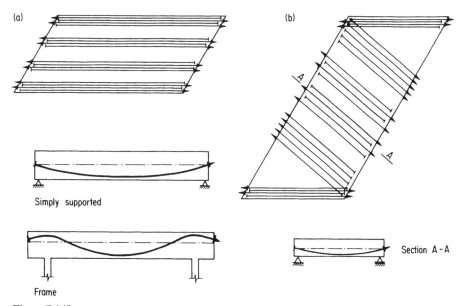

Figure 7.148
Tendon arrangements for single-span slabs: a, narrow; b, wide

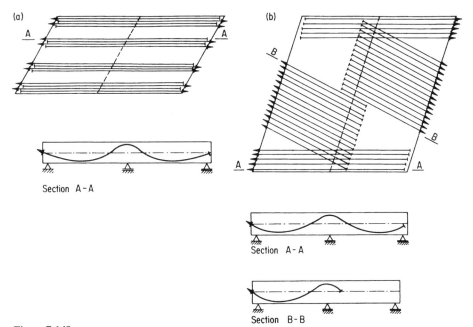

Figure 7.149
Tendon arrangements for multiple-span slabs: a, narrow; b, wide

Minimum mild reinforcing steel must be provided over the entire slab surface. This reinforcement is necessary for crack control and may be considered in the calculation of ultimate flexural resistance. The bars are normally arranged in an orthogonal pattern, parallel and perpendicular to the bridge axis for narrow or long slabs, and parallel and perpendicular to the support axis for short slabs.

The edges of the slab are strengthened with bars parallel to the edge and closed stirrups (figs. 7.150 a, b). The stirrups are designed for the superimposed shear flow due to vertical shear and twist according to the principles outlined in Section 4.4.2. They must be properly spliced with the transverse reinforcement of the slab.

A top layer of reinforcement is normally required in the corners of the slab that form an obtuse angle. This reinforcement can be detailed as shown in figure 7.150 c.

The calculation of the steel stresses under service conditions is difficult since the redistribution of sectional forces due to cracking is often substantial, even at low load levels. If the cracking moment under dead load and prestressing is exceeded by a significant amount in individual regions of the slab, it may be necessary to provide additional prestressing.

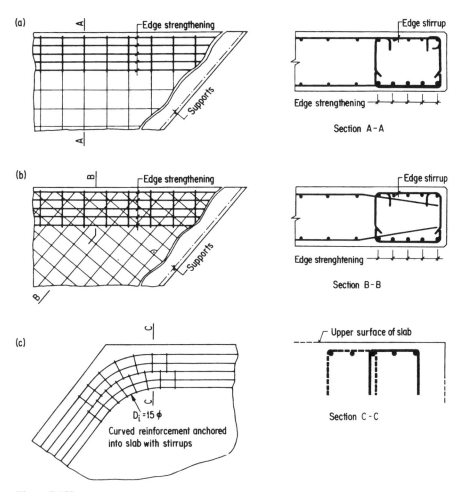

Figure 7.150
Reinforcement layout for skew slabs: a, edge reinforcement for narrow slabs; b, edge reinforcement for wide slabs; c, upper reinforcement in obtuse-angle corners

7.10 Cable-Stayed Bridges

7.10.1 Conceptual Design

a) General Considerations

The longest-spanning prestressed concrete bridges are supported by inclined steel cables. This arrangement extends the economic range of concrete bridges well beyond the limits of conventional girder bridges. Although arches are also statically feasible for relatively long spans, they are now only economical in exceptional situations.

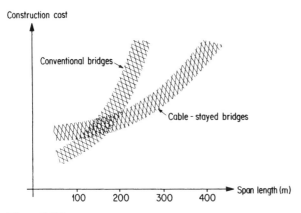

Figure 7.151
Cost comparison: conventional bridges and cable-stayed bridges

Although cable-stayed bridges have traditionally been preferred for long spans, they can be competitive with arches and conventional girders for spans as small as 150 m (fig. 7.151). They can also be appropriate for spans less than 150 m when clearance under the bridge is severely limited; the traditional alternatives – arches with suspended roadway, through trusses, and through girders – are expensive and visually inferior.

Cable-stayed bridges are generally uneconomical when the ratio of span length to bridge width is less than 10:1.

b) Cable System

Arranging the cables into two planes maximizes the torsional stiffness of the structural system and thus optimizes static and dynamic behaviour. Arrangements using one or three planes may nevertheless be preferable for bridges carrying divided highways. When only one plane of cables is provided, the torsional stiffness of the system is furnished by the girder alone. A torsionally stiff box girder is therefore required in such cases.

Figure 7.152 shows the most common cable patterns in elevation. The *fan* pattern (fig. 7.152a), which concentrates the cable anchorages at the top of the towers, is the most efficient. The detailing of the anchors, however, can be difficult and costly. The *harp* pattern (fig. 7.152b), simplifies detailing of anchors at both ends of the cables, but is the least efficient. The *half-fan* pattern (fig. 7.152c), in which the anchors are distributed over the upper portion of the tower only, is a popular intermediate solution.

The lower cable anchors should be spaced relatively close together. The spacing should be small enough to allow the girder to be built in cantilever construction

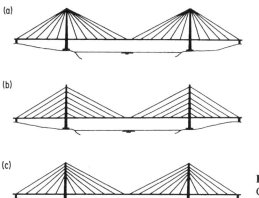

Figure 7.152
Cable patterns: a, fan; b, harp; c, half-fan

without temporary stays. For slender, flexible girders, anchor spacing is normally equal to segment length, which typically varies between 6 and 8 metres. For stiffer girders, anchor spacing can be an integer multiple of erection segment length. Small-diameter cables should be avoided, since the cost of corrosion protection, anchors, and erection is high relative to the cost of cable steel. The lower cable anchors should therefore not be spaced closer than 6 m apart. The chosen spacing should permit the replacement of an individual cable without falsework and ensure that the rupture of a cable, for example due to a vehicle fire, will not lead to collapse. The safety of the structure with one cable removed can be checked using a reduced factor of safety.

Backstays are cables that connect the tips of towers to rigid points at the ends of the stayed spans. They stiffen the system and reduce bending moments in towers and girder under the influence of partial live load. The participation of the cables, and hence the efficiency of the system, is thus enhanced by their use. The stress range in backstays is relatively high, and fatigue is thus often critical for their design. The stress in the backstays due to full live load on both side spans should not be less than 15 percent of tensile strength, to limit cable sag and the associated reduction in stiffness.

For aesthetic reasons, the surface formed by the cables should be plane or slightly curved. The anchors should therefore be arranged in a straight or a slightly curved line at the girder and at the tower. This requirement cannot be met when the anchors are concentrated at the top of the tower in a fan pattern; the crossing of the cables which results often produces a disturbing visual effect.

c) Towers

The main span should be supported symmetrically by two towers. Since supporting the main span by a single tower doubles the length of the cantilever, single towers should only be used in exceptional circumstances.

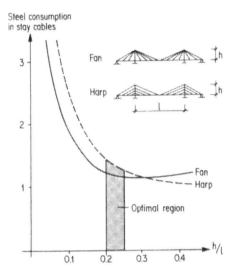

Figure 7.153
Effect of tower height on steel consumption in stay cables (adapted from Leonhardt and Zellner (1980))

Increasing tower height reduces the consumption of cable steel, but increases the costs of the tower itself. As shown in figure 7.153, the best compromise between these two costs results from when tower height is between 20 and 25 percent of main span length. Shorter towers may, however, be used when warranted by aesthetic considerations.

When backstays are used, any differences in cable forces on either side of the tower due to unbalanced live load are transferred directly to the backstays. This prevents large bending moments in the towers. In such cases, the towers can be relatively flexible in the logitudinal direction; stiffness can be determined to ensure adequate stability against buckling.

Towers should, however, be stiff transverse to the axis of the bridge. As shown in figure 7.154, various forms are possible, depending on girder width, tower height, and bridge height. The greatest torsional stiffness about the longitudinal axis of the bridge is achieved using A-shaped towers, since tower, cables, and girder act as a space truss. The lowest flexural and torsional frequencies of systems with A-shaped towers are typically sufficiently far apart to ensure adequate vibrational behaviour (see Section 7.10.5).

Towers that are vertical in the plane of the longitudinal axis of the bridge are normally most economical, since the lever arm of the cantilever system at the base of tower base is maximized for a given tower height (fig. 7.155). When towers are inclined away from the main span, their height must be increased to maintain the lever arm; when towers are inclined towards the main span, stresses in the backstays are significantly increased. The construction of inclined towers always entails additional costs and difficulties.

The location of the towers, and hence the span arrangement, can be determined from the allowable minimum stress and stress range in the cables. The graph of

Figure 7.154
Typical shapes for towers

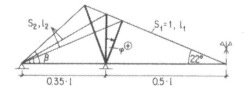

φ	β	S_2	$S_1 \cdot l_1 + S_2 \cdot l_2$
-20°	42°	0.8	0.92 · l
0°	30°	1.1	0.98 · l
+20°	23°	1.4	1.10 · l

Figure 7.155
Cable forces and consumption of cable steel
as a function of tower inclination

figure 7.156 gives upper bounds for the ratio of side spans to main span. This ratio
should be chosen as close as possible to the values obtained from the graph, which
minimize the length of the main span for a given bridge length. If the span lengths
are fixed from the outset and the most favourable span ratio cannot be chosen,
unloading of the main span cables under partial live loads must be prevented by
additional means. For short-span bridges, it may be sufficient to increase the
flexural stiffness of the girder, thus reducing the fraction of live load resisted by the
cables. Any transfer of load from cables to girder, however, reduces the overall
efficiency of the system. For longer spans, it may be preferable to use ballast or to
provide additional backstay cables to prevent horizontal displacements of the
tower tips.

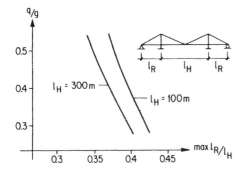

Figure 7.156
Recommended maximum ratio of side span
length to main span length (adapted from
Leonhardt and Zellner (1980))

d) Girder Cross-Section

The design of the girder cross-section must consider static behaviour in the transverse and longitudinal directions, construction methods, load-carrying mechanisms after the rupture of a cable, and dynamic behaviour.

Bending moments in the girder consist of two components: equilibrium moments, a function of the spacing between cable anchors, and compatibility moments, a function of the overall deformations of the system. When the cables have been properly prestressed, the equilibrium moments due to permanent load are those of a continuous beam rigidly supported at each of the anchor points. Compatibility moments due to dead load, caused by axial deformations of girder and cables, can be largely eliminated by a subsequent adjustment of the prestressing forces in the cables. Long-term shortening of the girder due to creep gradually moves the stresses due to permanent load towards the stresses of an idealized system cast simultaneously on falsework. Depending on the prestressing in the stay cables, up to two thirds of these stresses can be restored.

The system deformations induced by partial live loads are substantial, due to the flexibility of the cable system. The compatibility moments in the girder increase sharply with increasing girder stiffness (fig. 7.157). Although these moments unload the cables and tower, this results in no significant savings. The amount of girder reinforcement required cannot, however, be assessed of the basis of these moments alone. Girder reinforcement is also a function of the internal lever arm of the section, which increases with girder stiffness, and the transverse structural system.

Girders should normally be chosen as slender as possible, while maintaining appropriate transverse structural behaviour. Slenderness is limited by buckling stability and the integrity of the system in the event of the rupture of a cable. The ratio of girder depth to moment of inertia should be as large as possible for box and T-sections.

When two planes of cables are used, slabs with stiffened edges are most economical for bridge widths up to roughly 12 m; double-T girders with

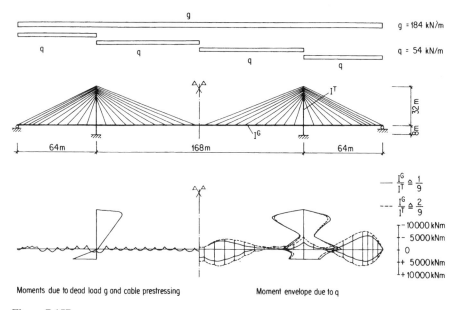

Figure 7.157
Effect of girder stiffness, I^G, on bending moments in the girder

diaphragms are most economical for greater widths (fig. 7.158). Box sections must be used for bridges with only one plane of cables to ensure adequate torsional stiffness (fig. 7.159).

e) Bearings and Joints

Bridges with two planes of cables normally require vertical supports only at the ends of the side spans. By eliminating vertical supports at the towers, negative moments in the girder can be substantially reduced.

Backstays normally produce uplift reactions at the ends of the side spans which cannot be compensated by ballast alone. Prestressed hold-down cables are thus required. The downward reactions at these locations can be increased by making the stayed spans continuous with the approach spans.

Fixing the girder against longitudinal displacement significantly increases the stability of the tower and is thus always recommended. The actual position of the fixed point is of secondary importance. Since cable-stayed bridges are flexible in the longitudinal direction, it can normally be located at the end of one of the side spans without difficulty, even for long-span bridges. For three-span bridges supported in this way, only one large expansion joint would be required. The fixed point can also, however, be located at one of the towers. Braking forces are thus

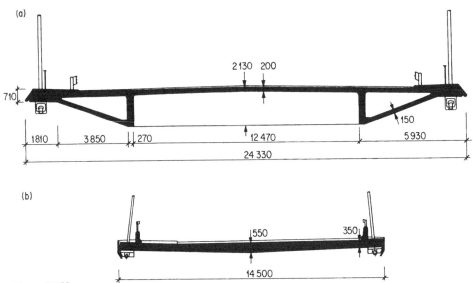

Figure 7.158
Cross-sections for bridges with two planes of cables (dimensions in mm): a, Pasco-Kennewick
Bridge, U.S.A., 1978; b, Diepoldsau Bridge, Switzerland, 1985

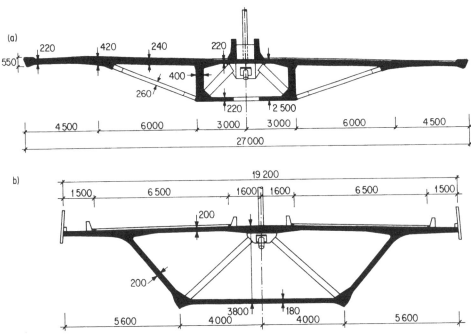

Figure 7.159
Cross-sections for bridges with one plane of cables (dimensions in mm): a, Chandoline Bridge,
Switzerland; b, Brotonne Bridge, France, 1977

carried to the tower foundation through bending. Since the axial force in the tower below the girder is normally large, only a small eccentricity will be produced.

Horizontal displacements transverse to the axis of the bridge due to wind and seismic loads can be restrained at the towers and at the ends of the side spans. At the towers, buttresses and neoprene bearings can be used for this purpose. A gap of 2 to 3 mm should be provided between buttress and bearing, to allow girder and tower to displace freely relative to each other in the vertical and longitudinal directions. Expansion joints at midspan of the main span are impractical, since they require complicated bearings (fixed vertically, free longitudinally) and induce unavoidable discontinuities in roadway grade due to long-term girder deflections and live load. Increasing the stiffness of the cables anchored near the joint can help to reduce, but will not eliminate, the rotation at the joint. Systems with expansion joints at midspan should thus only be considered for single-span bridges.

7.10.2 Cables and Anchorages

The choice of cable system is a function of tensile strength, modulus of elasticity, corrosion protection, fatigue resistance, anchor details, and erection procedure. The most commonly used types of stay cables are listed in table 7.2.

Table 7.2
Cable Types

Type	Typical System	$E_{C,0}$ (kN/mm^2)
Parallel bar	Dywidag	205
Parallel wire	BBRV	205
Parallel strand	VSL, Freyssinet	195
Locked-coil		165
Wire rope		150

The modulus of elasticity of stay cables, E_C, is a function of the material modulus, $E_{C,0}$, and sag. The influence of the latter factor is a function of stress, σ_C, and projected horizontal length, l_h. An idealized modulus of elasticity, $E_{C,i}$, which accounts for σ_C and l_h, can be defined as follows:

$$E_{C,i} = \frac{E_{C,0}}{1 + \dfrac{\gamma_C^2 \, l_h^2 \, E_{C,0}}{12\sigma_C^3}} \qquad\qquad (a)$$

The density of the cable, γ_C, is given by the following expression:

$$\gamma_C = \frac{G_C}{A_C \, l_h}$$

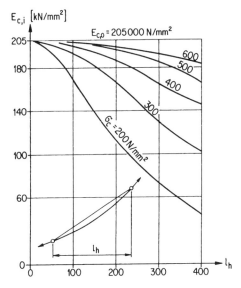

Figure 7.160
Idealized modulus of elasticity for stay cables
(adapted from Leonhardt and Zellner (1980))

where G_C is the total weight of the cable, including corrosion protection, and A_C is the area of steel in the cable. Equation (a) is graphed in figure 7.160 for an assumed $E_{C,0}$ of 205 kN/mm^2.

The value of E_C has a substantial influence on the overall stiffness of the structural system. Decreasing E_C increases both first order moments and second order moments, thus reducing the stability of the system. Since $E_{C,i}$ is a cubic function of σ_C, the stress in the cables should be as high as possible. The use of high-strength steel for stay cables is thus recommended, especially for long-span bridges.

Cables are the most susceptible component of the structural system to deterioration. Corrosion protection of the cables is thus of utmost importance; paint alone is insufficient.

Locked-coil cables and wire ropes are composed of galvanized wires. A zinc coating of roughly 100 g/m^2 is typical. Locked-coil cables are often coated with epoxy after fabrication. The durability thus obtained normally meets the standards prescribed for long-span highway bridges. Wire ropes, which cannot be reliably coated in this way, have a relatively low service life of approximately 50 years. They are therefore used only for small bridges, for instance pedestrian structures.

Parallel bar, wire, and strand cables are enclosed in steel or polyethylene ducts. The connection of the ducts to the anchors at either end must be watertight. After the cables have been prestressed, the ducts are carefully grouted. Polyethylene ducts have a lifetime of roughly 50 years; when properly maintained, however, they can last much longer. The entire cable must normally be replaced in the event of a failure in the corrosion protection mechanism.

When polyethylene ducts are used, cables can be fabricated under controlled conditions in the shop, complete with anchors, transported to the site on compact spools, and easily erected. Steel ducts normally use more expensive, corrosion-resistant materials and complicate erection.

The fatigue resistance of stay cables must be carefully checked. Although the fatigue resistance of stay cables and prestressing tendons in concrete are essentially identical, the design stress range of the former is significantly higher. Oscillating stresses are induced by cable vibrations due to wind and by live load. Wind-induced vibrations can be sharply reduced by dampers at the anchorages or by increasing the intrinsic damping of the cable itself. Bending of cables at the anchors must always be prevented.

No test results relating to the spectrum of live load stresses are available. The provisions contained in codes and standards apply primarily to steel bridges. Normal cables are primarily influenced by live loads applied in their immediate vicinity; backstays have a much larger region of influence. Fatigue stresses in backstays should therefore not be calculated for a single vehicle, even for highway bridges, but rather for a single truck of 400 kN and roughly 30 percent of the uniform design live load, q.

Extensive tests have shown that the fatigue resistance of stay cables is independent of the absolute magnitude of the fluctuating stresses, provided the upper limit, σ_u, is less than 50 percent of tensile strength (Birkenmaier 1980). Fatigue design is therefore based on a stress versus life diagram (S-N curve), in which the ordinate represents stress range, $\Delta\sigma$, and the abscissa represents the corresponding number of load cycles required to produce failure, N. Since the fatigue resistance of stay-cable anchors is normally greater than that of the cable proper, design S-N curves are based on those of a single wire, modified to account for the effects of group interaction in the cable. Thus, the design S-N curve for the cable, $\Delta\sigma_C(N)$, can be expressed as

$$\Delta\sigma_C(N) = \frac{1}{\gamma_1 \gamma_2} \Delta\sigma_W(N)$$

where $\Delta\sigma_W(N)$ denotes the 5 percent fractile of the single-wire S-N curve. The conversion factor, γ_1, accounts for group interaction, fabrication tolerances, and surface defects; it is normally equal to 1.3. The factor of safety, γ_2, can be taken as 1.25. The functions $\Delta\sigma_W(N)$ and $\Delta\sigma_C(N)$ are plotted for a typical wire/cable combination in figure 7.161.

The number of design load cycles, n, is a function of type of load and average daily truck traffic, and is generally specified in codes and standards. The values in table 7.3 can regarded as typical. Safety against failure in fatigue is ensured when the following inequality is satisfied:

$$\Delta\sigma_d \leqq \Delta\sigma_C(n) \tag{7.6}$$

where $\Delta\sigma_d$ is the computed stress range under design fatigue load.

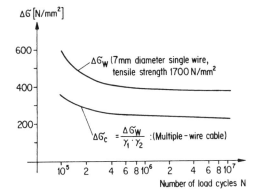

Figure 7.161
Comparison of allowable stress range in a
single wire and a multiple-wire cable (adapted from Birkenmaier (1980))

Table 7.3
Number of Design Load Cycles, n

Average Daily Truck Traffic	Concentrated Load Q	Uniform Load q
500	2 000 000	200 000
2 000	Number of cycles at which fatigue limit is reached	2 000 000

The fatigue resistance of backstays, in which fluctuating stresses are induced by
two load types, can be verified using an interaction equation. The specified
number of load cycles for design fatigue loads q and Q are denoted n_q and n_Q,
respectively. The corresponding computed stress ranges are likewise denoted $\Delta\sigma_q$
and $\Delta\sigma_Q$. Adequate safety against failure in fatigue is ensured when

$$\frac{n_q}{N(\Delta\sigma_q)} + \frac{n_Q}{N(\Delta\sigma_Q)} \leqq 1 \tag{7.7}$$

where $N(\Delta\sigma_q)$ and $N(\Delta\sigma_Q)$ are the maximum permissible number of load cycles for
the given stress range.

Special anchors are required to ensure that wire rupture in fatigue occurs in the
cable proper and not at the anchor; for this reason, conventional prestressing
anchors cannot be used for threaded bar, parallel wire, or parallel strand cables.
As shown in figure 7.162, parallel strand anchors transfer the cable force due to
prestressing and dead load to the bearing plate directly from the anchor head; the
force due to live load is transferred from the strands to the transition pipe in bond,
and then directly from the pipe to the bearing plate. Careful workmanship in
fanning out of the strands and in grouting are of primary importance to the fatigue
resistance of the cable. The details illustrated in figure 7.162 are also used to
anchor parallel bar cables.

The HiAm anchor, shown in figure 7.163 a, has been developed for parallel-wire
cables. The individual wired are anchored to a circular perforated plate by means

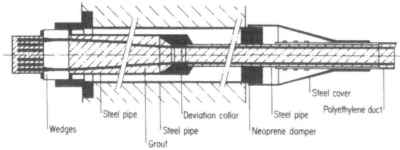

Figure 7.162
Anchor for parallel strand cables (VSL system)

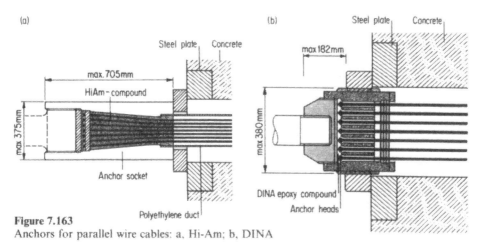

Figure 7.163
Anchors for parallel wire cables: a, Hi-Am; b, DINA

of cold-formed buttonheads; the plate bears against a steel socket filled with a mixture of epoxy and steel balls. Cables using HiAm anchors are completely prefabricated in the shop, thus ensuring a high degree of reliability. The DINA anchor, shown in figure 7.163 b, has also been used for parallel-wire cables.

Locked-coil cables are anchored using a steel socket. The wires at each end of the cable are "broomed" and inserted into the socket, which is then filled with molten zinc. Special attention must be given to the prevention of water penetration where the cable enters the socket.

Cables are normally prestressed from their lower end using hydraulic jacks. The upper anchorages should be arranged to permit easy replacement of the cables and should always be arranged symmetrically with respect to the longitudinal bridge axis. When a single plane of cables is used, the cables in the side spans can be doubled (fig. 7.164).

Cables can be anchored inside the tower when the tower dimensions are sufficiently large. The necessary clearance depends on the erection equipment used

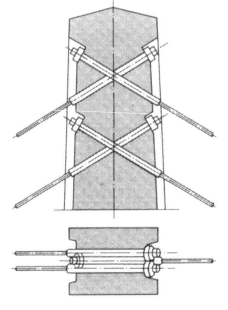

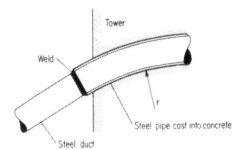

Figure 7.164
Suggested detail for anchorage of cables at tower (single plane of cables)

Figure 7.165
Connection of steel duct to steel pipe cast into tower concrete

and can be substantial. Transverse prestressing or structural steel members must be provided to resist the horizontal component of cable force.

Wire ropes can be carried over the tower with saddles. The radius must be at least 65 times the rope diameter or 600 times the wire diameter, whichever is greater.

Steel ducts, which can be used for threadbar or parallel strand cables, can be welded to a steel pipe which has been cast into the tower concrete (fig. 7.165). This arrangement enables the transmission of differential cable tensile forces to the tower through duct, weld, and pipe.

7.10.3 Analysis and Design

The sectional forces in the girder cannot be easily calculated due to the high degree of indeterminacy of the structural system and the high degree of interaction among the cables.

To prevent excessively high stresses under service conditions, the sectional forces due to the deformation of the system must be considered. It is therefore recommended to compute envelopes of the sectional forces using elastic methods. The redistribution of sectional forces can be considered to a limited extent by slight variations of the stiffness of tower and girder. In such cases, however, the response of the system under service conditions must always be investigated.

Two-dimensional models can normally be used to calculate the sectional forces. Second-order methods need only be used when dead load is greater than 5 percent of the critical load of the system, computed from the flexural stiffness of girder and tower at ultimate limit state.

Generally speaking, the towers are the most critical components of the system. The capacity of the system is practically exhausted with the formation of a mechanism in a tower, since the second-order moments increase sharply and the girder normally has only a small reserve of stability. Plastic hinges in the girder, however, are less dangerous, provided they do not endanger the stability of the towers. Hinges in the girder lead to only small increases in stress in the cables and towers when the girder cannot resist additional moments due to the deformation of the system.

Cables must resist static and dynamic actions. Although the design of normal cables is primarily controlled by static behaviour, the design of backstays is usually controlled by fatigue.

Cables are designed for static loads using working-stress methods. They can be considered safe when the maximum stress due to dead load plus live load is less than or equal to a specified allowable stress:

$$\sigma_{C,\max}(g+q) \leqq 0.45 f_{Ct} \cong 0.5 f_{Cy}$$

where f_{Ct} and f_{Cy} denote, respectively, the minimum tensile strength and nominal yield stress of the steel. The factor of safety, 0.45, also takes into account secondary flexural stresses that are not normally considered in the analysis.

Safety against fatigue failure is verified according to equation 7.6. Equation 7.7 is used when fatigue stresses are induced by two distinct load types.

Special consideration must be given to the load factors used for dead load and cable prestressing, to ensure that sections designed at ultimate limit state will also have adequate behaviour under service conditions. The permanent load moment under service conditions, M, is defined by the following sum:

$$M = M_g + M_{CP}$$

where M_g is the moment due to dead load and M_{CP} is the moment due to stay cable prestressing. Prestressing forces in the cables are chosen so that M approximates the moment diagram of a continuous beam supported at each of the cable anchor points. For design at ultimate limit state, it would appear at first glance appropriate to proceed as in classical prestressed concrete design, treating M_{CP} as a redundant moment due to prestressing. The ultimate moments could thus be formulated as

$$M_d = \gamma_S\, M_g + 1.0\, M_{CP}$$

As reported by Saul and Svensson (1983), however, this approach can result in values of M_d that differ greatly from M, or which are even of opposite sign; reinforcement designed for M_d would thus be inadequate under service conditions.

This difficulty can be avoided by applying the same load factor to M_g and M_{CP}:

$$M_d = \gamma_S\, (M_g + M_{CP}) \tag{a}$$

This results in moments at ultimate limit state that are compatible with those under service conditions. The apparent inconsistency of factoring M_{CP} by γ_S is resolved when equation (a) is rewritten as

$$M_d = \gamma_S\, (M_g + 1.0\, M_{CP}) + (\gamma_S - 1.0)\, M_{CP}$$

where the moments due cable prestressing are now factored by 1.0 as in classical design and the term $(\gamma_S - 1.0)\, M_{CP}$ can be considered as a redistribution of moments, valid at ultimate limit state.

As in the design of conventional bridges, stresses due to temperature, shrinkage, and creep need not be considered at ultimate limit state. Furthermore, they are usually not relevant under service conditions, provided the cross-section is adequately reinforced for the control and distribution of cracks.

The behaviour of slender girders at ultimate limit state must be verified taking into account the geometrical and material nonlinearities of the structural system. Permanent load, the corresponding cable prestressing forces, and the critical live load must be increased progressively to the γ_S level, proceeding from the system deformations due to temperature, shrinkage, and creep. Material stiffness must be decreased by the factor $1/\gamma_R$, i.e., for a given steel or concrete strain, the corresponding stress from the stress-strain diagram is to be reduced by $1/\gamma_R$.

7.10.4 Stability

Failure in buckling can be initiated in either the girder or the tower, since permanent load induces large compressive forces in both.

The girder is stabilized by the cable system and its own dead weight. Its buckling shape is a function of its slenderness ratio, support conditions, and cable arrangement, and thus cannot be easily determined.

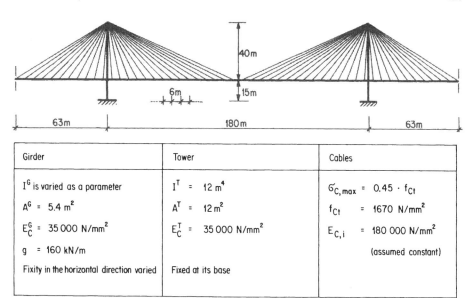

Figure 7.166
Bridge used for investigation of stability

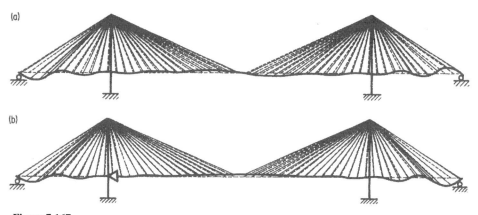

Figure 7.167
First buckling mode for bridge with very slender girder, $I^G = 0.02$ m^4: a, girder not fixed horizontally, $\lambda_1 = 3.88$; b, girder fixed horizontally at tower, $\lambda_1 = 3.88$

Figures 7.167 through 7.171 illustrate buckling shapes for the idealized cable-stayed bridge depicted in figure 7.166. Several different conditions of restraint of horizontal displacements were considered. The moment of inertia of the girder, I^G, was chosen as the parameter for the study; all other section and material properties were maintained constant. The first eigenvalue, λ_1, given for each case, relates the critical vertical buckling load and girder dead load as follows:

$$g_{crit} = \lambda_1 \, g$$

(a)

(b)

Figure 7.168
First buckling mode for bridge with slender girder, $I^G = 0.30$ m^4: a, girder not fixed horizontally, $\lambda_1 = 13.35$; b, girder fixed horizontally at tower, $\lambda_1 = 13.35$

(a)

(b)

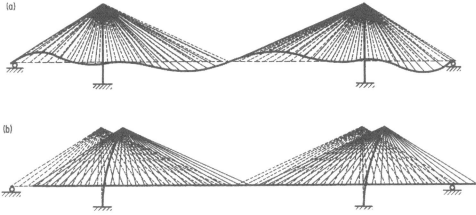

Figure 7.169
Higher antisymmetrical buckling modes, $I^G = 0.30$ m^4, girder not fixed horizontally: a, $\lambda_2 = 13.81$; b, $\lambda_3 = 13.99$

Figure 7.170
Buckling of tower, $I^G = 0.35$ m^4 (less than critical value), girder not fixed horizontally, $\lambda_1 = 13.99$

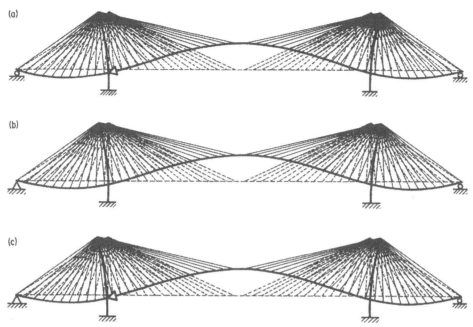

Figure 7.171
First buckling mode, $I^G = 10.00$ m^4, girder fixed horizontally; a, girder fixed at tower, $\lambda_1 = 42.34$; b, girder fixed at end of bridge, $\lambda_1 = 42.35$; c, girder fixed at tower and at end of bridge, $\lambda_1 = 42.79$

The bucking shapes corresponding to a very slender girder ($I^G = 0.02$ m^4) are given in figure 7.167. The buckling behaviour is independent of horizontal girder restraint and is characterized by short waves in the side spans. These are the result of the large horizontal component of the force in the backstays and the relative flexibility of the outer cables.

As girder stiffness is increased, longer waves and horizontal displacement at the tip of the tower are induced. Figure 7.168 shows the first buckling mode for $I^G = 0.30$ m^4, which is symmetric about the midspan axis, and is also independent of the horizontal restraint of the girder. If no horizontal restraint is given by the girder supports, antisymmetrical buckling shapes are produced by the higher buckling modes (fig. 7.169). The third mode shown is characterized by buckling in the tower only.

The horizontal support conditions begin to influence the mode of buckling after I^G reaches a critical value, approximately equal to 0.35 for the idealized bridge of these examples. When $I^G = 0.35$ and horizontal restraint is provided at one tower or at one end of the bridge, the bridge buckles roughly as shown in figure 7.168 b. When $I^G = 0.35$ and no horizontal restraint is provided, the bridge buckles as shown in figure 7.170, characterized by bending in the tower only and a horizontal rigid-body displacement of the girder. The buckling load is independent of girder stiffness when $I^G > 0.35$ and no horizontal restraint is provided.

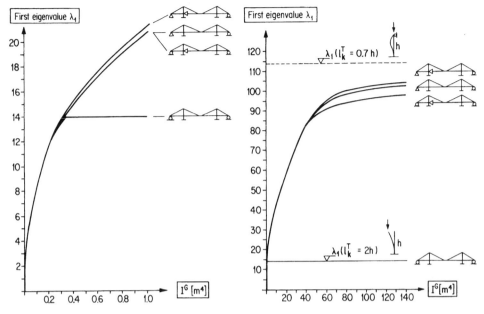

Figure 7.172
Influence of support conditions on stability

Figure 7.171 illustrates buckling figures for three bridges with $I^G = 10.0$ m^4, all of which are horizontally restrained at one of more of the supports. Horizontal restraint enables the joint participation of girder and column in the buckling resistance of the structure and thus significantly increases the buckling load (fig. 7.172).

From these examples, it can be concluded that premature buckling of the tower must be prevented by restraining the girder against horizontal displacement, and premature buckling of the girder must be prevented by providing a minimum girder stiffness, which will be a function of span length.

7.10.5 Dynamic Behaviour

Multiple-cable bridges with slender girders are considerably less sensitive to vibration than conventional girder bridges or suspension bridges. Due to the nonlinear stress-strain behaviour of stay cables, the frequency of free vibrations is a function of amplitude. Increases in amplitude due to near resonant vibration thus tend to move the natural frequency of the structure away from the frequency of excitation. In addition, the natural frequencies of the lower modes differ by only a small amount (fig. 7.173). Strong vibrations are therefore often excited in several different modes; the resulting vibrations largely counterbalance each other when superimposed.

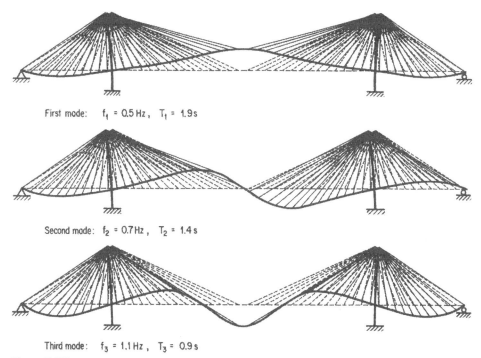

First mode: $f_1 = 0.5\,Hz$, $T_1 = 1.9\,s$

Second mode: $f_2 = 0.7\,Hz$, $T_2 = 1.4\,s$

Third mode: $f_3 = 1.1\,Hz$, $T_3 = 0.9\,s$

Figure 7.173
Typical natural frequencies and mode shapes for cable-stayed bridges

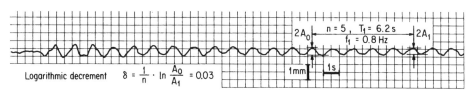

Logarithmic decrement $\delta = \dfrac{1}{n} \cdot \ln \dfrac{A_0}{A_1} = 0.03$ $2A_0$ $n = 5$, $T_1 = 6.2\,s$ $2A_1$ $f_1 = 0.8\,Hz$ 1mm   1s

Figure 7.174
Free vibration of the Diepoldsau Bridge, Switzerland (deflections measured at midspan with laser; bridge excited into vibration by impact of a 5 kN sandbag falling from a height of 1 m)

The damping of cable-stayed bridges is similar in magnitude to the damping of conventional girder bridges. Measurements of vibrations on the Diepoldsau Bridge in Switzerland yielded a logarithmic decrement of 0.03 (fig. 7.174). Similar model tests for the cable-stayed bridge over the Rio Paraná in Argentina yielded a logarithmic decrement of 0.15 (Walther et al. 1985).

Vibrations induced by vehicular traffic are small. As stated in Section 3.2.1, the excitation frequencies induced by truck wheel loads lie between 2 and 4 Hz (truck body vibration) and between 8 and 12 Hz (axle vibration). Since the lowest natural frequencies of cable-stayed bridges are normally less than 1 Hz, the likelihood of

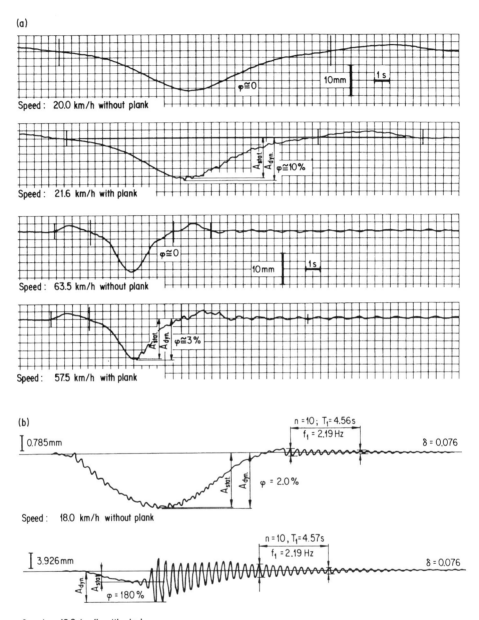

Figure 7.175
Dynamic increment, ϕ, computed from deflections at midspan: a, Diepoldsau Bridge (cable-stayed), vehicle weight 250 kN; b, Balma Bridge (conventional girder), vehicle weight 160 kN

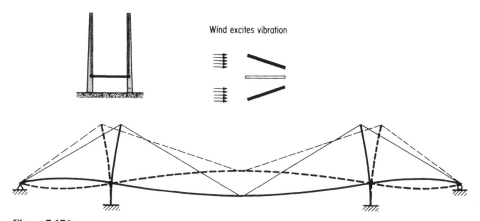

Figure 7.176
Flutter: coupled flexural and torsional vibration

resonant vibration due to traffic is small. The dynamic increment of live-load girder deflection, ϕ, is considerably smaller for cable-stayed bridges than for conventional girder bridges. For the Diepoldsau Bridge, the maximum value of ϕ induced by a single truck running over a plank on the deck was roughly 10 percent (Walther et al. 1985). For girder bridges, however, ϕ can reach values of up to 180 percent (fig. 7.175). The combination of a relatively low natural frequency and low dynamic increment results in vibrations due to live load that cause no unpleasant psychological effects on pedestrians.

Wind-induced vibrations are primarily the result of periodic gusts or vortex shedding. For cable-stayed bridges, the first mode shape is often similar to the deflected shape of the structure due to wind loads. Vibrations of dangerously large amplitude are highly unlikely, however, since resonant vibrations are hindered by the change in natural frequency resulting from increasing amplitude. Flutter due to simultaneous bending and torsional vibrations must, however, be considered (fig. 7.176). Bridges with H-shaped towers and slender girders are particularly sensitive in this regard, since their fundamental flexural and torsional frequencies are practically equal.

7.10.6 Construction

Cable-stayed bridges are normally built using the balanced-cantilever method. Precast segmental construction is well suited for bridges over navigable waters; otherwise, the girder can be cast in place using form travellers. The use of conventional falsework may be economical for short, low-level bridges, since the all of the cables can be erected and prestressed in a single operation. This simplifies the work of the cable erector and leads to lower costs.

The profile of the girder immediately after closure should be the camber curve required to compensate deflections due to superimposed dead load, residual creep,

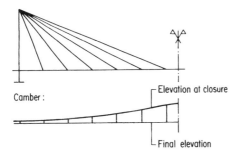

Camber :

Elevation at closure

Final elevation

Figure 7.177
Required camber at closure

and residual shrinkage (fig. 7.177). The moments in the girder should not differ greatly from the moments of a continuous beam supported at the cable anchor points. The cable forces, S_i^{ac}, and lengths, s_i, corresponding to this state can be easily calculated. (The superscript ac denotes *after closure*.) The vertical component of S_i^{ac} is essentially the self-weight of the suspended segment. The unbalanced horizontal force at the tower is equilibrated by backstay cables.

This state can be achieved by proper control of cable forces and lengths during cantilever construction or by a supplementary adjustment of the stays immediately before or after closure. The analytical complexity of the former option is generally preferable to the high cost of the latter. The following procedure can be used to control of the force and length of stay cables during cantilever construction:

The prestressing force in the ith cable, S_i^0, must satisfy the following equation:

$$S_i^0 - \frac{EA_i}{s_i} \delta_i \sin \alpha_i - \frac{EA_i}{s_i} \Delta l_i^G \cos \alpha_i = S_i^{ac} \qquad (a)$$

where S_i^{ac} has been previously computed. The second term on the left-hand side of this equation is the change in cable force due to δ_i, the vertical deflection of point i before continuity at midspan. The third term is the change in force due to Δl_i^G, the horizontal displacement of point i. The calculation of Δl_i^G must account for creep and shrinkage in the girder. Since δ_i and Δl_i^G are functions of S_i^0, equation (a) must be solved iteratively.

After the ith cable has been prestressed, the forces in the other cables can be approximated as follows:

$$S_{i-1} \cong S_{i-1}^{ac} + \frac{G}{\sin \alpha_{i-1}}$$

where G denotes the weight of the traveller , and

$$S_{i-k} \cong S_{i-k}^{ac}$$

for $k \neq i$.

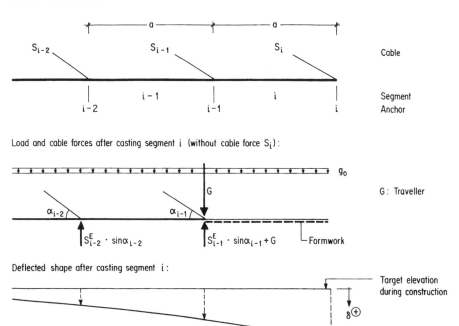

Load and cable forces after casting segment i (without cable force S_i):

Deflected shape after casting segment i:

Figure 7.178
Deflected shape after casting segment i

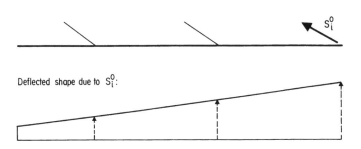

Deflected shape due to S_i^0:

Superposition of deflected shapes due to S_i^0 and due to weight of segment i (fig. 178):

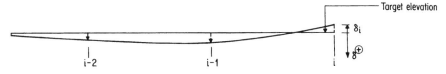

Figure 7.179
Deflected shape after prestressing cable i with force S_i^0

Deflected shape due to $-g_0$:

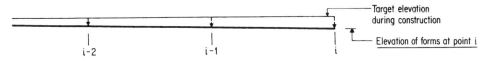

Superposition of the deflected shapes due to $-g_0$ and weight of segment i (fig.178)

Figure 7.180
Elevation of forms at point i

Form camber at point i is determined to compensate for elastic deformation of the traveller during casting of segment i and deformation of the girder due to S_i^0 (figs. 7.178, 7.179, and 7.180).

References

Birkenmaier, M. 1980. Fatigue Resistant Tendons for Cable-Stayed Construction. *IABSE Proceedings* P-30/80: 65–79.

Leonhardt, F. and W. Zellner. 1980. Cable-Stayed Bridges. *IABSE Surveys* S-13/80: 21–48.

Saul, R. and H. Svensson. 1983. Zur Behandlung des Lastfalls „ständige Last" beim Tragsicherheitsnachweis von Schrägkablebrücken (On the treatment of the load case "permanent load" in the check of safety in cable-stayed bridges). *Der Bauingenieur* 58 (1983): 329–335.

Walther, R., B. Houriet, W. Isler, P. Moïa. 1985. *Ponts haubanés* (Cable-stayed bridges). Lausanne: Presses polytechniques romandes.

8 Analysis and Design of Bridge Substructures

8.1 Piers

8.1.1 General Ideas

Piers have a profound influence on the appearance of bridges. Their primary aesthetic function, which increases in importance with increasing number of spans, is to enhance transparency. The influence piers have on the economy of bridges is much smaller in comparison. Provided their average height is less than about 30 m, piers will normally make up less than 5 percent of the total construction cost. Additional costs incurred to achieve aesthetically pleasing pier cross-sections will therefore be slight. Conversely, it will rarely be possible to justify the choice of poorly shaped piers on the basis of substantial cost savings. It is thus preferable to select the cross-section dimensions based on aesthetic considerations rather than economic criteria.

Transparency of the structure is maximized when pier width is small relative to superstructure width. Single piers are normally preferable in this regard, except in the case of low, wide bridges where two slender columns should be used. An acceptable degree of transparency is difficult to achieve when the upper ends of piers are widened into hammerheads. Guidelines for pier design to improve transparency are proposed in Section 2.3.

Construction is considerably simplified by the use of a constant cross-section. A parabolic profile, corresponding to the lateral load bending moment diagram, may nevertheless be more appropriate for very tall piers. Identical section dimensions should be used for all piers, regardless of variation in height.

The shorter of the two principal cross-section dimensions is chosen to keep the slenderness ratio of the tallest pier within reasonable limits. (The slenderness ratio is equal to the effective length of the pier, l_k, divided by its radius of gyration, i. Further discussion of l_k can be found in Section 8.1.2.) The slenderness ratio is normally greatest immediately prior to construction of the superstructure, when the completed pier stands freely. Slenderness ratios of up to 220 can be tolerated under these circumstances, provided the pier loads consist only of self-weight and wind on the pier itself. The corresponding slenderness ratios in the completed

structure would be 55 for columns fixed at both ends and 77 for columns fixed at
the base and hinged at the top, assuming the piers are restrained against sway. The
piers of cantilever-constructed bridges are highly stressed during construction by
the unbalanced girder weight. This situation can be alleviated by increasing the
section dimensions to reduce the stresses. It may be more practical, however, to
increase the allowable load by reducing the critical length through the use of
temporary cables or other means.

Aesthetic considerations normally lead to piers of relatively slender proportions.
The consequences of slenderness must be considered in the analysis and design of
the reinforcement.

8.1.2 Second-Order Analysis of Slender Reinforced Concrete Columns

a) Fundamental Considerations

The ultimate load of slender reinforced concrete columns can be reliably
calculated only when nonlinear behaviour is taken into account. Nonlinearity
results from changes in system geometry induced by the loads (geometrical
nonlinearity) and from cracking of concrete and plastification of concrete and
steel (material nonlinearity).

The calculation of sectional forces according to classical first-order theory is based
on equations of equilibrium formulated for the geometry of the undeformed
structure (fig. 8.1 a). This approach underestimates the actual sectional forces,
which must satisfy the equilibrium conditions of the deformed system geometry
(fig. 8.1 b). Methods of structural analysis in which the equilibrium equations are
formulated for the geometry of the deformed structure are called *second-order
methods*.

Assuming that material behaviour is linear and elastic, an exact second-order
analysis would consist of solving a fourth-order differential equation for the

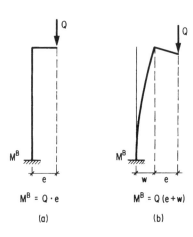

Figure 8.1
Moment at the base of an eccentrically-loaded
column: a, from first-order theory; b, from second-
order theory

member deflections, w. Sectional forces could then be obtained from the appropriate derivatives of w. The solution of the differential equation is only practical, however, for members of constant flexural stiffness EI. The structural response of members with variable EI is more easily calculated using approximate methods.

As a result of material nonlinearity, the flexural stiffness of a reinforced concrete section is a nonlinear function of the sectional forces M and N. A second-order analysis in which the functional relationship between EI, M, and N is rigorously formulated at each point would however be too complex for use in bridge design. Calculations can be considerably simplified by assuming constant flexural stiffness. Conservative yet reasonably accurate results are obtained when the stiffness EI_R corresponding to the sectional forces at the critical section is assigned to each point along the member. The second-order analysis can then proceed under the assumption of linear elastic material behaviour. Since the actual flexural stiffness will be greater than or equal to EI_R at all points, the deformations will be overestimated, resulting in a safe lower bound for the ultimate load.

b) Differential Equation of the Elastic Curve

The calculation of sectional forces is based on the solution of the differential equation of the elastic curve. The equation is formulated in this section for the general case of a column with arbitrary end conditions, subjected to an axial load Q and a lateral load q. The x-axis is located along the longitudinal axis of an idealized straight column. The origin of coordinates is located at the column's upper end (fig. 8.2). Two components of lateral deformation are considered: initial eccentricities resulting from inaccuracies in construction, $w_0(x)$; and unknown displacements produced by loads and imposed deformations, $w(x)$. The total deformation $w_{tot}(x)$ is thus:

$$w_{tot}(x) = w_0(x) + w(x) \tag{8.1}$$

The *initial geometry* of the structure is defined as the initial deformation $w_0(x)$.

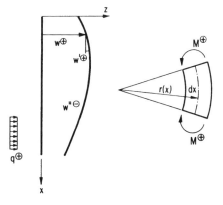

Figure 8.2
Notation and sign convention

As in first-order theory, the differential equation of the unknown displacements $w(x)$ is derived from moment-curvature relations. The geometrical relationship between displacement and radius of curvature, $r(x)$, is given by the following expression:

$$w''(x) = \frac{d^2 w(x)}{dx^2} \cong \frac{1}{r(x)} \tag{a}$$

Under the assumption that plane sections remain plane after deformation and that material behaviour is linear and elastic, equilibrium of a differential element requires that

$$\frac{1}{r} dx = -\frac{M}{EI} dx \tag{b}$$

The moment-curvature relationship is obtained by substitution of equation (b) into equation (a):

$$w''(x) = -\frac{M(x)}{EI} \tag{c}$$

or, equivalently,

$$EI = |Mr|$$

Taking two derivatives of equation (c) yields the fourth-order differential equation of the elastic curve $w(x)$ which, assuming EI is constant, is written as

$$w^{iv}(x) + \frac{M''(x)}{EI} = 0 \tag{d}$$

The bending moment $M(x)$ can be expressed as the sum of two components, $M_1(x)$ and $M_2(x)$. M_1 is defined by the equation of equilibrium of the initial system geometry. When the initial eccentricity $w_0(x)$ is equal to zero at all points along the member, the familiar equation of equilibrium for a straight beam

$$M_1'' = -q$$

is obtained.

The component M_2 is equal to the product of the axial force Q and the lateral displacement, w, plus any additional redundant moments necessary for compatibility in a statically indeterminate system:

$$M_2(x) = a_1 + a_2 x + Qw(x)$$
$$M_2'' = Qw''$$

The second derivative of the total bending moment can therefore be expressed as

$$M'' = M_1'' + M_2'' = M_1'' + Qw'' \tag{e}$$

The differential equation of the elastic curve is obtained by substituting equation (e) into equation (d):

$$w^{iv}(x) + \frac{Q}{EI} w''(x) = -\frac{M_1''(x)}{EI} \tag{8.2}$$

Its general solution is the sum of the general solution to the homogeneous equation

$$w^{iv}(x) + \frac{Q}{EI} w''(x) = 0 \tag{8.3}$$

and a particular solution to the inhomogeneous equation (8.2). The general solution of the homogeneous equation is

$$w_H = C_1 \sin \kappa x + C_2 \cos \kappa x + C_3 x + C_4 \tag{8.4}$$

where C_1, C_2, C_3, and C_4 are constants of integration and

$$\kappa = \sqrt{\frac{Q}{EI}} \tag{8.5}$$

The particular solution to the inhomogeneous equation is denoted w^P. It can be defined piecewise if M_1'' is discontinuous. The general solution to equation (8.2) is thus

$$w(x) = w_H(x) + w_P(x) = C_1 \sin \kappa x + C_2 \cos \kappa x + C_3 x + C_4 + w^P(x) \tag{8.6}$$

The constants C_i are determined from the boundary conditions.

The sectional forces can be evaluated from the appropriate derivatives of the elastic curve

$$M = -EIw'' \tag{8.7}$$

$$V = -EIw''' \tag{8.8}$$

c) *Closed-Form Solutions*

Closed-form solutions of equation (8.2) can be obtained for several cases of interest in the design of slender columns, three of which are presented in the following examples. The solution $w(x)$ is used to calculate sectional forces according to equations (8.7) and (8.8).

The *critical load* of the column, denoted Q_E, corresponds to the lowest eigenvalue of the differential equation. Deflections of indeterminate magnitude will occur when an axial load of magnitude Q_E is applied. Under these conditions, the system is said to have buckled. The critical load will be calculated for the columns of examples 8.2 and 8.3.

Example 8.1:
Cantilever column with uniform lateral load and concentrated loads at its tip

The system and loadings are shown in figure 8.3. The initial eccentricity, w_0, is assumed equal to zero at all locations along the column. The unknown displacement at the top of the column, $w(0)$, is denoted w^T. The basic statical relationships of the problem are as follows:

$$M_1 = M_0 - Hx - \frac{qx^2}{2}, \quad M_1'' = -q$$

$$M_2 = Q(w(x) - w^T), \quad M_2'' = Qw''(x)$$

$$M = M_1 + M_2 = M_0 - Hx - \frac{qx^2}{2} + Q(w(x) - w^T) = -EIw''$$

$$V = M' = -H - qx + Qw'(x) = -EIw'''$$

The differential equation of the elastic curve is obtained by substituting $M_1'' = -q$ into equation (8.2):

$$w^{iv} + \frac{Q}{EI} w'' = \frac{q}{EI} \tag{f}$$

The corresponding homogeneous equation has general solution

$$w_H = C_1 \sin \kappa x + C_2 \cos \kappa x + C_3 x + C_4, \qquad \text{(equation (8.4))}$$

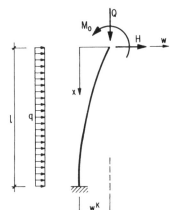

Figure 8.3
System of example 8.1

where $\kappa = \sqrt{(Q/EI)}$. A particular solution of the form $w^P = cx^2$ is assumed for the inhomogeneous equation. Substitution into equation (f) yields

$$w^P = \frac{q}{2Q} x^2 \qquad (g)$$

The general solution to equation (f) and the first four derivatives of the solution are thus

$$w = C_1 \sin\kappa x + C_2 \cos\kappa x + C_3 x + C_4 + \frac{q}{2Q} x^2$$

$$w' = C_1 \kappa \cos\kappa x - C_2 \kappa \sin\kappa x + C_3 + \frac{q}{Q} x$$

$$w'' = -C_1 \kappa^2 \sin\kappa x - C_2 \kappa^2 \cos\kappa x + \frac{q}{Q}$$

$$w''' = -C_1 \kappa^3 \cos\kappa x + C_2 \kappa^3 \sin\kappa x$$

$$w^{iv} = C_1 \kappa^4 \sin\kappa x + C_2 \kappa^4 \cos\kappa x$$

The constants of integration are solved from the boundary conditions:

At $x = 0$: $M(0) = -EIw''(0) = M_0$, $\qquad$ or $\quad w''(0) = -\dfrac{M_0}{EI}$

$$V(0) = -EIw'''(0) = -H + Qw'(0), \quad \text{or} \quad w'''(0) = (H - Qw'(0))/EI$$

At $x = l$: $w(l) = 0$

$\qquad w'(l) = 0$

The following constants of integration are thus obtained:

$$C_1 = \frac{1}{Q\kappa \cos\kappa l} \left[\left(M_0 + \frac{q}{Q} EI \right) \kappa \sin\kappa l - H - ql \right]$$

$$C_2 = \frac{1}{Q} \left(M_0 + \frac{q}{Q} EI \right)$$

$$C_3 = \frac{H}{Q}$$

$$C_4 = -\frac{\tan\kappa l}{Q\kappa} \left[\left(M_0 + \frac{q}{Q} EI \right) \kappa \sin\kappa l - H - ql \right]$$
$$\quad -\frac{1}{Q} \left(M_0 + \frac{q}{Q} EI \right) \cos\kappa l - \frac{Hl}{Q} - \frac{ql^2}{2Q}$$

Substituting these constants and the particular solution (equation (g)) into equation (8.6) yields the complete solution. The sectional forces in the column are obtained from equations (8.7) and (8.8). At the base of the column ($x = l$):

$$M(l) = \frac{1}{\cos\kappa l} \left[M_0 + \frac{q}{Q} EI - (H + ql) \frac{\sin\kappa l}{\kappa} \right] - \frac{q}{Q} EI$$

$$V(l) = -(H + ql)$$

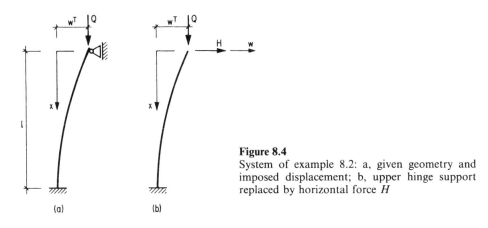

Figure 8.4
System of example 8.2: a, given geometry and imposed displacement; b, upper hinge support replaced by horizontal force H

Example 8.2:
Horizontal restraining force in a column with lower end fixed and upper end hinged, subjected to an imposed displacement

The hinge at the top of the column is replaced by a horizontal force H (fig. 8.4). This makes possible an equivalent formulation of the problem in terms of H, rather than in terms of the given horizontal displacement w^T. The initial eccentricity w_0 is equal to zero at all points along the column. The basic statical relationships are as follows:

$$M_1 = -Hx, \qquad\qquad M_1'' = 0$$
$$M_2 = Q(w(x) - w^T), \quad M_2'' = Qw''$$
$$M = M_1 + M_2 = -Hx + Q(w(x) - w^T) = -EIw''$$
$$V = M' = -H + Qw' = -EIw'''$$

Since $M_1''(x) = 0$ for all x, the differential equation of the elastic curve (equation (8.2)) reduces to

$$w^{iv} + \frac{Q}{EI}\, w'' = 0 \tag{h}$$

Since this equation is itself homogeneous, only w_H is considered:

$$w = w_H = C_1 \sin \kappa x + C_2 \cos \kappa x + C_3 x + C_4, \qquad \text{(equation (8.4))}$$

where $\kappa = \sqrt{(Q/EI)}$. The boundary conditions are as follows:

At $x = 0$: $V(0) = -EIw'''(0) = -H + Qw'(0)$, or $w'''(0) = (H - Qw'(0))/EI$

$\qquad\qquad M(0) = -EIw''(0) = 0$, $\qquad\qquad$ or $w''(0) = 0$

At $x = l$: $w(l) = 0$

$\qquad\qquad w'(l) = 0$

These boundary conditions yield the following constants of integration:

$$C_1 = \frac{-H}{Q\kappa\cos\kappa l}$$

$$C_2 = 0$$

$$C_3 = \frac{H}{Q}$$

$$C_4 = \frac{H}{Q}\left(\frac{\tan\kappa l}{\kappa} - l\right)$$

The complete solution to equation (h), in terms of H, is therefore

$$w(x) = -\frac{H}{Q}\left(\frac{\tan\kappa x}{\kappa x} - x\right) + \frac{H}{Q}\left(\frac{\tan\kappa l}{\kappa} - l\right)$$

The value of H is obtained by setting $w(0) = w^T$ in the above equation:

$$H = \frac{Q\kappa}{\tan\kappa l - \kappa l}\, w^T \qquad\qquad\qquad (i)$$

Equation (i) can be used to calculate the critical load Q_E of a column with lower end fixed and upper end free, since the horizontal restraining force at its tip will vanish when the buckling load is reached. $H = 0$ implies that then $\tan\kappa l = \pm\infty$, and hence

$$\kappa l = n\frac{\pi}{2}, \qquad \text{for any positive integer } n$$

The critical load Q_E corresponds to $n = 1$:

$$\kappa l = l\sqrt{\frac{Q}{EI}} = \frac{\pi}{2},$$

$$Q = \frac{\pi^2 EI}{4l^2} = Q_E$$

Example 8.3:
Buckling load of a column with lower end fixed and upper end hinged

The structural system is shown in figure 8.5, where H denotes the unknown horizontal reaction at the top of the column. The column has no initial eccentricity. The basic statical relationships are

$$M_1(x) = 0, \qquad\qquad M_1''(x) = 0$$
$$M_2(x) = Hx + Qw(x), \qquad M_2''(x) = Qw''(x)$$

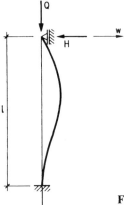

Figure 8.5
System of example 8.3

Since $M_1'' = 0$, the differential equation of the elastic curve is homogeneous

$$w^{iv} + \frac{Q}{EI}\, w'' = 0$$

Its general solution is

$$w = w_H = C_1 \sin \kappa x + C_2 \cos \kappa x + C_3 x + C_4 \qquad \text{(equation (8.4))}$$

where $\kappa = \sqrt{(Q/EI)}$. The constants of integration are obtained from the boundary conditions:

At $x = 0$: $w(0) = 0$

   $M(0) = -EIw''(0) = 0,$ or $w''(0) = 0$

At $x = l$: $w(l) = 0$

   $w'(l) = 0$

which can be expressed as the following system of equations:

$$C_2 + C_4 = 0$$
$$C_2\, \kappa^2 = 0$$
$$C_1 \sin \kappa l + C_2 \cos \kappa l + C_3 l + C_4 l = 0$$
$$C_1\, \kappa \cos \kappa l - C_2\, \kappa \sin \kappa l + C_3 = 0$$

The critical load Q_E corresponds to the nontrivial solutions of this system of equations. Since C_2 and C_4 vanish, Q_E can be obtained by setting the determinant of the matrix of coefficients of C_1 and C_3 equal to zero:

$$\sin \kappa l - \kappa l \cos \kappa l = 0 \quad \text{or} \quad \kappa l = \tan \kappa l$$

which has as its solution

$$\kappa l = 1.43\pi$$

The critical load is obtained by replacing κ by $\sqrt{(Q/EI)}$ in the above equation, yielding

$$Q = \frac{2.04\pi^2 EI}{l^2} = Q_E$$

Critical loads are commonly expressed in the form $Q_E = \pi^2 EI/l_k^2$, where l_k is defined as the *effective length* of the column. Thus, for the column of example 8.3, $l_k = l/\sqrt{2.04} = 0.7l$. The quantity l_k is a function of end conditions and column stiffness; critical lengths for commonly-used column types can be found in figure 6.5.

d) Approximate Calculations: Vianello's Method

Figure 8.6 shows an axially loaded column, hinged at both ends, with initial eccentricity $w_0(x)$. (The maximum value of $w_0(x)$ is denoted $w_{0,max}$.) Application of the load Q results in first-order moments M_0, which are a function of this initial geometry:

$$M_0(x) = Q w_0(x)$$

M_0 induces additional displacements w_1, which can be calculated according to first-order theory using the method of virtual work:

$$w_1(x) = \int \frac{M_0(s)}{EI} \bar{M}_x(s)\,ds = \int \frac{Q w_0(s)}{EI} \bar{M}_x(s)\,ds \qquad (j)$$

where $\bar{M}_x(s)$ are the moments obtained when a unit lateral load is applied at point x. Assuming EI is constant, the above equation can be rewritten as

$$w_1(x) = \frac{Q}{EI} \int w_0(s)\,\bar{M}_x(s)\,ds \qquad (8.9)$$

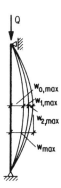

Figure 8.6
Incremental deflections of an axially-loaded column with initial eccentricity w_0

In a similar manner, the additional deflections due to $M_1 = Q w_1$, calculated using first-order theory, are

$$w_2(x) = \frac{Q}{EI} \int w_1(s) \, \bar{M}_x(s) \, ds \tag{k}$$

Additional terms $w_i(x)$ are obtained analogously. The total deflection $w(x)$ can thus be expressed as an infinite series

$$w(x) = w_0(x) + w_1(x) + w_2(x) + \ldots \tag{l}$$

If the additional displacement w_1 is proportional to the initial displacement w_0, then equation (8.9) can be rewritten as

$$w_1(x) = \frac{Q}{EI} \int w_0(s) \, \bar{M}_x(s) \, ds = \alpha w_0(x) = \frac{Q}{EI} K w_0(x)$$

where $\alpha = QK/EI$, and $K w_0(x) = \int w_0(s) \, \bar{M}_x(s) \, ds$. Since K is a function only of the initial displacement, it will be constant; it is conveniently expressed as $K = l^2/c$, for a constant c. It follows that

$$w_1(x) = \frac{Q l^2}{c EI} w_0(x) = \alpha w_0(x) \tag{m}$$

$$\alpha = \frac{Q l^2}{c EI} \tag{n}$$

Equation (k) thus becomes

$$w_2(x) = \alpha^2 w_0(x)$$

and equation (l) is transformed into

$$w(x) = w_0(x) (1 + \alpha + \alpha^2 + \ldots)$$

If $0 < \alpha < 1$, this infinite series can be summed; the total deflection will be

$$w(x) = w_0(x) \frac{1}{1 - \alpha} \tag{o}$$

If $\alpha = 1$, the deflections $w(x)$ will be infinite. The condition $\alpha = 1$ thus corresponds to the critical load Q_E. Since $w(x)$ is proportional to $w_0(x)$ for all values of α, it follows that $w_0(x)$ must be proportional to the buckled shape of the member. The critical load is thus given by

$$Q_E = \frac{c EI}{l^2}$$

obtained by setting $\alpha = 1$ in equation (n). Expressed in terms of Q_E, equation (n) becomes

$$\alpha = \frac{Q}{Q_E}$$

Equations (m) and (o) can likewise be rewritten as

$$w_1(x) = \frac{Q}{Q_E} w_0(x) \tag{8.10}$$

$$w(x) = w_0(x) \frac{1}{1 - \dfrac{Q}{Q_E}} \tag{8.11}$$

respectively. [1] Although derived here for the particular case of a statically determinate column with constant flexural stiffness, these equations are general, valid for both statically indeterminate columns and columns with variable EI. They are exact provided $w_0(x)$ is proportional to the buckled shape of the member.

When the applied load Q is equal to the critical load Q_E, equation (8.10) reduces to

$$w_1(x) = w_0(x) \tag{8.12}$$

Hence, when $Q = Q_E$, the additional displacement w_1 due to $M_0 = Qw_0$ will equal the initial deformation. This observation leads to a procedure for calculating Q_E: given $w_0(x)$, $w_1(x)$ is calculated as a function of Q using equation (8.9). The two expressions are equated and solved for Q, which will be equal to Q_E.

The procedure is suitable for approximate calculations. The actual buckled shape can be replaced by a curve w_0 of similar shape that satisfies the same boundary conditions. The calculation is then carried out as described above. When the buckled shape of the member is not obvious, the elastic curve produced by a lateral load can be assumed for w_0. The final result is relatively insensitive to the choice of load; either a concentrated or a uniform load can therefore be used. This technique is particularly helpful in the case of columns of variable stiffness.

Although equation (8.11) was derived for the case of an initial eccentricity w_0, it is also valid in the more general case of deformations induced by the combined

---

[1] It is possible to express the quotient Q/Q_E in terms of axial force N. This is often necessary for consistency, when the context is clearly one of sectional forces and not of external loads. In these cases, the critical load is simply rewritten N_E, whose value is identical to Q_E in all cases where axial load is concentrated at the ends of the column. The parameter α can thus be equivalently defined as

$$\alpha = \frac{N}{N_E}$$

action of initial eccentricity, load, and imposed displacement. The deflection w_1 is defined as the first-order displacement due to the total first-order moment by rewriting equation (j) as

$$w_1(x) = \int \frac{M_1(s)}{EI} \bar{M}_x(s)\, ds$$

where M_1 is the sum of the first-order moment due to initial eccentricity and the first-order moment due to loads and imposed displacements:

$$M_1 = M_1[w_0] + M_1[g, q, \ldots]$$

The additional displacement w_2 is calculated using equation (k); higher terms are defined analogously. The sum $w_1(x) + w_2(x) + w_3(x) + \ldots$ is denoted $w(x)$, the total second-order displacement due to M_1. If w_1 is proportional to the buckled shape, then equation (8.11) can be rewritten as

$$w(x) = w_1(x)\, \frac{1}{1 - \dfrac{Q}{Q_E}} \tag{8.13}$$

The total deformation can now be formulated using equation (8.1):

$$w_{tot}(x) = w_0(x) + w(x)$$

where $w_0(x)$ is the initial eccentricity and $w(x)$ is as defined in equation (8.13).

Equation (8.13) can be used for approximate calculations by relaxing the condition that $w_1(x)$ be proportional to the buckled shape. Provided $w_1(x)$ satisfies the same boundary conditions, and is reasonably similar to the buckled shape, results of acceptable accuracy can be obtained.

Example 8.4:
Cantilever column with uniform lateral load and concentrated loads at its tip

The system is shown in figure 8.7. The deflection at the tip of the column due to first-order moments, w_1^T, is

$$w_1^T = \frac{q l^4}{8\,EI} + \frac{H l^3}{3\,EI} - \frac{M_0 l^2}{2\,EI} = \frac{l^2}{EI}\left(\frac{q l^2}{8} + \frac{H l}{3} - \frac{M_0}{2}\right)$$

The critical load of a cantilever column, derived in example 8.2, is

$$Q_E = \frac{\pi^2 EI}{4 l^2}$$

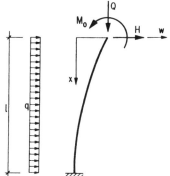

Figure 8.7
System of example 8.4

The second-order deflection at the tip can now be obtained from equation (8.13):

$$w^T = \frac{l^2}{EI}\left(\frac{ql^2}{8} + \frac{Hl}{3} - \frac{M_0}{2}\right)\frac{1}{1 - \dfrac{4Ql^2}{\pi^2 EI}}$$

The corresponding second-order moment at the base of the column is

$$M^B = -\left(\frac{ql^2}{2} + Hl - M_0 + Qw^T\right)$$

This approximate solution agrees closely with the exact solution calculated from the differential equation in example 8.1. The difference between the two solutions is less than 0.5 percent at all load levels; the agreement is so good that the two solutions cannot be distinguished in figure 8.8.

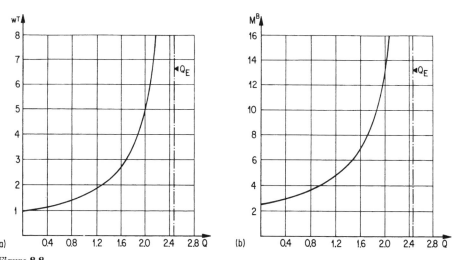

Figure 8.8
Results of example 8.4, plotted as a function of axial load Q, for $M_0 = -1, H = 1, q = 1, l = 1$, and $EI = 1$: a, deflection at top of column w^T; b, moment at base of column M^B

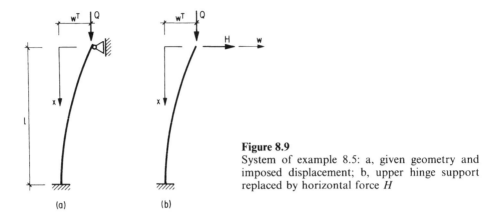

Figure 8.9
System of example 8.5: a, given geometry and
imposed displacement; b, upper hinge support
replaced by horizontal force H

Example 8.5:
Horizontal restraining force in a column with lower end fixed and upper end hinged,
subjected to an imposed displacement

The system is shown in figure 8.9 a. As in example 8.2, the hinged column with
imposed displacement w^T is replaced by a cantilever column with lateral load H at
its upper end (fig. 8.9 b). The first-order deflection at the tip of the column
produced by H is

$$w_1(0) = \frac{Hl^3}{3EI}$$

The second-order deflection at the tip is calculated from equation (8.13), given the
critical load of a cantilever column $Q_E = \pi^2 EI/4l^2$:

$$w(0) = \frac{Hl^3}{3EI}\left(\frac{1}{1 - \dfrac{4Ql^2}{\pi^2 EI}}\right)$$

This equation is rearranged to yield the following expression for H:

$$H = \frac{3EI}{l^3} w(0)\left(1 - \frac{4Ql^2}{\pi^2 EI}\right)$$

The given tip displacement w^T is now substituted for $w(0)$, thus completing the
solution. The corresponding moment at the base of the column,
$M^B = -(Hl + Qw^T)$, is given by the equation

$$M^B = -\left[\frac{3EI}{l^2} w^T\left(1 - \frac{4Ql^2}{\pi^2 EI}\right) - Qw^T\right]$$

The approximate and exact solutions are compared in figure 8.10.

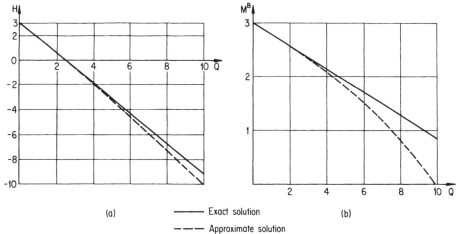

(a)　　　　　——— Exact solution　　　　　(b)
　　　　　———— Approximate solution

Figure 8.10
Results of example 8.5, plotted as a function of axial load Q, for $w^T = 1$, $l = 1$, and $EI = 1$: a, horizontal reaction H; b, moment at base of column M^B

Example 8.6:
Critical load of a column with lower end fixed and upper end hinged

The system and axial load are shown in figure 8.11a. Nodes 0 through 6 are equally spaced over the height of the column.

Equation (8.10) will be used to solve for Q_E. The true buckled shape is approximated by w_0, the deflected shape due to a horizontal load $H = 1$ applied at node 2 (figs. 8.11b, c). The initial deflection w_0 and the additional deflection w_1, produced by the first-order moments due to w_0, will be equated at node 2.

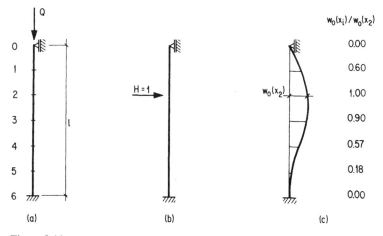

(a)　　　　　　(b)　　　　　　(c)

Figure 8.11
System of example 8.6: a, geometry and axial load; b, transverse load; c, deflections w_0

Table 8.1
Calculation of $w_1(x_2)$ Using Simpson's Rule

Node	$\dfrac{M}{Q_E w_0(x_2)}$	$\dfrac{\bar{M}}{l}$	n	$\dfrac{M\bar{M}n}{Q_E l w_0(x_2)}$
0	0.000	0.000	1	0.000
1	0.631	0.086	4	0.218
2	1.000	0.173	2	0.346
3	0.928	0.093	4	0.344
4	0.575	0.012	2	0.014
5	0.184	−0.068	4	−0.050
6	0.000	−0.148	1	0.000
Total	$\dfrac{M\bar{M}n}{Q_E l w_0(x_2)}$			0.872

The deflection $w_1(x_2)$ is calculated analogously to equation (8.9) using the method of virtual work. Because the system is statically indeterminate, computational effort can be reduced by considering the real displacements and curvatures for a statically determinate primary system, and the virtual loads and moments for the given statically indeterminate system of figure 8.11. The quantity M/EI thus denotes the real curvature due to the moment $M = Q_E w_0$; $\bar{M}$ denotes the virtual moment due to the load $H = 1$ at node 2. The integral

$$w_1(x_2) = \int_0^l \frac{M}{EI} \bar{M} dx$$

is evaluated numerically using Simpson's rule with the help of table 8.1. Thus, for $\Delta x = l/6$,

$$w_1(x_2) = \frac{\Delta x}{3} \sum \frac{M}{EI} \bar{M} n = \frac{\Delta x}{3} \frac{0.872 Q_E l}{EI} w_0(x_2) = 0.0485 \frac{Q_E l^2}{EI} w_0(x_2)$$

The critical load is obtained from the condition $w_1(x_2) = w_0(x_2)$ (equation (8.12)):

$$Q_E = 20.6 \frac{EI}{l^2}$$

This approximate solution agrees to within 2 percent of the exact solution obtained from the differential equation

$$Q_E = \frac{\pi^2 EI}{(0.7l)^2} = 20.2 \frac{EI}{l^2}$$

8.1.3 Calculation of Ultimate Resistance for Flexure and Axial Force

a) Fundamentals

The ultimate resistance of a section subjected to bending moment M and axial force N is defined by specific combinations of these two sectional forces, which can be represented simply and clearly by means of an *M-N* interaction diagram. The calculation of the ultimate sectional force combinations (M_R, N_R) is based on the following assumptions:

1. Plane sections remain plane after deformation
2. The tensile strength of concrete can be neglected
3. Idealized stress-strain diagrams can be used for both steel and concrete

As discussed in Section 4.3.1, material strengths can be defined in several different ways. In the context of *M-N* interaction diagrams, it is customary to use specified design values for the strength of both concrete and steel, f_c and f_{sy}, respectively.

The stress-strain behaviour of concrete can be idealized in several ways. The diagram given in the CEB-FIP Model Code (Comité Euro-International du Béton 1977) is commonly used. The stress-strain curve consists of a parabolic segment for strains between 0 and $0.67\varepsilon_{cu}$ and a constant segment for strains between $0.67\varepsilon_{cu}$ and $1.0\varepsilon_{cu}$. (ε_{cu} denotes the ultimate compressive strain.) A simplified version of this diagram, also shown in figure 8.12a, consists of two constant segments. Concrete stress is assumed equal to zero for strains less than $0.2\varepsilon_{cu}$, and equal to f_c for strains greater than $0.2\varepsilon_{cu}$. Use of this simplified diagram will result in a considerable reduction in computational effort without significant loss in accuracy.

The stress-strain diagram of reinforcing steel is normally idealized as bilinear (fig. 8.12b). It is characterized by linear increase in stress up to the yield strain ε_{sy}, beyond which the steel stress is constant and equal to f_{sy}. The specified maximum permissible steel strain, $\varepsilon_{s,max}$, normally lies between 0.005 and 0.007.

b) Interaction Diagrams for Ultimate Sectional Forces

The ultimate limit state of the cross-section is achieved when the extreme fibre compressive strain reaches ε_{cu}, or when the strain in the outer layer of reinforcing

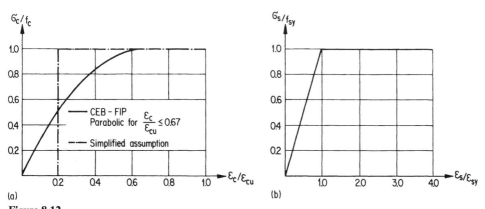

Figure 8.12
Idealized stress-strain diagrams: a, concrete; b, reinforcing steel

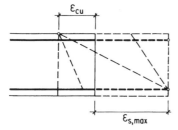

Figure 8.13
States of strain at ultimate limit state; effective cross-section resistance

steel reaches $\varepsilon_{s,\,max}$ (fig. 8.13). These ultimate states of strain define the *effective resistance* of the cross-section. A point on the *M-N* interaction diagram is calculated by assuming one of these ultimate states of strain. The corresponding stresses are then calculated from the material stress-strain diagrams and integrated to obtain the ultimate sectional force combination $(M_R,\ N_R)$.

Large curvatures are associated with the ultimate strains defining the effective cross-section resistance. This implies that the ultimate limit state of a slender member will be accompanied by large deflections w and large moments Qw. The maximum axial load Q which can be sustained by the system thus deformed will be lower than the actual ultimate load, which is reached at a lower curvature and with lower Qw moments. It is therefore preferable to define the ultimate resistance using reduced states of strain, obtained by limiting the maximum strain in the outer layers of reinforcing steel to the yield strain:

$$\varepsilon_{s,\,max} = \varepsilon_{sy}$$

$(\varepsilon_{sy} = 0.0022$ for $f_{sy} = 460$ N/mm^2.) The cross-section resistance defined by the reduced states of strain (fig. 8.14) will be called the *reduced resistance*. Although it is somewhat lower than the effective resistance, it will result in a higher ultimate load of the member, due to the smaller curvatures and Qw moments.

These concepts are illustrated in figure 8.15 for the simple case of an eccentrically loaded cantilever column. The axial load Q is increased from zero until failure of the member occurs. The interaction diagrams of the effective resistance (R) and the reduced resistance (R_{red}) at the base of the column are given in figure 8.15a.

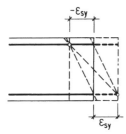

Figure 8.14
States of strain at ultimate limit state; reduced cross-section resistance

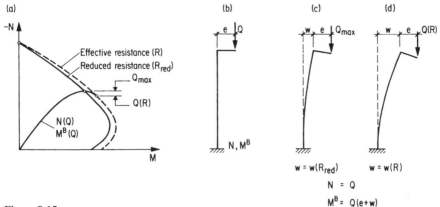

Figure 8.15
Ultimate load of an eccentrically loaded column: a, interaction diagrams; b, system and load; c, ultimate deformations (reduced resistance); d, ultimate deformations (effective resistance)

The sectional forces $M(Q)$, $N(Q)$ produced by the increasing axial load Q are plotted on the same set of axes. The $M(Q)$, $N(Q)$ curve reaches a maximum at its point of intersection with the interaction curve of the reduced resistance. The corresponding axial load, denoted Q_{max}, is the ultimate load of the system. Beyond this point, increases in strain are only possible by reducing the load. The effective cross-section resistance can thus only be reached for an axial load $Q(R)$ less than Q_{max}.

The four characteristic states of strain shown in figures 8.16 through 8.19 are normally sufficient to generate interaction diagrams. Further computational effort can be saved by using the rectangular stress block of the simplified concrete stress-strain diagram of figure 8.12 a. An interaction diagram calculated using these simplifications is compared with the corresponding "exact" diagram in figure 8.20. It is apparent that any error is small and on the conservative side.

The interaction diagrams of the effective resistance model and the reduced resistance model typically differ by only a few percent. The states of strain of figure 8.13 can thus be used to calculate the cross-section resistance, even for slender members, provided deformations and second-order moments are computed using

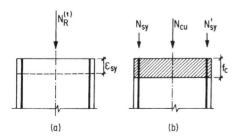

Figure 8.16
State of strain 1: a, sectional forces and strains; b, concrete stresses and resultant forces in concrete and steel

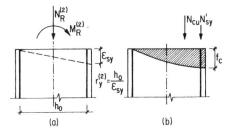

Figure 8.17
State of strain 2: a, sectional forces and
strains; b, concrete stresses and resultant
forces in concrete and steel

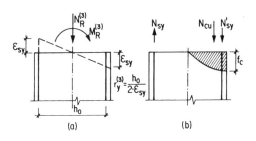

Figure 8.18
State of strain 3: a, sectional forces and
strains; b, concrete stresses and resultant
forces in concrete and steel

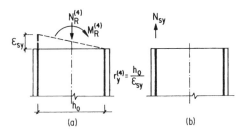

Figure 8.19
State of strain 4: a, sectional forces and
strains; b, resultant forces in steel

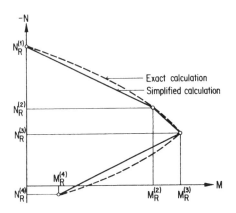

Figure 8.20
Approximate and exact M-N interaction
diagrams

the reduced states of strain. The slightly unconservative ultimate sectional forces thus obtained can be tolerated since the calculation of deflections and second-order moments, which normally disregards the stiffness of the concrete between cracks, is sufficiently conservative.

Charts and diagrams to assist in the calculation of ultimate cross-section resistance have been published for many commonly used cross-section types (Menn et al. 1977; Walther 1977). Interaction diagrams for rectangular and circular columns are given in the Appendix. Most published interaction diagrams are given in terms of nondimensional quantities to extend their range of application. Commonly used dimensionless expressions are as follows:

Axial force: $\quad n_R = \dfrac{N_R}{bhf_c}$ $\hspace{4cm}$ (8.14)

Moment: $\quad m_R = \dfrac{M_R}{bh^2 f_c}$ $\hspace{4cm}$ (8.15)

Interaction diagrams are normally published for specific arrangements of reinforcement. The diagrams given in the Appendix, for example, assume that the reinforcement of a rectangular section is located only at the extreme tensile (A_s) and compressive (A'_s) fibres, and is arranged symmetrically about the axis of the cross-section ($A_s = A'_s$). The mechanical reinforcement ratio is defined in terms of $A_{s,\text{tot}} = A_s + A'_s$:

$$\omega_{\text{tot}} = \frac{A_{s,\text{tot}}}{bh}\frac{f_{sy}}{f_c} \hspace{4cm} (8.16)$$

c) Interaction Diagrams for Prestressed Compression Members

The ultimate resistance of a prestressed compression member can be calculated with reasonable accuracy using a simplified method. It is assumed that the prestressing steel is located on the flexural tension side of the section. Due to its pre-strain, the prestressing steel and mild steel located at the same depth will yield roughly simultaneously (fig. 8.21).

Mild steel and prestressing steel can thus be considered as equal in the calculation of the interaction diagram. Both can be assumed to yield at an increment of strain equal to ε_{sy}. As shown in figure 8.22, the axial force and moment due to initial prestress, P and M_P, produce only a translation of the origin of the diagram.

Yielding of prestressing steel on the compression side of the section, however, will always be preceded by crushing of the concrete. When only a small amount of prestressed compression reinforcement has been provided, the usual ultimate states of strain of figure 8.13 or figure 8.14 can be used.

The total bending moment due to loads and prestressing must be considered in calculating the deformations of prestressed compression members.

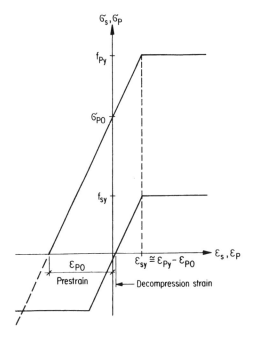

Figure 8.21
Stress-strain behaviour of mild steel and prestressing steel in a prestressed compression member

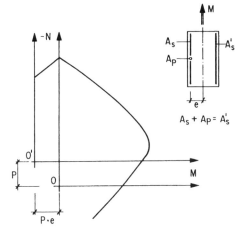

Figure 8.22
Interaction diagrams of a non-prestressed member (origin O), and of a prestressed member (origin O')

8.1.4 Flexural Stiffness of Reinforced Concrete Sections

The flexural stiffness of a reinforced concrete section is a function of the sectional forces M and N. This is a consequence of the nonlinear stress-strain behaviour of concrete and reinforcing steel, and of the negligible tensile strength of concrete. Flexural stiffness is also a function of duration of loading, due to creep in concrete.

The influence of cracking and nonlinear stress-strain behaviour on flexural stiffness can be observed in the relationship between bending moment, M, and

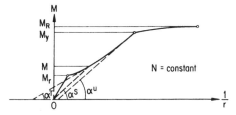

Figure 8.23
Moment-curvature diagram of a reinforced concrete section for initial loading, assuming constant axial force

curvature, $1/r$, in a previously uncracked section. A typical moment-curvature relationship is illustrated in figure 8.23, in which M is increased from zero to the ultimate moment M_R while the axial force N is held constant for all values of M. The loads are assumed to be of short duration; the effects of creep can thus be neglected.

The diagram is nonlinear and is characterized by three distinct segments separated by discontinuities in slope. The first extends from the origin to the cracking moment M_r. The section is uncracked and its behaviour is essentially linear in this region. The second segment extends from M_r to M_y, defined as the moment at which yielding in the outer layer of reinforcing steel begins. The third segment extends from M_y to M_R, the ultimate moment of the section. Moments M_y and M_R correspond to the states of strain of figures 8.14 and 8.13, respectively.

Flexural stiffness can be defined in several ways. Figure 8.23 shows three possibilities. The *tangent stiffness* EI^T is equal to the slope of the moment-curvature diagram for a given moment M. It can be expressed as

$$EI^T = \left| \frac{dM}{d(1/r)} \right| = \tan \alpha^T$$

The *secant stiffness* EI^S is equal to the slope of a line passing through the origin and intersecting the moment-curvature diagram at a given point $(1/r, M)$:

$$EI^S = \left| \frac{M}{1/r} \right| = |Mr| = \tan \alpha^S$$

Both EI^T and EI^S are defined for initial loading only. When the section is unloaded, or when load is reapplied to a previously loaded section, the resulting moment-curvature diagram will assume a more linear character; EI^T and EI^S will be approximately equal for all values of M. In this case, the two stiffnesses are denoted collectively EI^U:

$$EI^U = \tan \alpha^U$$

The use of EI^S, the secant stiffness derived from the moment-curvature diagram of initial loading, will be conservative and will lead to reasonably accurate deflection

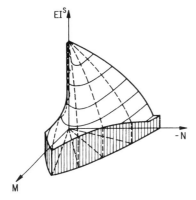

EI^S

$-N$

M

Figure 8.24
Secant stiffness EI^S as a function of M and N

calculations. The value of EI^S can be determined simply and directly from the moment M and the state of strain. Figure 8.24 shows the variation of secant stiffness with the sectional forces M and N; the values of EI^S along the shaded boundary of the surface correspond to the ultimate sectional forces.

It was proposed in Section 8.1.3 that the calculation of deformations and second-order moments in slender columns be based on the reduced cross-section resistance, in which the ultimate limit state of the section is reached when the strain at the outer layer of reinforcement is equal to ε_{sy}. The corresponding *reduced* secant stiffness is denoted EI_y:

$$EI_y = r_y M_y$$

where r_y and M_y are the radius of curvature and moment, respectively, obtained from one of the reduced ultimate states of strain.

The role of the reduced stiffness EI_y can be visualized using the moment-curvature diagram of figure 8.23. It is apparent that the small increase in moment from M_y to M_R results in a large increase in curvature. Increasing the moment beyond M_y thus results in large increases in second-order moments but only negligible increases in cross-section resistance. It is therefore reasonable to define the capacity of slender compression members in terms of M_y, and to calculate deformations at ultimate limit state using the corresponding stiffness EI_y.

Axial force N is plotted against EI_y in figure 8.25. Each point on the curve is obtained from a state of strain in which $\varepsilon_{s,\,max} = \varepsilon_{sy}$. The curve can be quickly sketched using the four characteristic states of strain of figures 8.16 through 8.19. The corresponding radii of curvature are as follows:

State of stress 1: $r_y^{(1)} = \infty$

State of stress 2: $r_y^{(2)} = \dfrac{h_0}{\varepsilon_{sy}}$

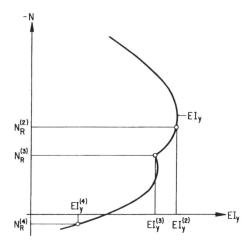

Figure 8.25
Relation between reduced stiffness EI_y
and axial force N

State of stress 3: $r_y^{(3)} = \dfrac{h_0}{2\varepsilon_{sy}}$

State of stress 4: $r_y^{(4)} = \dfrac{h_0}{\varepsilon_{sy}}$

The calculation of member deformations is considerably simplified by assuming constant flexural stiffness along the length of the member, equal to EI_y at the critical section. The virtual work integral confirms that the greatest contribution to the displacement will be from locations where M is large and EI is small. At these points, the actual stiffness will be close to the minimum stiffness EI_y. At the remaining locations in the member, where M is small and EI is large, the actual stiffness will be considerably greater than EI_y. The contribution of these points to the total displacement will be much less, due to the smaller bending moments. The end result will thus be relatively insensitive to the larger error in EI at these locations. Deformations calculated in this way will thus be reasonably accurate. Because a lower bound of member stiffness is used at all locations, any error will be on the conservative side.

An exact calculation of the deformations and second-order moments may nevertheless be worthwhile for very tall bridge piers. An iterative procedure can be used for this purpose. The second-order moments $M(x)$ are first calculated assuming constant flexural stiffness EI_y. The stiffness of the uncracked homogeneous section is then assigned to all locations where $M(x) < M_r$. At all other locations, the radius of curvature $r(x)$ corresponding to $M(x)$ is calculated. The flexural stiffness in the cracked portion of the member is then obtained from the equation

$$EI(x) = |M(x)r(x)|$$

Table 8.2
Reduction of Steel Strain Due to the Participation of Concrete Between
Cracks ($\varrho_{tot} = A_{s, tot}/A_c$)

	Average Steel Strain ε_{sm}	
	$\varrho_{tot} \leq 0.6\%$	$\varrho_{tot} > 0.6\%$
$M_r \leq M \leq 2M_r$	$0.7\varepsilon_s^{II}(M)$	$0.8\varepsilon_s^{II}(M)$
$2M_r \leq M \leq M_y$	$0.9\varepsilon_s^{II}(M)$	$1.0\varepsilon_s^{II}(M)$

(The contribution of the concrete between the cracks can be considered in the calculation of $r(x)$ by means of an average steel strain ε_{sm} as given in table 8.2.) The calculation of the second-order moments is then repeated using $EI(x)$.

Creep in concrete results in progressive increase in strain at constant stress. The time-varying strain produced by the sustained concrete stress σ_c is given by

$$\varepsilon_c(t) = \varepsilon_{c, el}(1 + \phi(t))$$

at time t, and

$$\varepsilon_{c, \infty} = \varepsilon_{c, el}(1 + \phi_\infty)$$

after many years. The quantity $\varepsilon_{c, el}$ denotes the strain corresponding to σ_c in the stress-strain diagram for short-term loading (fig. 8.12 a). The effect of creep can be visualized as a stretching of the stress-strain diagram in the horizontal direction by a factor of $(1 + \phi)$. It is apparent that concrete will be able to sustain strains in excess of ε_{cu} under long-term loads.

Points on the interaction diagram of a cross-section subjected to sustained load can be computed using normal procedures, provided the concrete stresses are calculated from the stress-strain diagram stretched by the factor $(1 + \phi)$. Since the ultimate long-term strain in the concrete is greater than ε_{cu}, the use of the strain diagrams of figures 8.13 and 8.14 will lead to an underestimate of the cross-section resistance. The difference between the short-term and long-term stresses for strains near ε_{cu} will, however, be relatively small. The cross-section resistance under sustained load obtained from the states of strain of figure 8.13 will therefore closely approximate the actual resistance. On the other hand, the difference between the short-term and long-term stresses corresponding to ε_{sy} will be considerably greater. Use of the strain diagrams of figure 8.14 will thus result in sectional forces which are significantly smaller than the actual resistance under sustained load.

It is possible to define new states of strain that lead to a realistic value for the ultimate long-term load. They must be sufficiently high to maximize cross-section resistance, yet must also limit second-order effects. A rigorous calculation of such an "optimal" strain distribution for sustained loads is, however, impractical. The effect of creep is primarily one of increasing member deformations, while leaving the cross-section resistance essentially unchanged. Calculations can thus be simplified by using the reduced cross-section resistance, obtained from the strain limitation $\varepsilon_{s,\,max} \leqq \varepsilon_{sy}$ and the short-term stress-strain diagram, for both short-term and long-term loads. Long-term deformations at ultimate limit state must be calculated, however, using a reduced flexural stiffness to account for the effects of creep. The reduced stiffness is defined as follows:

$$EI_y(\phi) = k_\phi\, EI_y$$

The simplification is, in all cases, conservative.

The reduction factor k_ϕ is a function of the creep coefficient ϕ and of axial force. Graphs of k_ϕ as a function of ϕ and the nondimensional ratio n_R/n_{R0} are given in the Appendix for both circular and rectangular sections. The quantity n_{R0} is a nondimensional expression for the ultimate axial force in pure compression, and is equal to $1 + \omega_{tot}$. Quantities ω_{tot} and n_R are defined in equations (8.16) and (8.14), respectively. For the typical case $\phi = 2$ and $n_R/n_{R0} = 0.3$, k_ϕ lies between 0.5 and 0.6.

The deformations due to sustained and short-term loads can be combined by idealizing the former as a short-term effect. The procedure will be formulated using the following notation:

G: sustained load

Q: short-term load

N_E: critical compression

The first-order displacement due to sustained load, $w_1(G)$, is computed using the flexural stiffness $k_\phi\, EI_y$. The corresponding second-order displacement is obtained from equation (8.13):

$$w(G) = w_1(G)\,\frac{1}{1 - \dfrac{N(G)}{N_E(\phi)}} \tag{8.17}$$

where the critical compression has also been reduced due to effects of creep:

$$N_E(\phi) = \frac{\pi^2\, k_\phi\, EI_y}{l_k^2}$$

The idealized short-term displacement $\bar{w}_1(G)$ is defined as the first-order deflection that would produce $w(G)$ in the absence of creep:

$$w(G) = \bar{w}_1(G)\,\frac{1}{1 - \dfrac{N(G)}{N_E}} \tag{a}$$

where $N_E = \pi^2 EI_y/l_k^2$ is calculated using the flexural stiffness for short-term loads. The following expression for $\bar{w}_1(G)$ is obtained by rearranging equation (a):

$$\bar{w}_1(G) = w(G)\left(1 - \frac{N(G)}{N_E}\right) \tag{8.18}$$

The total deformation due to the combined effects of sustained load G and short-term load Q can now be calculated:

$$w = (\bar{w}_1(G) + w_1(Q)) \frac{1}{1 - \dfrac{N(G) + N(Q)}{N_E(\phi)}} \tag{8.19}$$

where $w_1(Q)$ is the first-order displacement due to short-term load.

Example 8.7:
Concentrically loaded column with initial eccentricity w_0, subjected to long-term load G and short-term load Q

The system, loadings, and critical loads are shown in figure 8.26. The calculation of the total lateral deflection, w_{tot}, proceeds as follows:

$$w_1(G) = w_0 \frac{N(G)}{N_E(\phi)} \qquad\qquad \text{(Equation (8.10))}$$

$$w(G) = w_0 \frac{N(G)}{N_E(\phi)} \frac{1}{1 - \dfrac{N(G)}{N_E(\phi)}} = 1.0\,w_0 \qquad\qquad \text{(Equation (8.17))}$$

$$\bar{w}_1(G) = w(G)\left(1 - \frac{N(G)}{N_E}\right) = 0.76\,w_0 \qquad\qquad \text{(Equation (8.18))}$$

$$w_1(Q) = w_0 \frac{N(Q)}{N_E} = 0.08\,w_0$$

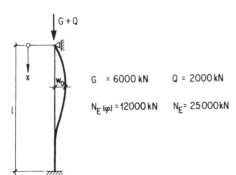

G   = 6000 kN   Q = 2000 kN

$N_{E\,(\phi)}$ = 12000 kN N_E = 25000 kN

Figure 8.26
System of example 8.7

$$w(G+Q) = (\bar{w}_1(G) + w_1(Q)) \frac{1}{1 - \dfrac{N(G) + N(Q)}{N_E}} = 1.23\,w_0$$

(Equation (8.19))

$$w_{tot} = w_0 + w(G+Q) = 2.23\,w_0$$

(Equation (8.1))

8.1.5 Imposed Deformations

Redundant support reactions and sectional forces are induced in statically in-determinate structures by imposed deformations due to self-equilibrating stresses or support displacements. Because they exist merely to satisfy compatibility, these reactions and sectional forces are proportional to the stiffness of the system, decreasing as the structure's deformability increases as a result of cracking, material plastification, and creep.

Imposed deformations are of little significance to the behaviour of non-slender members at ultimate limit state. Provided the system is sufficiently ductile, the compatibility conditions that define the accompanying redundant forces need no longer be satisfied. In slender compression members, however, imposed deformations increase the second-order moments and must therefore be con-sidered at ultimate limit state.

Cracking induced by imposed deformations must always be checked under service conditions. This is best accomplished using the geometrical methods described in Section 4.8.4, since the accompanying nonlinear behaviour cannot be adequately dealt with by statical methods.

The following example shows how imposed deformations can be considered in the design of slender columns at ultimate limit state, and investigates their influence on the second-order moments.

Example 8.8:
Concentrically loaded column with initial eccentricity, with and without an imposed displacement at its upper end

(In this example, square brackets will be used to denote the action causing the displacement, while parentheses will be used to denote the point at which the function is evaluated. Thus, $w[w_0](x)$ will be the displacement at point x due to initial eccentricity w_0.)

1. Imposed displacement not considered:

The system and loadings are shown in figure 8.27. The maximum initial eccentricity is denoted $w_{0,max}$. The total deflection due the initial eccentricity is calculated as follows:

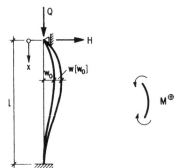

Figure 8.27
System of example 8.8; no imposed displacement

$$w_1[w_0](x) = w_0(x)\frac{Q}{Q_E} \qquad\qquad \text{(Equation (8.10))}$$

$$w[w_0](x) = w_1[w_0](x)\,\frac{1}{1-\dfrac{Q}{Q_E}} = w_0(x)\frac{Q}{Q_E}\,\frac{1}{1-\dfrac{Q}{Q_E}} = w_0(x)\frac{Q}{Q_E-Q}$$

$$\text{(Equation (8.13))}$$

$$w_\text{tot}(x) = w_0(x)+w[w_0](x)=w_0(x)\left(1+\frac{Q}{Q_E-Q}\right) = w_0(x)\,\frac{1}{1-\dfrac{Q}{Q_E}}$$

$$\text{(a) (Equation (8.1))}$$

The second-order moments are thus

$$M(x)= Qw_\text{tot}(x) - H[w_0]\,x$$

where $H[w_0]$ is the statically redundant horizontal reaction at the top of the column. Its value can be obtained from the boundary condition $w_\text{tot}(0) = 0$:

$$w_\text{tot}(0) = \int \frac{M}{EI}\,\bar{M}\,dx = 0,$$

where $\bar{M}(x)$ is the cantilever moment due to a unit lateral load at the top of the column:

$$\bar{M}(x) = -x \qquad\qquad\qquad\qquad\qquad\qquad \text{(b)}$$

Assuming constant flexural stiffness, the maximum moment is

$$M_\text{max} = 0.733\,Qw_{0,\text{max}}\,\frac{1}{1-\dfrac{Q}{Q_E}}$$

and is located at roughly the upper third point of the column.

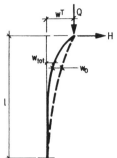

Figure 8.28
System of example 8.8 with imposed displacement w^T

2. Imposed displacement considered:

The same column, with identical axial load and initial eccentricity w_0, is now subjected to an imposed displacement $w(0) = w^T$ at its upper end (fig. 8.28). The quantities w_0 and w^T are of opposite sign. Equation (8.1) is again used to formulate the total deflection:

$$w_{tot}(x) = w_0(x) + w(x) = w_0(x) + w[w_0](x) + w[w^T](x) \qquad (c)$$

where $w(x)$ has been expressed as the sum of a deflection due to w_0 and a deflection due to w^T. From equation (a),

$$w_0(x) + w[w_0](x) = w_0(x) \; \frac{1}{1 - \dfrac{Q}{Q_E}}$$

The solution of $w[w^T](x)$ begins, as usual, by expressing it as a function of first-order deflections using equation (8.13):

$$w[w^T](x) = w_1[w^T](x) \; \frac{1}{1 - \dfrac{Q}{Q_E}}$$

$w_1[w^T]$ is the first-order deflection due to the moment

$$M_1(w^T) = -H_0 x,$$

where

$$H_0 = \frac{3EI}{l^3} w^T$$

The solution of $w[w^T](x)$ is thus complete, and can be substituted into equation (c) to obtain the total second-order deflection. The second-order moments are now computed:

$$M[w_0, w^T](x) = -Q(w_{tot}(0) - w_{tot}(x)) - Hx$$
$$= -Q(w_{tot}(0) - w_0(x) - w[w_0](x) - w[w^T](x)) - Hx$$

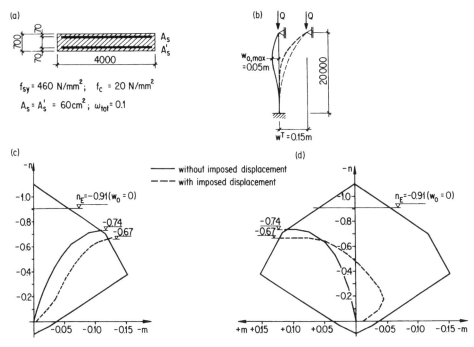

Figure 8.29
Ultimate load of the column of example 8.8 with and without imposed displacement (dimensions in mm): a, section properties; b, load and imposed displacement; c, response at upper third point; d, response at base

The quantity H is the statically redundant horizontal reaction whose value is obtained from the boundary condition $w_{tot}(0) = w^T$ at the top of the column:

$$w^T = \int \frac{M[w_0, w^T]}{EI} \bar{M} dx$$

where $\bar{M}(x)$ is as given in equation (b).

As shown in figure 8.29, the imposed deformation results in increased bending moments and a lower ultimate load relative to the column without the imposed deformation.

8.1.6 Design at Ultimate Limit State

The familiar condition

$$S_d = S(\gamma_S Q) \leq \frac{1}{\gamma_R} R \tag{8.20}$$

can also be used in the design of slender compression members at ultimate limit state. The load factors, denoted γ_S, are defined in design codes and standards. The value of γ_S for dead load is normally larger when its effect is unfavourable, and smaller when its effect is favourable. The values of γ_S for all other loads do not normally vary. The computed cross-section resistance is reduced by the resistance factor γ_R, which is applied equally to both coordinates of the interaction diagram: $(M_R/\gamma_R, N_R/\gamma_R)$.

Statistical variation in the flexural stiffness of slender columns can be accounted for by calculating deformations using a design value of stiffness EI_d:

$$EI_d = \frac{EI_y}{\gamma_R}$$

Many published interaction diagrams express the cross-section resistance in terms of unfactored moments and axial forces. Factoring the sectional forces by $\gamma = \gamma_R \gamma_S$ simplifies the use of these diagrams, while maintaining the specified margin of safety. Inequality (8.20) is thus transformed into

$$S^* = \gamma_R\,S(\gamma_S\,Q) \leq R \quad \text{or} \quad S^* = S(\gamma_S\,\gamma_R\,Q) \leq R \tag{8.21}$$

Equivalence of the two structural safety criteria requires that flexural stiffness not be factored by $1/\gamma_R$ when inequality (8.21) is used.

Two equivalent statements of the safety criterion can thus be formulated for a given axial load Q, first-order moment M_1, and first-order displacement w_1. In both statements, the parameter η is used to relate the geometrical eccentricity of the member to the statical eccentricity of the axial load at the critical section (fig. 8.30).

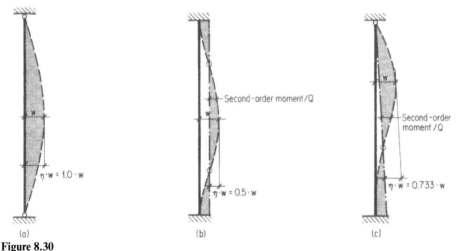

Figure 8.30
Maximum eccentricity ηw_0 of axial force for various end conditions: a, hinged-hinged; b, fixed-fixed; c, hinged-fixed

1. Loads factored by γ_S, resistance factored by $1/\gamma_R$, stiffness factored by $1/\gamma_R$ (inequality (8.20)):

$$N_d = \gamma_S Q \leq \frac{1}{\gamma_R} N_R$$

$$M_d = \gamma_S M_1 + \gamma_S Q \eta w_{1,d} \frac{1}{1 - \dfrac{\gamma_S Q}{Q_{E,d}}} \leq \frac{1}{\gamma_R} M_R$$

The first-order deflections and critical load are calculated using the factored flexural stiffness EI_y/γ_R

$$w_{1,d} = \frac{\gamma_S Q}{EI_y/\gamma_R} \alpha l^3,$$

and

$$Q_{E,d} = \frac{\pi^2 EI_y/\gamma_R}{l_k^2}$$

(The coefficient α is a function of the first-order moments and system geometry and is obtained from the virtual-work integral.)

2. Loads factored by $\gamma = \gamma_R \gamma_S$, resistance and stiffness not factored (inequality (8.21)):

$$N^* = \gamma_R \gamma_S Q \leq N_R$$

$$M^* = \gamma_R \gamma_S M_1 + \gamma_R \gamma_S Q \eta w_1^* \frac{1}{1 - \dfrac{\gamma_R \gamma_S Q}{Q_E}} \leq M_R$$

where the first-order deflections and critical load are calculated using the unfactored flexural stiffness EI_y:

$$w_1^* = \frac{\gamma_R \gamma_S Q}{EI_y} \alpha l^3,$$

and

$$Q_E = \frac{\pi^2 EI_y}{l_k^2}$$

(α is defined as above.)

8.1.7 Use of Design Aids

a) Nondimensional Ratios

As stated in Section 8.1.3, published interaction diagrams normally express the ultimate sectional forces in terms of nondimensional ratios. The dimensionless

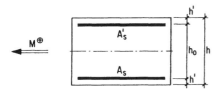

Figure 8.31
Symbols used for rectangular cross-sections

quantities used in the diagrams given in the Appendix are defined as follows for rectangular and circular cross-sections:

1. Rectangular Sections:

   The symbols used are given in figure 8.31. It is assumed that the section is symmetrically reinforced, i.e., $A_s = A'_s$.

   1. Axial force:

   $$n_R = \frac{N_R}{bhf_c}$$

   2. Moment:

   $$m_R = \frac{M_R}{bh^2 f_c}$$

   3. Flexural stiffness:

   $$ei_y = \frac{EI_y}{bh^3 f_c} = \frac{r_y}{h} m_y$$

   4. Radius of curvature:

   $$\frac{r_y}{h}$$

   5. Mechanical reinforcement ratio:

   $$\omega_{tot} = \frac{A_s + A'_s}{bh} \frac{f_{sy}}{f_c}$$

2. Circular Sections:

   The symbols used are given in figure 8.32. The area of reinforcement, A_s, is assumed to be uniformly distributed around the circumference.

   1. Axial force:

   $$n_R = \frac{4 N_R}{\pi h^2 f_c}$$

   2. Moment:

   $$m_R = \frac{4 M_R}{\pi h^3 f_c}$$

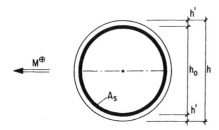

Figure 8.32
Symbols used for circular cross-sections

3. Flexural stiffness:
$$ei_y = \frac{4EI_y}{\pi h^4 f_c} = \frac{r_y}{h} m_y$$

4. Radius of curvature:
$$\frac{r_y}{h}$$

5. Mechanical reinforcement ratio:
$$\omega_{tot} = \frac{4A_s}{\pi h^2} \frac{f_{sy}}{f_c}$$

The parameter r_y/h is a nondimensional expression for the radius of curvature corresponding to the plastic moment m_y. Values of r_y/h can also be obtained from the interaction diagrams in the Appendix. The dimensionless flexural stiffness ei_y, used to calculate the second-order moments, can be computed from

$$ei_y = \frac{r_y}{h} m_y \cong \frac{r_y}{h} m_R \tag{8.22}$$

b) Design Procedures

As stated in Section 8.1.1, column dimensions will normally have been chosen on the basis of aesthetic criteria. The design of slender bridge columns for adequate structural safety is thus primarily concerned with determining the amount of reinforcement to be provided. Two design procedures can be used for this purpose:

1. Design review:

(1) The section dimensions and axial force $n^* = \gamma_R n_d$ are given. An estimate is made of the required reinforcement ratio, ω.

(2) The ultimate moment m_R and radius of curvature r_y/h corresponding to n^* and ω are obtained from the interaction diagram. The flexural stiffness ei_y is calculated using equation (8.22).

(3) The second-order displacements and second-order moment m^* are calculated. The safety of the cross-section is adequate for the assumed reinforcement ratio provided $m^* \leq m_R$.

(4) Another reinforcement ratio is chosen and the procedure is repeated if $m^* > m_R$ (design inadequate), or if m^* is much less than m_R (design overly conservative). In the latter case, it is incorrect to use the reinforcement ratio corresponding to the point (n^*, m^*), which will always be insufficient.

The procedure is conservative since the deformations of the system are calculated using the flexural stiffness ei_y corresponding to the ultimate cross-section resistance (n_R, m_R), which will be less than the stiffness corresponding to the actual design sectional forces (n^*, m^*).

2. Direct design:

A rapidly converging iterative procedure is derived here for the simple case of an axially loaded column with initial eccentricity $w_0(x)$. The maximum value of $w_0(x)$ is denoted $w_{0,max}$. The second-order design moment, expressed in nondimensional form, is given by the expression

$$m^* = n^* \frac{\eta w_{0,max}}{h} \frac{1}{1 - \dfrac{n^*}{n_E}} \tag{8.23}$$

(The parameter η is defined in Section 8.1.6.) Equation (8.22) is used to rewrite the critical load in terms of m_R and r_y/h:

$$n_E = \frac{\pi^2 ei_y}{(l_k/h)^2} = \frac{\pi^2 m_R(r_y/h)}{(l_k/h)^2} \tag{a}$$

The following expression is obtained by setting $m_R = m^*$ and substituting equation (a) into equation (8.23):

$$m^* = n^* \left(\frac{\eta w_{0,max}}{h} + \frac{(l_k/h)^2}{\pi^2(r_y/h)} \right) \tag{8.24}$$

The moment m^* can be computed directly from n^* using this expression.

The direct design procedure can thus be formulated as follows:

(1) The section dimensions and axial force $n^* = \gamma_R n_d$ are given. The reinforcement ratio ω is assumed equal to 0.2.

(2) The radius of curvature r_y/h corresponding to n^* and ω is obtained from the interaction diagram.

(3) The moment m^* is obtained from equation (8.24).

(4) New values of ω and r_y/h corresponding to (n^*, m^*) are obtained from the interaction diagram. The procedure is repeated until the difference between two successive values of m^* is sufficiently small.

Example 8.9:
Concentrically-loaded rectangular column with initial eccentricity w_0

The system geometry, loading Q, and support conditions are shown in figure 8.33. The initial eccentricity is assumed proportional to the buckled shape of the column. Its maximum value, in nondimensional form, is

$$\frac{w_{0,max}}{h} = \frac{l_k}{300 h}$$

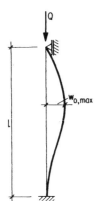

Figure 8.33
System of example 8.9

The relevant geometrical information and material parameters are as follows:

$$
\begin{array}{lll}
l = 30.0\,\text{m} & w_{0,\max} = 0.070\,\text{m} & N = 20000\,\text{kN} \\
l_k = 21.0\,\text{m} & \eta = 0.733 & f_c = 24\,\text{N/mm}^2 \\
h = 0.7\,\text{m} & h'/h = 0.05 & f_{sy} = 460\,\text{N/mm}^2 \\
b = 4.0\,\text{m} & l_k/h = 30.0 & \gamma = \gamma_R\,\gamma_S = 1.8
\end{array}
$$

The interaction diagram is taken from the Appendix for a rectangular section, $h'/h = 0.05$, $\omega = 0.00$ to 0.50, $\Delta\omega = 0.050$. The nondimensional axial load is

$$
n^* = \frac{\gamma N}{bhf_c} = 0.536
$$

The reinforcement will be designed first by the design review procedure, and then by the direct design procedure.

1. Design review:

(1) Assume $\omega_{\text{tot}} = 0.25$.

(2) $m_R = 0.20$

   $\dfrac{r_y}{h} = 275$

   $ei_y = \dfrac{r_y}{h}\,m_R = 55$

(3) $m^* = n^* \dfrac{\eta\,w_{0,\max}}{h}\,\dfrac{1}{1 - \dfrac{n^*}{n_E}} = 0.35$ (Equation (8.23))

(4) Since $m^* > m_R$, the assumed reinforcement ratio is unacceptable.

(1)   Assume $\omega_{tot} = 0.35$.

(2) $m_R = 0.24$

$$\frac{r_y}{h} = 265$$

$$ei_y = \frac{r_y}{h}\, m_R = 63.6$$

(3) $m^* = n^* \dfrac{\eta\, w_{0,\max}}{h} \dfrac{1}{1 - \dfrac{n^*}{n_E}} = 0.17$

(4)   Since $m^* < m_R$, the design is acceptable.

2. Direct Design:

Equation (8.24) is adapted to the given parameters by substituting the values of n^*, η, $w_{0,\max}$, h, and l_k/h. The following expression is obtained:

$$m^* = 0.0393 + \frac{48.9}{r_y/h}$$

The steps of the iterative calculation are presented in table 8.3.

Table 8.3
Solution of Example 8.9 by Direct Design ($n^* = 0.536$)

Iteration	ω	$\dfrac{r_y}{h}$	$m^* = 0.0393 + \dfrac{48.9}{r_y/h}$
1	0.20	280	0.214
2	0.27	275	0.217
3	0.30	270	0.220
4	0.30	270	0.220

8.1.8 Special Cases

a) Columns During Construction

Bridge columns are normally free-standing prior to construction of the super-structure. In this state, the effective length l_k is equal to twice the height l provided the column is fixed to a rigid base. As a result, the slenderness ratio l_k/i of the free-standing column will be higher than l_k/i for the same column in the completed structure. Values of l_k/i of up to 240 are possible for free-standing columns.

The stresses in the free-standing column are induced by its self-weight g and wind load q_w. The critical distributed load of a cantilever column, $(gl)_E$, is equal to

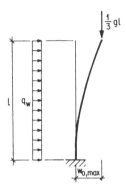

Figure 8.34
Statical model for the calculation of second-order effects in a free-standing column during construction

about three times the critical load Q_E applied at the tip (Timoshenko and Gere 1961:105). Second-order deflections and moments can therefore be calculated using an equivalent tip load equal to $gl/3$. The statical model shown in figure 8.34 must be modified as necessary to account for any elastic rotations at the base of the column.

Dead load can have either a favourable or an unfavourable effect on behaviour at ultimate limit state. The design must be checked using the formulas of Section 8.1.6 with two different load factors γ_S for dead load. The total safety factor for dead load $\gamma_G = \gamma_R \gamma_S$ may therefore be different than the total safety factor for wind, γ_w.

The second-order deflection at the top of the free-standing column, neglecting any rotation of the foundation, is given by the expression

$$(w^T)^* = w_{0,\text{max}} + \left(\frac{\gamma_w q_w l^4}{8 EI_y} + \frac{\gamma_G gl/3}{Q_E} w_{0,\text{max}}\right) \frac{1}{1 - \dfrac{\gamma_G gl/3}{Q_E}}$$

(The initial displacement, $w_{0,\text{max}}$, is obtained from the relevant design standard and is normally left unfactored.) The corresponding ultimate sectional forces at the base of the column are

$$N^* = -\gamma_G gl$$

$$M^* = -\frac{\gamma_w q_w l^2}{2} - \gamma_G \frac{gl}{3} (w^T)^*$$

b) *Expansion Bearings*

Superstructure displacements due to prestressing, shrinkage, temperature, and other actions produce frictional forces in expansion bearings, which must be considered in the column design. Properly detailed and maintained bearings with

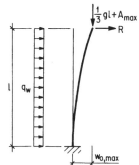

Figure 8.35
Statical model for the calculation of second-order effects in a column
with expansion bearings at the top

steel rollers or PTFE sliding surfaces produce frictional forces equal to roughly 5 percent of the vertical dead load reaction from the superstructure. The loads to be considered in the design of the column thus consist of the total self-weight of the column gl, the maximum superstructure reaction A_{max}, bearing friction, R, and wind q_w (fig. 8.35).

The deflection of the column tip at ultimate limit state, neglecting foundation rotations, is

$$(w^T)^* = w_{0,\max} + \left(\frac{\gamma q_w l^4}{8 EI_y} + \frac{\gamma R l^3}{3 EI_y} + \frac{\gamma(gl/3 + A_{\max})}{Q_E} w_{0,\max}\right)$$

$$\cdot \frac{1}{1 - \dfrac{\gamma(gl/3 + A_{\max})}{Q_E}} \tag{a}$$

(The γ factors may not all be equal.) The ultimate sectional forces at the base of the column are thus

$$N^* = -\gamma(gl + A_{\max})$$

$$M^* = -\frac{\gamma q_w l^2}{2} - \gamma R l - \gamma(gl/3 + A_{\max})(w^T)^*$$

The computed column tip deflection (equation (a)) may be greater than the maximum possible superstructure displacement u^G. In such a case, the frictional force R acts in the opposite direction, resisting the tip displacement and thus reducing the design moments. The value of R must therefore be limited to a conservatively small multiple of the superstructure dead load reaction A_g:

$$R < \alpha A_g$$

where α lies between 0.005 and 0.010, depending on bearing type. Alternatively, it can be assumed that the displacement u^G is imposed at the column tip with the bearings blocked (fig. 8.36). The design would then be adequate at ultimate

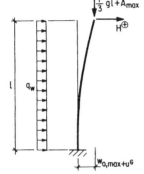

Figure 8.36
Statical model for the calculation of second-order effects in a column with blocked bearings and imposed superstructure displacement u^G

limit state provided the inequalities of Section 8.1.6 are satisfied. The resulting horizontal force at the column tip H, required for compatibility, must act opposite to the displacement, and satisfy

$$|H^*| < \alpha A_g$$

8.1.9 Flexible Systems

a) General Remarks

In the context of column design, structural systems are defined as *flexible* when longitudinal forces due to wind, earthquake, braking, and other actions are transferred from superstructure to foundation through the columns, or *fixed* when the transfer is made directly, normally by means of fixed bearings at an abutment. The portions of a bridge between internal hinges, and thus disconnected from the abutments, are normally designed as flexible systems.

The inherent stability of fixed systems is relatively high, since all columns whose tips cannot displace relative to the superstructure are effectively restrained against sway in the longitudinal direction. Their effective length is thus substantially reduced. The stability of flexible systems, however, can only be provided by columns that are not braced against sway, and whose effective length is correspondingly greater. For these reasons, fixed systems are normally more economical, except in the case of very long bridges.

Due to the inherent inaccuracy in the calculation of the neutral point for superstructure displacements, the detailing of bearings for flexible systems requires particular attention. This topic is discussed further in Chapter 6.

b) Global Stability of Flexible Systems

Flexible structural systems must have sufficient longitudinal stiffness to ensure that deformations and second-order moments are not excessive. The global stability of the bridge provides an effective measure in this regard. The second-order longitudinal displacement of the superstructure u^G can be expressed approximately as

$$u^G = u_1^G \frac{1}{1 - \dfrac{1}{\gamma_E}} \tag{8.25}$$

where u_1^G is the first-order superstructure displacement and γ_E is the inherent factor of safety against buckling of the system. The second-order moments M resulting from u^G can likewise be expressed as

$$M = M_1 \frac{u^G}{u_1^G} = M_1 \frac{1}{1 - \dfrac{1}{\gamma_E}} \tag{8.26}$$

where M_1 is the first-order moment.

The factor γ_E should be greater than 3.0. The global stability of the system is often checked, however, prior to designing the column reinforcement. In such a case, γ_E is calculated assuming the specified minimum reinforcement in all columns. The stiffness of the system can be considered adequate provided γ_E calculated in this way is greater than 2.0. It is assumed that the column reinforcement will be increased during final design, thus raising γ_E to an acceptable level. If the global stability is checked after the final design has been completed, γ_E should not be less than 3.0.

The buckling load of the system is reached when, for an arbitrary longitudinal displacement of the superstructure u^G, the sum of the horizontal forces H_i transferred from the superstructure to the columns is equal to zero. A procedure to calculate γ_E can be based on this condition. The system buckling load is expressed as the product of the column loads Q and the unknown safety factor γ_E. The forces H_i are likewise expressed as functions of the buckling load $\gamma_E Q$ and the superstructure displacement u^G. The horizontal forces are then added and their sum equated to zero:

$$\sum H_i = 0 \tag{a}$$

This equation can then be solved for γ_E.

Expressions for the horizontal forces H_i are derived below for three types of columns. Since the loads are increased by a global safety factor γ_E, the column flexural stiffness EI_y is left unfactored. A value of $220 h M_R$ can be used as an initial lower-bound estimate of EI_y, corresponding to minimum column reinforcement. The sign convention is given in figure 8.37.

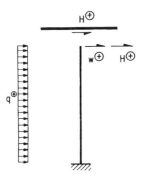

Figure 8.37
Sign convention for flexible systems

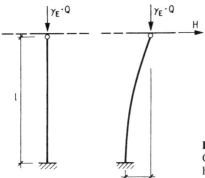

Figure 8.38
Column with lower end fixed and upper end hinged:
horizontal force H corresponding to system buckling
load

1. Lower end fixed, upper end hinged (fig. 8.38):

   Displacement at the top of the column:

   $$w^T = w_1^T \frac{1}{1 - \dfrac{\gamma_E Q}{Q_E}}$$

   where

   $$w_1^T = \frac{H l^3}{3 E I_y} \quad \text{and} \quad Q_E = \frac{\pi^2 E I_y}{4 l^2}$$

   Horizontal force at the top of the column:

   $$H = \frac{3 E I_y}{l^3} w^T \left(1 - \frac{\gamma_E Q}{Q_E}\right) = \frac{12 w^T}{\pi^2 l} (Q_E - \gamma_E Q)$$

2. Both ends fixed (fig. 8.39):

   Displacement at the top of the column:

   $$w^T = w_1^T \frac{1}{1 - \dfrac{\gamma_E Q}{Q_E}}$$

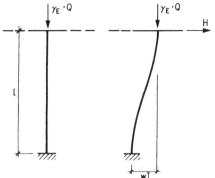

Figure 8.39
Column with both ends fixed: horizontal force H corresponding to system buckling load

where

$$w_1^T = \frac{Hl^3}{12\,EI_y} \quad \text{and} \quad Q_E = \frac{\pi^2\,EI_y}{l^2}$$

Horizontal force at the top of the column:

$$H = \frac{12\,EI_y}{l^3}\,w^T\left(1 - \frac{\gamma_E\,Q}{Q_E}\right) = \frac{12\,w^T}{\pi^2\,l}\,(Q_E - \gamma_E\,Q)$$

Equal expressions for the horizontal force H are obtained for both column types. The different support conditions therefore produce only a change in the critical load of the column.

3. Both ends hinged (fig. 8.40):

Horizontal force at the top of the column:

$$H = -\frac{w^T}{l}\,\gamma_E\,Q$$

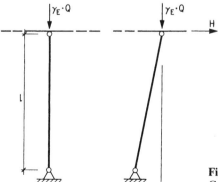

Figure 8.40
Column with both ends hinged: horizontal force H corresponding to system buckling load

Since all displacements at the top of the column must be equal ($w_i^T = u^G$, for all i), the condition of equation (a) can be written

$$\frac{12}{\pi^2} u^G \left[\sum_i \frac{Q_{E,i}}{l_i} - \gamma_E \left(\sum_i \frac{Q_i}{l_i} + \frac{\pi^2}{12} \sum_j \frac{Q_j}{l_j} \right) \right] = 0 \tag{b}$$

where the index i denotes the statically indeterminate columns which stabilize the system (types 1 and 2 above), while j denotes the statically determinate columns (type 3 above). The quantity $Q_{E,i}$ is the critical load of column i. Equation (b) can be solved for γ_E:

$$\gamma_E = \frac{\sum_i \dfrac{Q_{E,i}}{l_i}}{\sum_i \dfrac{Q_i}{l_i} + \dfrac{\pi^2}{12} \sum_j \dfrac{Q_j}{l_j}}$$

c) Calculation of the Ultimate Load of the System

The buckling load derived above is not a reliable indicator of the ultimate load of the system. Under the influence of wind loads, longitudinal forces in the superstructure, and bearing eccentricity, the ultimate capacity at critical sections will be reached for loads substantially less than the buckling load.

An exact calculation of the ultimate load of a flexible system would account for the superstructure construction sequence, as well as material parameters and climatic and geotechnical factors. Such a calculation would be very complex analytically. It would moreover be based on many parameters which are subject to considerable statistical variability or are difficult to determine at the design stage. The large computational effort required would thus not result in increased accuracy. It is therefore reasonable to use a simplified procedure.

It will be assumed that the ultimate load of the system has been reached at the formation of the first plastic hinge, that is, when the ultimate resistance has been reached at any one section of the system. This assumption results in a lower bound for the ultimate load, since the formation of a single plastic hinge will not constitute a failure mechanism unless the system is statically determinate. Due to second-order effects, however, only a limited increase in load is normally possible after the first plastic hinge has been formed. The procedure is therefore a conservative and reasonably accurate indicator of the true ultimate load of the system.

The safety of each column section at ultimate limit state must therefore be checked according to the inequalities of Section 8.1.6, which will be expressed here in the following form:

$$S^*(M^*, N^*) \leqq R \tag{Inequality (8.21)}$$

When this inequality is satisfied at all locations, the ultimate load of the system will be greater than the factored design loads.

The sectional forces at ultimate limit state $S*$ must be calculated considering three types of lateral actions: (1) imposed column tip displacements due to prestressing, shrinkage, and temperature change in the superstructure; (2) lateral wind loads q_w, applied directly to the column; and (3) horizontal forces H acting at the column tip. The second-order displacements due to each of these actions are combined and used to compute the second-order moments.

The imposed column tip displacement due to prestressing, shrinkage, and temperature change in the superstructure is denoted w_G^T. This displacement can normally be computed using the flexural stiffness of the homogeneous, uncracked column sections. The superstructure construction sequence must be accounted for in the calculation. It can be assumed that column stiffness remains constant during the entire construction period. The initial eccentricity $w_0(x)$, whose sign is chosen to produce the most unfavourable effect, is combined with w_G^T resulting in the total initial deformation, w_I. This deflection will closely approximate the column's buckled shape. The relation can be formulated at the top of the column as follows:

$$w_I^T = w_G^T + w_0^T$$

The superstructure acts as a rigid link among the columns. Any column tip displacement in addition to w_I^T must therefore be equal for all columns of the bridge. These additional displacements, denoted w^T, are produced by wind load on the columns and the horizontal force H applied at the column tips, and include the second-order displacements due to the initial deformation w_I^T.

The unknown horizontal forces can be calculated by expressing the force at column i, H_i, as a function of the tip displacement w^T. The equilibrium condition of the superstructure is used to solve for w^T (fig. 8.41):

$$H_{\text{tot}}^* - \sum H_i^* = 0 \tag{c}$$

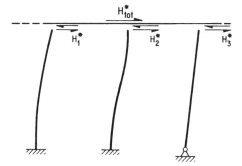

Figure 8.41
Horizontal forces H_i^* transferred from super-structure to columns at ultimate load

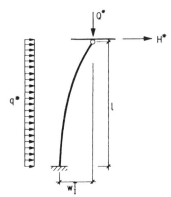

Figure 8.42
Column with lower end fixed and upper end hinged:
ultimate load of the system

where H^*_{tot} is the total horizontal load (factored by $\gamma = \gamma_R \gamma_S$) originating from the superstructure and includes the minimum frictional force at all expansion bearings, applied in opposite direction to the loads.

The expressions for H_i are derived here for three common column types. Loads are factored by $\gamma = \gamma_R \gamma_S$; the column flexural stiffness EI_y is left unfactored:

1. Lower end fixed, upper end hinged (fig. 8.42):

First-order displacements at column tip:

$$w_1 = w_1[H^*] + w_1[q^*] + w_1[w_I^T]$$

where

$$w_1^T[H^*] = \frac{H^* l^3}{3 EI_y}$$

$$w_1^T[q^*] = \frac{q^* l^4}{8 EI_y}$$

$$w_1^T[w_I^T] = \frac{Q^*}{Q_E} w_I^T$$

(Q is equal to the vertical reaction from the superstructure plus approximately one third of the total column self-weight.)

Second-order displacement at column tip:

$$w^T = \left(\frac{H^* l^3}{3 EI_y} + \frac{q^* l^4}{8 EI_y} + \frac{Q^*}{Q_E} w_I^T \right) \frac{1}{1 - \dfrac{Q^*}{Q_E}}$$

where

$$Q_E = \frac{\pi^2 EI_y}{4 l^2}$$

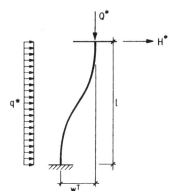

Figure 8.43
Column with both ends fixed: ultimate load of the system

Horizontal force at column tip:

$$H^* = w^T \frac{12}{\pi^2} \left(\frac{Q_E}{l} - \frac{Q^*}{l} \right) - \frac{3q^*l}{8} - \frac{12Q^*}{\pi^2 l} w_I^T \qquad \text{(d)}$$

2. Fixed at both ends (fig. 8.43):

First-order displacements at column tip:

$$w_1^T(H^*) = \frac{H^* l^3}{12 EI_y}$$

$$w_1^T(q^*) = \frac{q^* l^4}{24 EI_y}$$

$$w_1^T(w_I^T) = \frac{Q^*}{Q_E} w_I^T$$

(Q is equal to the vertical reaction from the superstructure plus approximately one half of the total column self-weight.)

Second-order displacement at column tip:

$$w^T = \left(\frac{H^* l^3}{12 EI_y} + \frac{q^* l^4}{24 EI_y} + \frac{Q^*}{Q_E} w_I^T \right) \frac{1}{1 - \dfrac{Q^*}{Q_E}}$$

where

$$Q_E = \frac{\pi^2 EI_y}{l^2}$$

Horizontal force at column tip:

$$H^* = w^T \frac{12}{\pi^2} \left(\frac{Q_E}{l} - \frac{Q^*}{l} \right) - \frac{q^*l}{2} - \frac{12Q^*}{\pi^2 l} w_I^T \qquad \text{(e)}$$

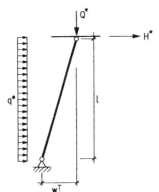

Figure 8.44
Column with both ends hinged: ultimate load of the system

3. Both ends hinged (fig. 8.44):

Horizontal force at column tip:

$$H^* = -\frac{Q^*}{l}(w_I^T + w^T) - \frac{q^* l}{2} \tag{f}$$

The horizontal forces H_i calculated from equations (d), (e), and (f) are substituted into equation (c). The value of w^T obtained from equation (c) is then substituted back into equations (d), (e), and (f) to obtain the value of H_i for each column.

The ultimate sectional forces can now be computed and checked against the available cross-section resistance. If it is determined from the sectional forces that certain columns or substantial portions thereof are still in the uncracked state, the calculation can be repeated using the stiffness of the uncracked homogeneous section in these regions. This can lead to a significant increase in the ultimate load of the system in some cases.

d) Inclined-Leg Frame Bridges

Inclined-leg frame bridges behave as flexible two-dimensional frames. The transfer of longitudinal horizontal forces from the superstructure to the supports induces bending in the superstructure. The superstructure moment envelope must therefore be calculated by superimposing the effects of the vertical and the horizontal forces.

Longitudinal displacements and the resulting bending moments are magnified by second-order effects according to equations (8.25) and (8.26). The magnitude of the increase will be a function of γ_E, the inherent factor of safety against buckling of the system.

The critical load case for buckling of the system consists of dead load plus live load applied to one half of the bridge (fig. 8.45a). If the structure is symmetrical, the

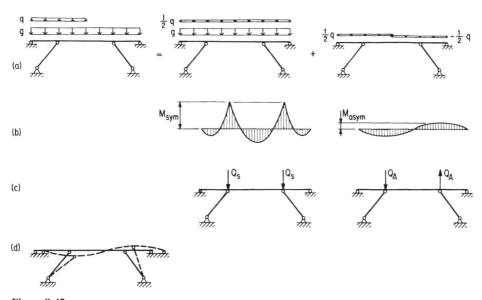

Figure 8.45
Inclined-leg frame bridge: a, symmetric and antisymmetric load components; b, moment diagrams; c, simplified load components; d, buckled shape

loading can be expressed as the sum of a symmetrical component and an antisymmetrical component. The superstructure moment diagrams produced by each component are shown in figure 8.45 b. The calculation of γ_E can be simplified through the use of concentrated loads Q^S and Q^A (fig. 8.45 c). Instability is induced by compression which is a result of the symmetrical component alone. The buckling load of the system is therefore expressed as a multiple of Q^S:
$Q_E = \gamma_E Q^S$.

A rigorous calculation of the buckling load of the system can be made using the stiffness method, taking into account axial deformation. A system of equations is established for the joint rotations and displacements of the structure loaded with $\gamma_E Q^S$. The buckling load is obtained by setting the determinant of the system equal to zero.

In simple cases, the buckling load can be calculated directly using Vianello's method, presented in Section 8.1.2. Assuming an initial displacement δ_0 corresponding to the buckled shape of the structure, the buckling load is reached when the first-order displacement δ_1 is equal to δ_0:

$$\delta_1 (\gamma_E Q^S, \delta_0) = \delta_0 \qquad (8.27)$$

The possible formation of plastic hinges in the girder at ultimate limit state must be considered in computing the moments M_0 due to δ_0 and $\gamma_E Q^S$. When the girder is designed for the moment diagrams of figure 8.45 b, without considering moment redistribution, M_0 can be calculated using normal procedures, as outlined in the following example:

Example 8.10:
Inclined-leg frame without plastic hinge

The system and loadings are shown in figure 8.46. The initial displacement, consists of the rigid-body rotation of the columns through angle δ_0 and the corresponding deflections of the superstructure. The resulting bending moment diagram (M_0) is shown in figure 8.47. The additional rotation of the columns, δ_1, is computed using the method of virtual work:

$$\delta_1 = \int_0^{l/2} \frac{M_0}{EI_y} \bar{M} \, dx = \frac{1}{3EI_y} \gamma_E Q^S \frac{\delta_0}{\tan \alpha} \frac{(ac/2)^2}{b(a+c/2)}$$

The virtual moments $\bar{M}$ are given in figure 8.48. The factor of safety against buckling is obtained from equation (8.27):

$$\gamma_E = \frac{3EI_y b(a+c/2)\tan\alpha}{Q^S(ac/2)^2}$$

(a)

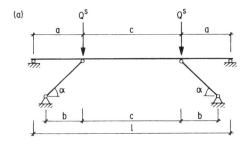

(b)

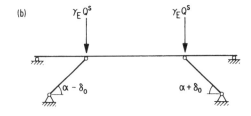

Figure 8.46
System of example 8.10: a, geometry and loading; b, initial displacement δ_0

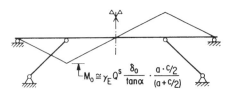

Figure 8.47
Moments M_0 due to $\gamma_E Q^S$ and δ_0

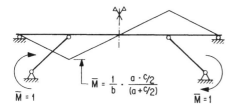

Figure 8.48
Moments $\bar{M}$ due to a unit rotation at the base of each of the inclined columns

The moments M_0 are proportional to the moments due to the concentrated antisymmetrical loads Q^A. When the girder is designed to form a plastic hinge at the support before the ultimate load is reached, M_0 must be altered accordingly. In the following example, it is assumed that the plastic moment at the supports is equal to the design moment due to symmetrical load $(g + q/2)$. The antisymmetrical load of figure 8.45c will therefore cause no increase in moment at the right support. The moments M_0 can thus be considered to act on a hinged system.

Example 8.11:
Inclined-leg frame with plastic hinge

The system and loadings are shown in figure 8.49. The girder is hinged at the right inclined support. The initial deflection δ_0 is identical to the previous example (fig. 8.46). The bending moments M_0 due to δ_0 and $\gamma_E Q^S$ are shown in figure 8.50. The method of virtual work is again used to solve for δ_1:

$$\delta_1 = \int_0^{l/2} \frac{M_0}{EI} \bar{M}\, dx = \frac{1}{6EI_y} Q^S \frac{\delta_0 (ac)^2 (a+c)}{2b \tan \alpha (a + c/2)^2}$$

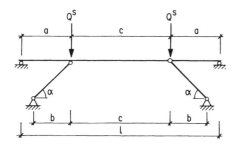

Figure 8.49
System of example 8.11

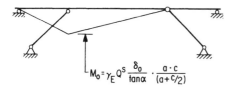

Figure 8.50
Moments M_0 due to $\gamma_E Q^S$ and δ_0

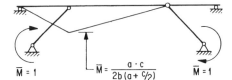

$\bar{M} = 1$ $\bar{M} = \dfrac{a \cdot c}{2b\left(a + c/2\right)}$ $\bar{M} = 1$

Figure 8.51
Moments $\bar{M}$ due to a unit rotation at the base of each of the inclined columns

(The virtual moments $\bar{M}$ are given in figure 8.51.) The value of γ_E is obtained from equation (8.27):

$$\gamma_E = \frac{6\,EI_y\,b\,(a + c/2)^2 \tan\alpha}{Q^S (ac)^2 \,(a + c)}$$

8.2 Foundations

8.2.1 General Remarks

Foundations are a relatively expensive component of bridges. Foundation costs can vary from 30 to 80 percent of the superstructure cost, depending on the topographical, geotechnical and hydrological conditions. The foundation system and its interaction with the superstructure also have a profound effect on the overall structural behaviour of bridges. The choice of structural system and the careful analysis and design of its components are therefore just as important for the substructure as for the superstructure.

The foundation design must provide an adequate margin of safety against the occurrence of the following limit states: loss of bearing capacity, overturning, excessive settlement, and slope stability. It must be based on a thorough description of the geotechnical and hydrological conditions, obtained from borings, piezometer readings and, where necessary, measurements of long-term soil displacements. The substantial statistical variability of all geotechnical parameters must be considered. Laboratory tests are the primary means of identifying the range of variation of the parameters.

8.2.2 Spread Footings

The simplest and most economical foundation type is the spread footing. These foundations can be used when a competent soil or rock layer is present relatively close to the surface. The depth of the bottom of the footing must lie below the maximum depth of frost penetration; the upper surface of the footing must be located at least 0.5 m below finished grade.

Footing dimensions must be selected to prevent excessive settlements, over-turning of the structure, and bearing pressures in excess of the available capacity.

Stability against overturning is of particular importance for columns during construction and for columns that are not restrained against sway. In addition, slope stability must be considered in the design of footings located in inclined terrain.

Foundation settlements are calculated for the total dead load reaction. When appropriate, settlements can be calculated according to elastic theory using published influence charts (Terzaghi 1943; additional charts from various sources can be found in Klöckner, Engelhardt, and Schmidt 1982). The allowable foundation settlement is usually chosen to limit cracking in the superstructure or to limit the maintenance required for subsequent bearing adjustment. It is often practical to reduce potential settlements through deep vibration, injections, or other densification measures.

Resistance against overturning must be checked for the most critical load combination. The limit state is defined as a bearing pressure at the edge of the footing equal to the bearing capacity of the underlying material. All loads that induce overturning are increased by the factor γ_S. Dead load is multiplied by a smaller factor when it resists overturning. The allowable soil pressure along the edge of the footing can be obtained from the stress-deformation diagram of the soil. It is reduced by an appropriately chosen resistance factor. Adequate stability is ensured when the maximum factored stress at the footing edge is less than the factored allowable soil pressure. Second-order effects resulting from the deformations of both column and soil must be considered.

The bearing capacity of the soil is rarely of controlling significance in the calculation of footing dimensions, which are normally determined from the allowable settlements. Slope stability may, however, be critical in inclined terrain. The factored resistance of the soil against sliding must be greater than the maximum shear stress due to factored foundation loads at the critical surface.

The footing slab is designed for two load cases: maximum vertical load, and maximum eccentricity of soil pressure. Its thickness at the column face should be chosen to eliminate the need for shear reinforcement. This will normally be the case when: the resultant of the soil pressure outside the column perimeter is directed to the column by a compression diagonal inclined at roughly 45°, and the compression diagonal resulting from soil pressure at the edge of the footing is inclined at roughly 30° (fig. 8.52). Any sloping of the top surface of the footing should be made without top forms for reasons of economy and should not hinder the proper consolidation of the concrete. Truss models can be used in the design of the reinforcement. The reinforcing steel must be properly anchored at the edges of the footing.

Spread footings that are not located at the immediate surface are normally built in sloped or shored excavations. Soldier piles and timber lagging can be used for shoring when the excavation is in the dry. Otherwise, sheet pile walls can be used in sand or gravel; drilled concrete pile walls are recommended in coarser material. Deep footings are built as shafts or cofferdams.

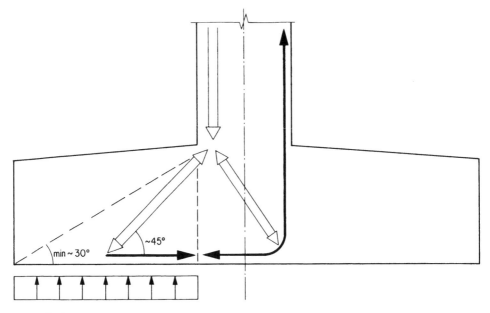

Figure 8.52
Spread footings: geometrical requirements for no shear reinforcement

When spread footings are not appropriate for the given geotechnical conditions, deep foundations such as shafts, cofferdams, or piles must be used.

8.2.3 Shaft Foundations

Shaft foundations have the following advantages:

1. The soil is visible and fully accessible for inspection over the entire depth of the foundation
2. The definitive depth of the foundation can be established without difficulty during construction, to correspond with the actual soil conditions encountered
3. At least a portion of the reaction from the bridge can be compensated by the weight of the material removed from the excavation. This makes shafts especially suitable for foundations in unstable soils, particularly in inclined terrain

Shaft foundations are best suited for construction in cohesive soils where the water table lies below the bottom of the foundation. They can also be built, however, in granular soils when shoring, injections, or other appropriate measures are taken. Provided the permeability of the soil is relatively low, the water table can be lowered below the bottom of the footing using wellpoints or other means.

The shape and dimensions of the shaft cross-section can be freely chosen. The diameter should nevertheless be large enough to permit mechanized excavation. A circular section is normally chosen, except in inclined terrain where elliptical sections can resist the non-uniform soil pressure more efficiently. Varying the interior cross-section with depth is not recommended, since suspended formwork of constant diameter is normally used to cast the walls. The footing can be made slightly wider than the shaft, but only under appropriate conditions.

Shaft foundations can be divided into two classes, according to the manner in which they transfer the reaction from the superstructure to the ground.

1. *Bearing shafts* transfer the entire column load to the soil through the footing at the bottom of the shaft (fig. 8.53). The walls serve primarily as stay-in-place shoring for the excavation. When required, they can also protect the columns against lateral pressure due to long-term soil displacements. The load on the bearing layer is in all cases reduced by the weight of the excavated material. Connecting the column to the top of the shaft fixes it solidly into the upper surface of the ground and reduces its effective length.
2. *Friction shafts* transfer a significant portion of the column load to the soil through skin friction. The column normally rests on a thick concrete cap on top of the shaft (fig. 8.54), or is connected to the shaft along its entire depth (fig. 8.55).

Drainage must be provided in shafts that have been designed to relieve part of the load on the bearing layer. It is recommended to dimension the drainage pipes very conservatively, to reduce the risk of blockage through silting.

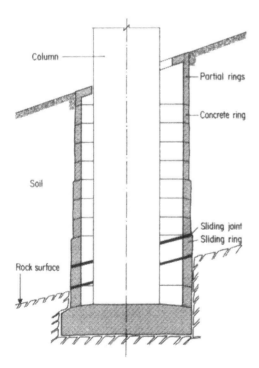

Figure 8.53
Bearing shaft for soil subject to horizontal displacements

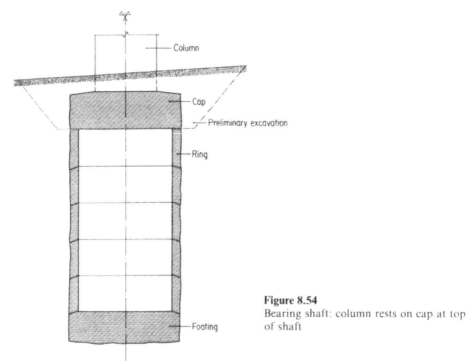

Figure 8.54
Bearing shaft: column rests on cap at top of shaft

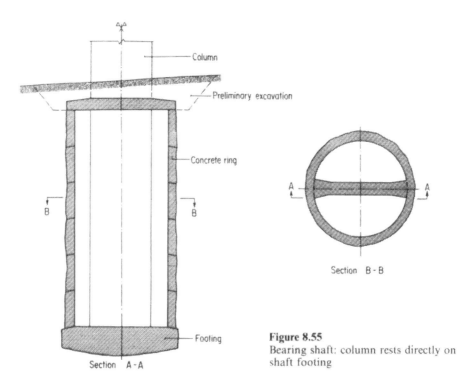

Section A - A

Section B - B

Figure 8.55
Bearing shaft: column rests directly on shaft footing

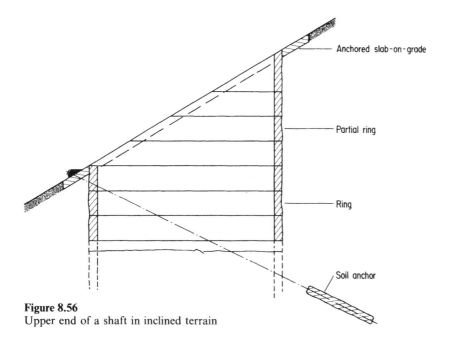

Anchored slab-on-grade

Partial ring

Ring

Soil anchor

Figure 8.56
Upper end of a shaft in inclined terrain

Shafts are constructed by the underpinning method. The height of the rings varies
between 0.8 and 1.5 m, depending on the cohesion of the soil. Construction of the
upper partial rings required in steeply sloping terrain is facilitated by first casting a
0.3 m thick concrete slab on grade around the excavation, and anchoring this slab
into the soil (fig. 8.56).

In the absence of soil movements, the load on the shaft walls is derived from the
horizontal pressure in the at-rest condition, reduced to account for the beneficial
effects of arch action around the shaft. An expression for the lateral soil pressure
on the shaft, σ^R, is derived below for a shaft with radius r, and soil of unit weight γ
and friction angle ϕ. The inclination angle of the soil surface is denoted β.

In the at-rest condition, the vertical and horizontal soil pressures are as follows:

$$\sigma_v = \gamma z$$
$$\sigma_h = K_0 \, \sigma_v$$

where K_0, the coefficient of lateral stress at rest, can be expressed approximately as
a function of the coefficient of active stress K_a

$$K_0 = K_a \, [1 + \sin(\phi - \beta)]$$

where

$$K_a = \cos^2 \phi / \{1 + \sqrt{[\sin \phi \sin(\phi - \beta)/\cos \beta]}\}^2$$

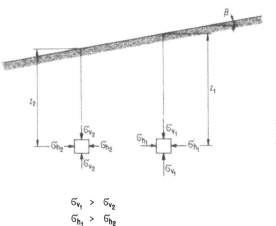

Figure 8.57
At-rest pressure in inclined terrain

The horizontal at-rest pressure is therefore

$$\sigma_h = K_a \left[1 + \sin(\phi - \beta) \right] \sigma_v \tag{a}$$

The at-rest pressure on the uphill side of a large-diameter shaft will be greater than the at-rest pressure on the downhill side (fig. 8.57). Passive pressure must therefore be mobilized on the downhill side to maintain equilibrium.

The unsupported excavation below the completed portion of the shaft is made possible by a local redistribution of stress known as arching. The load is transferred away from the excavated face in both the horizontal and the vertical directions. The vertical arch action disappears as the shaft construction progresses deeper. A significant portion of the horizontal arch action, however, remains after the shaft is completed, and contributes to the reduction of the load on the shaft. Equation (a) can thus be modified to take this beneficial effect into account. The lateral pressure acting on the uphill portion of the shaft, σ^R, is defined as a multiple of σ_h:

$$\sigma^R = A \sigma_h \tag{b}$$

where the reduction factor A is defined as follows (Lang and Huder 1985, 154–156):

$$A = \frac{1 - \exp\left(-\dfrac{z}{r} K_a \tan \phi\right)}{\dfrac{z}{r} \tan \phi} \tag{c}$$

The pressure acting on the shaft wall (uphill side) is obtained by substituting equations (a) and (c) into equation (b):

$$\sigma^R = K_a r \gamma \left[1 + \sin(\phi - \beta) \right] \frac{1 - \exp\left(-K_a \dfrac{z}{r} \tan \phi\right)}{\tan \phi}$$

an equal and opposite pressure is applied to the downhill side of the shaft for equilibrium.

Vertical skin friction

$$\tau^R = \tfrac{2}{3}\,\sigma^R \tan \phi$$

can be assumed for friction shafts. Settlement of the shaft is required before this skin friction can be fully mobilized.

The design of the shaft walls can be based on the model shown in figure 8.58. Second-order methods are normally required for the calculation of the sectional forces. The critical load N_E is given approximately by

$$N_E = N \frac{w_0}{w_1}$$

where N is the tangential force in the shaft ring, w_0 is the initial eccentricity, and w_1 is the first-order wall deflection due to N and w_0. The initial eccentricity is assumed proportional to the buckled shape, and can be approximated as the first-order deflections due to the applied loads. For shafts located in soils susceptible to horizontal displacements, a range of possible load combinations (q_1, q_2) exists.

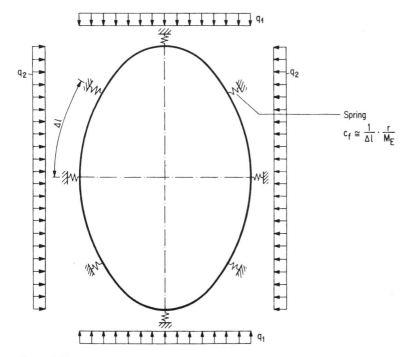

Figure 8.58
Statical model for the design of shaft walls

The flexibility of the spring, c_f, can be calculated approximately from the local shaft radius r and the constrained modulus of elasticity M_E obtained from an oedometer test:

$$\bar{c}_f \cong \frac{r}{M_E} \qquad \text{for continuous support, and}$$

$$c_f = \frac{\bar{c}_f}{\Delta l} \qquad \text{for discrete springs with spacing } \Delta l$$

Long-term movements in the soil should be prevented by providing a drainage system to stabilize the entire slope. Borings for the drainage can be located at the surface of the ground and inside the shafts.

When the construction of a drainage system is not practical, the shaft should be designed to absorb at least a portion of the soil movement. If a well-defined sliding surface can be identified in the soil, the shaft should be made discontinuous at the plane of intersection. This will permit the unrestrained movement of the upper portion of the shaft along with the soil mass. Two or three joints are normally sufficient (fig. 8.55). Rock wool pads, about 50 mm in thickness, can be used as joint material.

Excavation costs can be prohibitive when stable soil layers are located at great depths. In such cases, it may be more practical to found the shaft closer to the surface, excavating only what is necessary to compensate the column load. The system is then detailed so that any deformations induced in the structure by shaft displacements can be subsequently corrected. This can be accomplished by providing a bearing between the column base and the shaft cap, thus permitting the column to be jacked back into position after the shaft displaces.

Long-term movements in the soil can induce an increase or a decrease in the initial ring pressure in the shaft walls, depending on the geometry of the sliding surface. Since this geometry cannot normally be clearly determined, the rings should be conservatively designed for an increase in pressure. A long-term program to monitor the shaft wall deformations should also be considered. A conservative shaft wall design is recommended even when the soil is stable, since shaft foundations are extremely difficult to rehabilitate.

8.2.4 Cofferdams

The classical methods of shaft foundation construction cannot be used below the water table. In such cases, cofferdams may be a practical solution. Cast-in-place concrete cofferdams are shown in figure 8.59. These can be used in sandy or gravelly soils without boulders or cobbles larger than 200 mm in diameter (This material can be excavated relatively easily underwater without blasting.) Steel sheet piling can also be used in these conditions. Cofferdams consisting of drilled concrete piles are best used in soils in which boulders are present. The ratio of

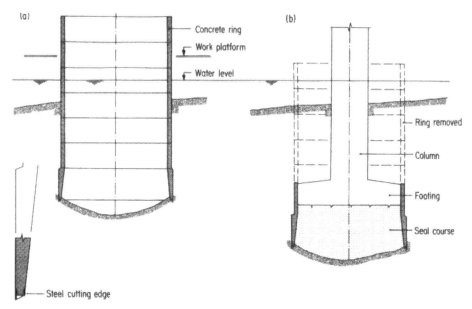

Figure 8.59
Cofferdam foundations: a, after completion of underwater excavation; b, final state after partial removal of cofferdam walls

perimeter to cross-sectional area, and hence the overall material consumption, can be minimized by selecting a circular cross-section.

The walls are constructed segmentally above the water table and continuously lowered by excavation underneath the previously completed portions. The lowest segment, which is provided with a strong steel cutting edge, has an exterior diameter 100 to 200 mm greater than the other segments.

The initial stages of the excavation can be undertaken in the dry as long as the intrusion of water is slight and there is no danger of soil liquefaction. Below a given depth, however, underwater excavation methods must be used. Skin friction, which is produced as the cofferdam is lowered, can be minimized by the injection of a bentonite slurry between the walls and the soil from the interior of the cofferdam. This measure is not appropriate in permeable soils, where the slurry may flow into the cofferdam from below. Wall friction is best reduced in these cases by stuffing straw between the walls and the soil.

After the cofferdam has been excavated to the desired depth, it is sealed with tremie concrete placed underwater. Any mud that settles to the bottom of the excavation must be removed before the seal is placed. Settlement should be minimized by casting the seal immediately after completion of the excavation. After the seal has hardened, the cofferdam can be dewatered. The hardened seal course provides sufficient resistance against hydrostatic uplift. The structural footing and the pier can then be cast using normal procedures. For piers located in rivers, the upper portion of the cofferdam can be removed to roughly 4 m below the river bed, depending on scour danger.

An alternate construction procedure is possible with sheet-pile and drilled-pile cofferdams, which can be sealed before excavation with soil injections. In this way, all excavation can be done in the dry.

8.2.5 Pile Foundations

A large number of different pile systems is available to match the wide variety of possible soil conditions and applications. Piles can be used for deep foundations whenever the soil is not susceptible to horizontal displacements and adequate access is available for the equipment required to place them.

a) Pile Types

Piles can be classified according to the following characteristics:

1. Production: Produced on-site or prefabricated
2. Placement: Driven or drilled
3. Load transfer: End-bearing or friction

Prefabricated piles can be made of timber, steel, or concrete. Their dimensions are limited by the equipment used to transport and place them. As a result, their load-carrying capacity is relatively small. The design service loads for prefabricated piles range between 100 and 1000 kN, depending on soil conditions and pile cross-section. Although timber piles are most often used for temporary structures, they can also be used for permanent structures provided they are permanently located below the water table. Their low cost renders them particularly suitable for the foundations of small bridges. Rolled H-sections are the most common steel pile type. They are appropriate for use as end-bearing piles. Concrete piles are normally more economical than steel piles. Hollow concrete sections produced by spinning have been used to minimize weight.

Piles can be produced on site from plain concrete, reinforced concrete, or injected mortar. Their diameter ranges from 0.5 m to 1.5 m. Because their length is practically unlimited, their full capacity can be utilized. The design service loads of commonly used in-situ piles range from 1000 to 10000 kN, depending on pile cross-section. The diameter of these piles can be locally reduced due to a collapse of the surrounding soil during placement of concrete. It is therefore recommended that in-situ piles be tested after construction.

Driven piles can be used when the soil contains no boulders or other impediments to driving. Piles can be driven by impact, vibration, or jacking. The surrounding soil is compacted by the displacement of soil and by vibration. The capacity of driven piles can be estimated from the resistance encountered during driving. The diameter and length of driven piles are limited. Pile driving is not recommended in sensitive soils where the accompanying vibrations may result in a rearrangement of the soil structure and possibly subsidence of the adjacent area.

For drilled piles, an equal volume of soil is removed before the pile is placed into the ground. A steel shell is normally first placed by twisting it back and forth about its longitudinal axis. The soil is excavated as the shell penetrates into the ground. Granular material can be brought to the surface using an auger. Boulders can be broken up by coring. The shell is withdrawn as the pile is concreted. Underwater concreting methods should be used for piles extending below the water table, since dewatered shafts can cave in as the casing is removed. A slurry wall can be used instead of a steel shell under certain circumstances.

As their name implies, friction piles transfer load from the structure to the soil primarily by skin friction. A high coefficient of friction between soil and pile is therefore essential for these piles to function properly. Bearing piles transfer the load primarily at the pile tip. A sound layer of rock or soil is required. In all cases, however, the load is actually transferred by a combination of friction and bearing.

b) Pile Systems

Friction piles are normally driven, since the skin friction is enhanced by the driving operation. Prefabricated piles or injected piles are suitable for relatively small distributed loads, and where short piles can be used. Concrete piles that are cast in place and then driven are suitable for large concentrated loads and when long piles are required. The bearing capacity at the tip can be increased by a bell-shaped end. Cast-in-place friction piles are used only in exceptional cases, for instance sensitive soils, soils with impediments to driving, very deep bearing layers, or very large concentrated loads. The pile capacity, which is determined from careful inspection of the excavated material, should be confirmed by a load test. This is normally expensive for large piles.

Pile group action is a problem particular to friction piles. The minimum pile spacing should be about three times the pile diameter. It can be assumed for the calculation of foundation settlements and carrying capacity that the pile group behaves as a spread footing of identical dimensions.

Cast-in-place concrete drilled piles are well suited for bearing on rock, since their tips can be properly keyed into sound material below the rock surface. Skin friction must be relied on for additional resistance in certain cases where the bearing layer is less than ideal. Cast-in-place driven piles may be more suitable in these situations. Bearing piles must always be checked for negative skin friction produced by settlement of the upper layers of soil. This phenomenon can considerably increase the load acting on the pile.

The upper ends of piles are normally connected to the base of the column by means of a pile cap. It is also possible, however, to design the piers as extensions of the piles. In such cases, increased deviations from the dimensions shown in the plans must be expected, due to the length and inclination of the piles.

If a column is supported by a single pile or one row of piles, column fixity must be provided through pile bending. This implies that the upper layers of soil must be capable of resisting the corresponding lateral pressures. The largest bending moments in the column-pile system occur under the foot of the column in the region of connection to the pile. This type of foundation is not recommended for columns that have expansion bearings at their upper end or for columns that stabilize flexibly supported bridges. Columns that are restrained against sway by a fixed superstructure can, however, be founded without undue problems on a single row of piles.

When piles are grouped into two or more rows, lateral loads are resisted primarily by axial forces. Pile bending will be small and will only occur when the horizontal reaction at the base of the column is resisted by vertical piles.

Prefabricated and injected piles required to resist horizontal loads are preferably battered. Battered piles are less practical to construct in cast-in-place concrete, due to difficulties in placing the reinforcement and concrete. When these small-diameter piles must be used vertically, the surrounding soil will contribute to the horizontal and flexural resistance. In such cases, the pile will behave as a beam on an elastic foundation.

References

Comité Euro-International du Béton. 1977. *Model Code for Concrete Structures*. Bulletin d'information no. 117-E. Paris: CEB.

Klöckner, W., K. Engelhardt, and H.-G. Schmidt. 1982. Gründungen (Foundations). In vol. 2 of *Beton-Kalender 1982*. Berlin and Munich: Verlag von Wilhelm Ernst & Sohn.

Lang, H.-J., and J. Huder. 1985. *Bodenmechanik und Grundbau* (Soil Mechanics and Foundation Engineering). 3rd ed. Berlin: Springer-Verlag.

Menn, C., J. Kammenhuber, U. Oelhafen, M. Grenacher, R. Bonomo, and L. Gruber. 1977. *Berechnung und Bemessung von Stützen und Stützensystemen* (Analysis and design of columns and column systems). Zurich: Eidgenössische Technische Hochschule Zürich, Institut für Baustatik und Konstruktion.

Terzagi, K. 1943. *Theoretical Soil Mechanics*. New York: John Wiley & Sons.

Timoshenko, S. P., and J. M. Gere. 1961. *Theory of Elastic Stability*. 2nd ed. New York: McGraw Hill.

Walther, R. 1977. *Design Charts for Reinforced Concrete Sections*. Lausanne: Ecole polytechnique fédérale de Lausanne, Chaire de béton armé de béton précontraint.

Appendix: Diagrams for the Design of Slender Columns

A1 Use of Diagrams

a) Fundamentals

The following diagrams can be used for the design of columns with rectangular or circular cross-section. They are valid for slender members and members stressed primarily in axial compression. The following notation is used:

1. Mechanical reinforcement ratio:

$$\omega = \frac{A_s + A_s'}{bh} \frac{f_{sy}}{f_c} \qquad \text{(rectangular sections)}$$

$$\omega = \frac{A_s}{\frac{\pi}{4} h^2} \frac{f_{sy}}{f_c} \qquad \text{(circular sections)}$$

2. Normalized sectional forces:

$$n_R = \frac{N_R}{bhf_c} \qquad m_R = \frac{M_R}{bh^2 f_c} \qquad \text{(rectangular sections)}$$

$$n_R = \frac{N_R}{\frac{\pi}{4} h^2 f_c} \qquad m_R = \frac{M_R}{\frac{\pi}{4} h^3 f_c} \qquad \text{(circular sections)}$$

where the design values of material strength, f_c and f_{sy}, are defined in Section 4.3.1.

b) Buckling Diagrams

These diagrams give normalized axial force, n_R, as a function of slenderness l_k/h. An initial eccentricity $w_0 = l_k/300$ has been assumed. The diagrams can be used for values of ω ranging from 0.1 to 1.2, $h'/h = 0.05$ or 0.10, and $\phi = 0.0$ (short-term load) and 2.0 (long-term load).

c) m_R-n_R Interaction Diagrams

The strain in the reinforcing steel has been limited to 0.0022, which corresponds to yield for f_{sy} – 460 N/mm². When bending moment is the predominant sectional force, the cross-section resistance determined in this way is smaller than the actual resistance computed assuming no limitation on steel strain, particularly for circular sections with low reinforcement ratios. This assumption is nevertheless reasonable for slender columns and leads to a more accurate calculation of the actual ultimate load.

In addition to the normalized cross-section resistance (m_R, n_R), the diagrams also give the corresponding normalized radius of curvature r_y/h, used in the calculation of flexural stiffness. The following relations apply:

$$ei_y = \frac{EI_y}{bh^3 f_c} \qquad \text{(rectangular sections)}$$

$$ei_y = \frac{EI_y}{\frac{\pi}{4} h^4 f_c} \qquad \text{(circular sections)}$$

The diagrams can be used for values of ω ranging from 0.0 to 0.5 and from 0.0 to 2.0, and $h'/h = 0.02$, 0.05, or 0.10.

d) k_ϕ-Diagrams for the Reduction of Stiffness Due to Creep

These values can be used to calculate the design value of flexural stiffness for long-term loading, $EI_y(\phi)$, computed as follows:

$$EI_y(\phi) = k_\phi EI_y$$

The factors k_ϕ are given for values of ϕ ranging from 0.0 to 3.0, as a function of the ratio n_R/n_{R0}, where $n_{R0} = 1 + \omega$ denotes ultimate axial force in the absence of bending.

A2 Notation

Notation :

Rectangular sections

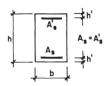

Normalized quantities:

$$n_R = \frac{N_R}{b \cdot h \cdot f_c} \qquad ei_y = \frac{EI_y}{b \cdot h^3 \cdot f_c}$$

$$m_R = \frac{M_R}{b \cdot h^2 \cdot f_c} \qquad \omega = \frac{A_s + A_s'}{b \cdot h} \cdot \frac{f_{sy}}{f_c}$$

Circular sections

Normalized quantities:

$$n_R = \frac{N_R}{\pi/4 \cdot h^2 \cdot f_c} \qquad ei_y = \frac{EI_y}{\pi/4 \cdot h^4 \cdot f_c}$$

$$m_R = \frac{M}{\pi/4 \cdot h^3 \cdot f_c} \qquad \omega = \frac{A_s}{\pi/4 \cdot h^2} \cdot \frac{f_{sy}}{f_c}$$

A3 Buckling Diagrams

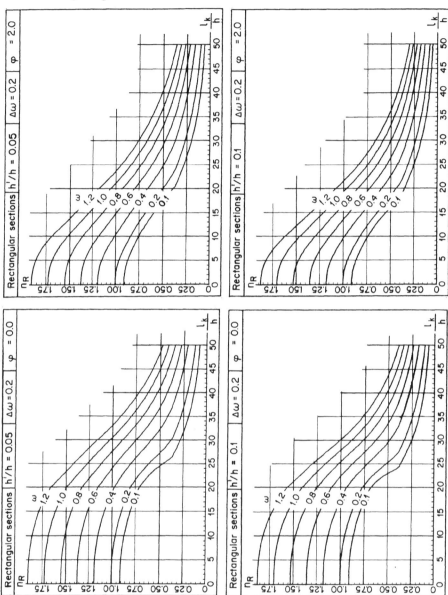

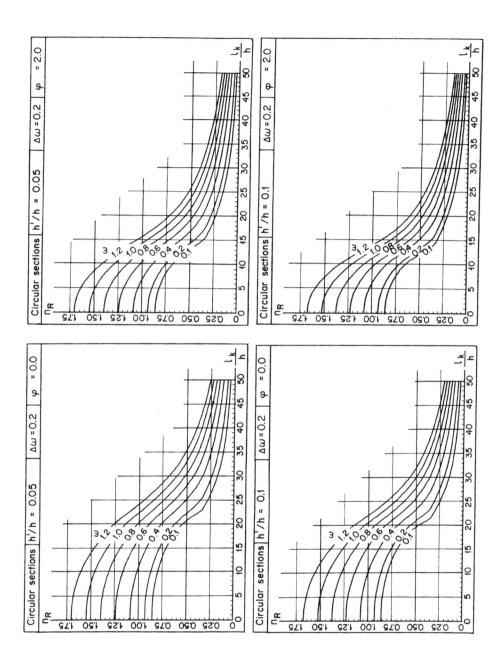

A4 m_R-n_R Interaction Diagrams

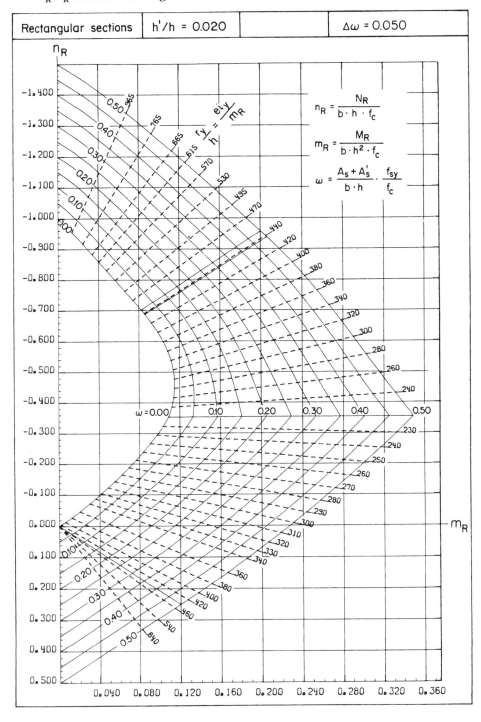

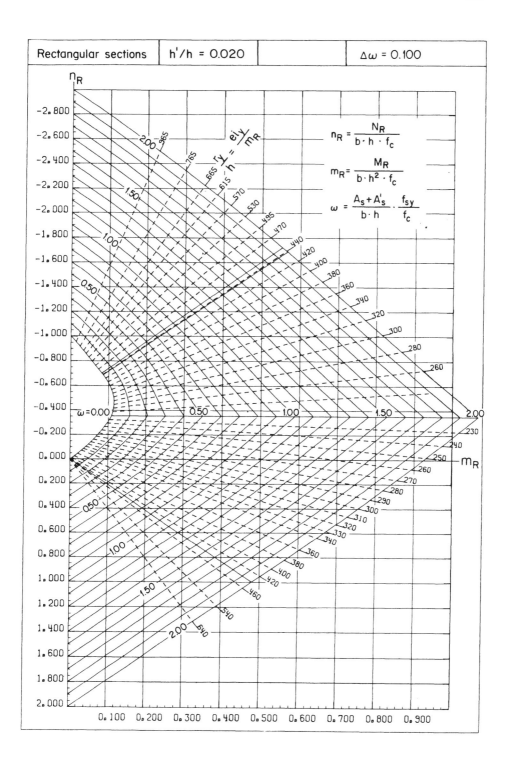

Rectangular sections   h'/h = 0.020   Δω = 0.100

$$n_R = \frac{N_R}{b \cdot h \cdot f_c}$$

$$m_R = \frac{M_R}{b \cdot h^2 \cdot f_c}$$

$$\omega = \frac{A_s + A'_s}{b \cdot h} \cdot \frac{f_{sy}}{f_c}$$

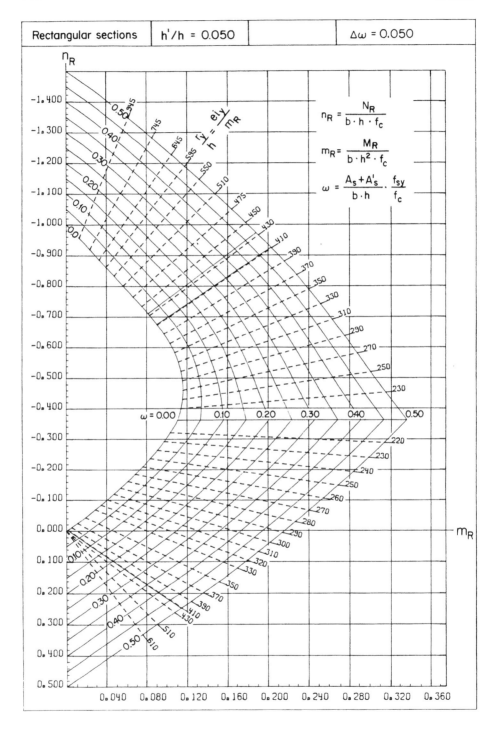

Rectangular sections | h'/h = 0.050 | | Δω = 0.050

$$n_R = \frac{N_R}{b \cdot h \cdot f_c}$$

$$m_R = \frac{M_R}{b \cdot h^2 \cdot f_c}$$

$$\omega = \frac{A_s + A'_s}{b \cdot h} \cdot \frac{f_{sy}}{f_c}$$

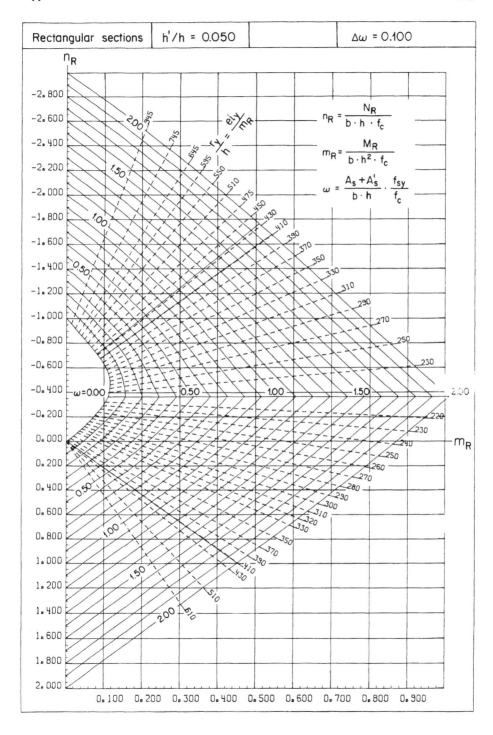

Rectangular sections | $h'/h = 0.050$ | | $\Delta\omega = 0.100$

$$n_R = \frac{N_R}{b \cdot h \cdot f_c}$$

$$m_R = \frac{M_R}{b \cdot h^2 \cdot f_c}$$

$$\omega = \frac{A_s + A_s'}{b \cdot h} \cdot \frac{f_{sy}}{f_c}$$

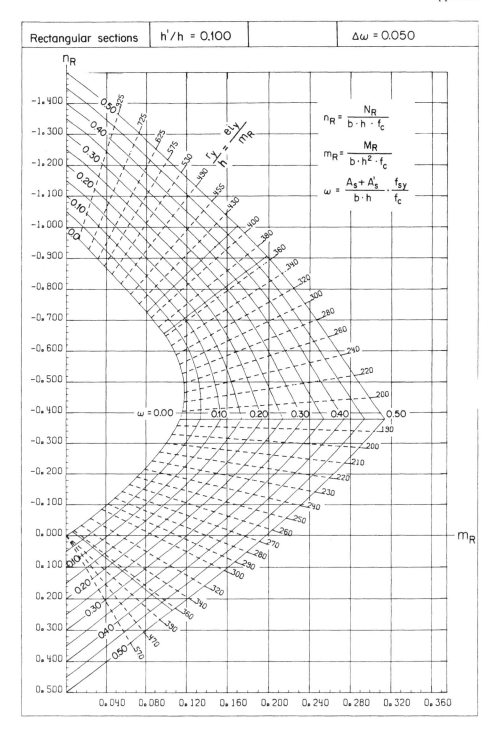

Rectangular sections | $h'/h = 0.100$ | | $\Delta\omega = 0.050$

$$n_R = \frac{N_R}{b \cdot h \cdot f_c}$$

$$m_R = \frac{M_R}{b \cdot h^2 \cdot f_c}$$

$$\omega = \frac{A_s + A_s'}{b \cdot h} \cdot \frac{f_{sy}}{f_c}$$

Rectangular sections	h'/h = 0.100		Δω = 0.100

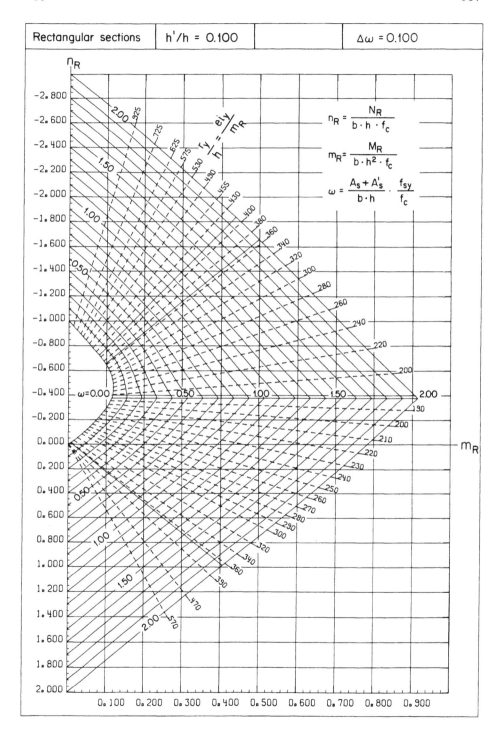

$$n_R = \frac{N_R}{b \cdot h \cdot f_c}$$

$$m_R = \frac{M_R}{b \cdot h^2 \cdot f_c}$$

$$\omega = \frac{A_s + A'_s}{b \cdot h} \cdot \frac{f_{sy}}{f_c}$$

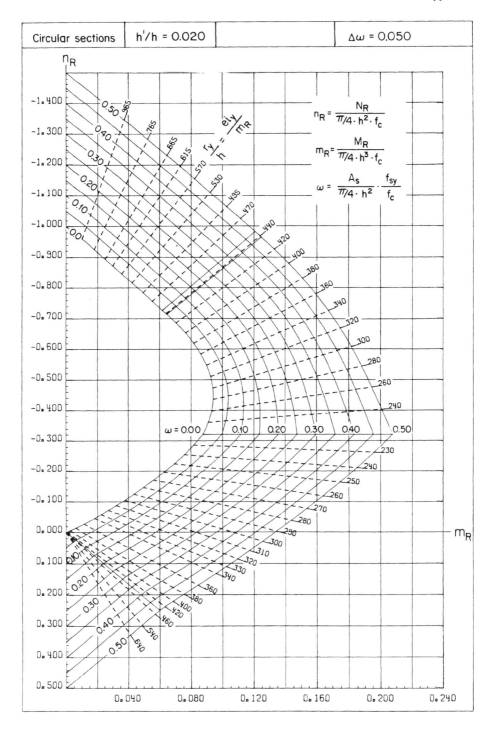

Circular sections   h'/h = 0.020   Δω = 0.050

$$n_R = \frac{N_R}{\pi/4 \cdot h^2 \cdot f_c}$$

$$m_R = \frac{M_R}{\pi/4 \cdot h^3 \cdot f_c}$$

$$\omega = \frac{A_s}{\pi/4 \cdot h^2} \cdot \frac{f_{sy}}{f_c}$$

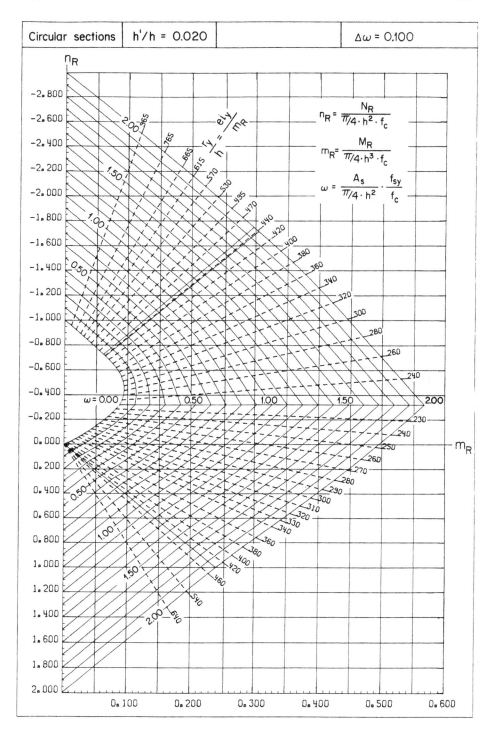

Circular sections | h'/h = 0.020 | Δω = 0.100

$$n_R = \frac{N_R}{\pi/4 \cdot h^2 \cdot f_c}$$

$$m_R = \frac{M_R}{\pi/4 \cdot h^3 \cdot f_c}$$

$$\omega = \frac{A_s}{\pi/4 \cdot h^2} \cdot \frac{f_{sy}}{f_c}$$

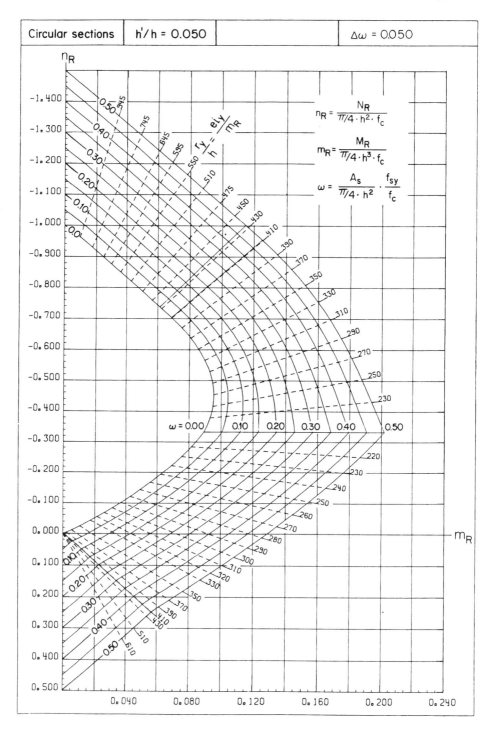

Circular sections   h'/h = 0.050   Δω = 0.050

$$n_R = \frac{N_R}{\pi/4 \cdot h^2 \cdot f_c}$$

$$m_R = \frac{M_R}{\pi/4 \cdot h^3 \cdot f_c}$$

$$\omega = \frac{A_s}{\pi/4 \cdot h^2} \cdot \frac{f_{sy}}{f_c}$$

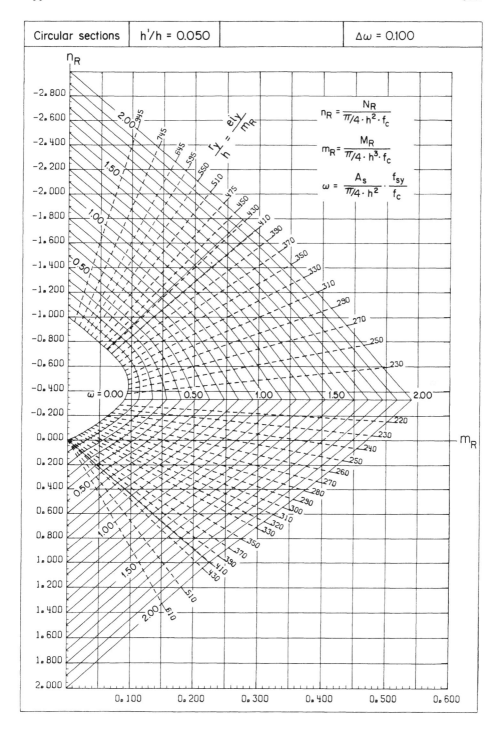

Circular sections | h'/h = 0.050 | | Δω = 0.100

$$n_R = \frac{N_R}{\pi/4 \cdot h^2 \cdot f_c}$$

$$m_R = \frac{M_R}{\pi/4 \cdot h^3 \cdot f_c}$$

$$\omega = \frac{A_s}{\pi/4 \cdot h^2} \cdot \frac{f_{sy}}{f_c}$$

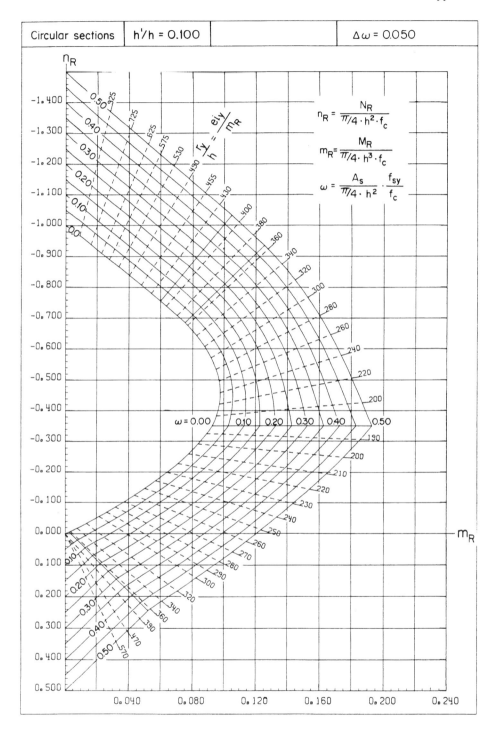

Circular sections | h'/h = 0.100 | Δω = 0.050

$$n_R = \frac{N_R}{\pi/4 \cdot h^2 \cdot f_c}$$

$$m_R = \frac{M_R}{\pi/4 \cdot h^3 \cdot f_c}$$

$$\omega = \frac{A_s}{\pi/4 \cdot h^2} \cdot \frac{f_{sy}}{f_c}$$

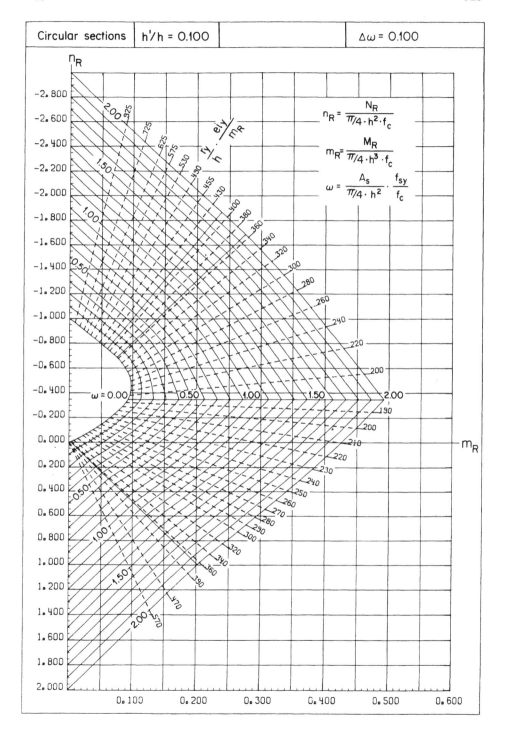

Circular sections | h'/h = 0.100 | | Δω = 0.100

$$n_R = \frac{N_R}{\pi/4 \cdot h^2 \cdot f_c}$$

$$m_R = \frac{M_R}{\pi/4 \cdot h^3 \cdot f_c}$$

$$\omega = \frac{A_s}{\pi/4 \cdot h^2} \cdot \frac{f_{sy}}{f_c}$$

A5 k_ϕ Diagrams for the Reduction of Stiffness Due to Creep

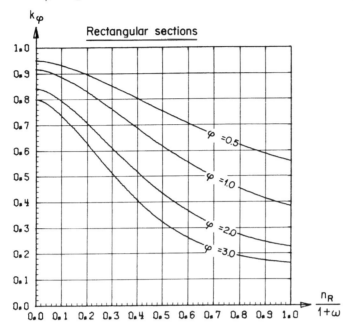

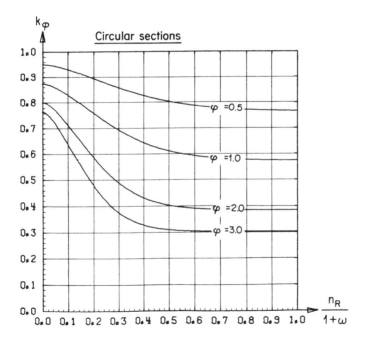

Index

Page numbers in *italics* refer to figures

37795592R00307

Made in the USA
Middletown, DE
06 December 2016